A-Z

HANDBOOK 4TH EDITION

Chemistry

Andrew Hunt

DIGITAL
EDITION

PHILIP ALLAN
UPDATES

Philip Allan Updates, an imprint of Hodder Education, an Hachette UK company, Market Place, Deddington, Oxfordshire OX15 0SE

Orders

Bookpoint Ltd, 130 Milton Park, Abingdon, Oxfordshire OX14 4SB
tel: 01235 827720
fax: 01235 400454
e-mail: uk.orders@bookpoint.co.uk

Lines are open 9.00 a.m.–5.00 p.m., Monday to Saturday, with a 24-hour message answering service. You can also order through the Philip Allan Updates website: www.philipallan.co.uk

ISBN 978-0-340-99100-8

First published 1998
Second edition 2000
Third edition 2003
Fourth edition 2009

Impression number 5 4 3 2 1
Year   2014   2013   2012   2011   2010   2009

Typeset by Macmillan, India.

Printed by Antony Rowe Ltd, Chippenham, Wiltshire.

Environmental information
Hachette UK's policy is to use papers that are natural, renewable and recyclable products and made from wood grown in sustainable forests. The logging and manufacturing processes are expected to conform to the environmental regulations of the country of origin.

Contents

A–Z entries 1

Appendices

    Chemistry revision lists 388

    Hints for exam success..403

    Examiners' terms 405

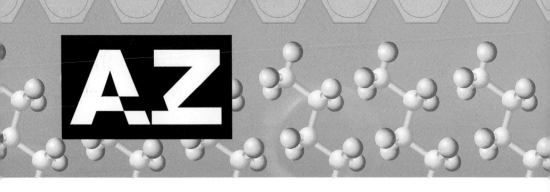

How to use this book

The *A–Z Chemistry Handbook* is an alphabetical textbook designed to help you at every stage of an advanced, pre-university chemistry course. Entries begin with one-sentence definitions which explain why the terms are important. These are followed by fuller explanations and examples together with diagrams, tables and equations. Words in light-bold italics are cross-references to other entries in the book.

You can use the book as a course companion and revision aid in conjunction with these specifications:

- English A-level courses from AQA, Edexcel and OCR (Chemistry and Salters Chemistry)
- CCEA Advanced Chemistry in Northern Ireland
- WJEC Advanced Chemistry in Wales
- SQA Higher and Advanced Higher Chemistry in Scotland
- Cambridge International: Advanced Chemistry and the Pre-U Certificate in Chemistry
- International Baccalaureate

Course companion

The handbook covers all the main ideas in chemistry specifications at this level. This concise, alphabetical book makes it easy for you to find the explanations, information and examples that you need when starting to study a new topic. Worked examples show you how to tackle the calculations.

Entries include many aspects of practical and applied chemistry, including both analytical chemistry and organic preparations. Furthermore, the book explains how chemists gather valid and reliable data and interpret their findings in terms of theories and models.

Revision aid

The appendices will help your revision. The appendix on pages 388–407 includes revision lists for all the important topics. The second appendix offers hints for exam success. The third appendix explains the terms that examiners use when setting questions.

In addition, the website that accompanies this book shows you exactly which topics you need to revise for each component of the chemistry course you are following.

A–Z Online

This new digital edition of the *A–Z Chemistry Handbook* includes free access to a supporting website and a free desktop widget to make searching for terms even quicker. Log on to **www.philipallan.co.uk/a-zonline** and create an account using the unique code provided on the inside front cover of this book.

Once you are logged on, you will be able to:

- search the entire database of terms in this handbook
- print revision lists specific to your exam board
- get expert advice from examiners on how to get an A* grade
- create a personal library of your favourite terms
- expand your vocabulary with our word of the week

You can also add the other *A–Z Handbooks (digital editions)* that you own to your personal library on A–Z Online.

Acknowledgements

Firstly, I thank the teachers and editors who advised on the first edition of this book: Geoffrey Barraclough, Rod Clough, Jeffrey Hancock and Ted Lister. Their wise comments helped me to develop the content and style of this handbook.

Secondly, I thank the team at Philip Allan Updates, especially Katie Blainey who has been the driving force behind this new generation of *A–Z Handbooks*.

Finally, I thank all the people that I have worked with when writing chemistry textbooks and while contributing to the Nuffield Advanced Chemistry, SATIS 16–19, Salters Advanced Chemistry and Science in Society projects. My involvement in these collaborations has clarified my understanding of chemical ideas and taught me much about the importance of chemistry in our lives.

Andrew Hunt

The publishers would like to thank the following for permission to reproduce illustrations in this book:

(page 72) from *Science for Understanding Tomorrow's World: Global Change for Education* with permission of the International Council for Science, 51, Bd. de Montmorency, 75016 Paris, France.

(page 80) from *Salters Advanced Chemistry, Chemical Storylines* by Salters, with permission of Heinemann Educational Publishers, a division of Reed Educational & Professional Publishing Ltd.

While every effort has been made to contact copyright holders, any omissions brought to our attention will be remedied in future printings.

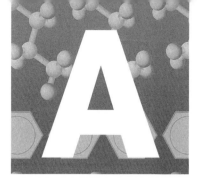

***A*~r~** is the symbol for *relative atomic mass*.

absolute zero (0 K) is the temperature at which atoms and molecules in crystals are effectively motionless. It is the lowest temperature on the absolute or *Kelvin* temperature scale. On the Celsius scale this temperature is $-273.15°C$.

Gases appear to behave as if they would have zero volume at absolute zero. This would be true of a theoretical *ideal gas* but in practice real gases turn to liquids and solids and occupy a definite volume before absolute zero is reached.

absorption of liquids and gases happens when a *fluid* soaks into the pores of a material, like water soaking into a sponge. Absorption should be carefully distinguished from *adsorption*.

absorption of radiation: when *electromagnetic radiation* passes through a material, some or all of the wavelengths of the radiation may be absorbed. When white light passes through a red filter, for example, all the wavelengths are absorbed except for red which passes through – so it looks red.

The gases in the *atmosphere* absorb most of the radiation reaching the Earth from the Sun. *Ozone in the stratosphere* helps to protect living things by absorbing harmful *ultraviolet radiation*.

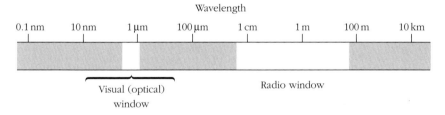

The bands of wavelengths in the electromagnetic spectrum of sunlight which can pass through the Earth's atmosphere to reach the surface of the Earth. It is as if there are two 'windows' letting through bands of radiation while all other wavelengths are shut out (by absorption).

Chemists use *spectroscopy* to study the absorption of radiation. From an *absorption spectrum* they can make deductions about the composition and structure of the sample.

absorption spectrum: a plot showing how strongly a sample absorbs radiation over a range of frequencies. Absorption spectra from *infrared spectroscopy* and *ultraviolet spectroscopy* give chemists valuable information about the composition and structure of chemicals.

accuracy of data is determined by how close a measured quantity is to the correct value. In chemical analysis the correct value is often not known and so chemists need to estimate the *measurement uncertainty* (see also *errors of measurement*).

acids are compounds with the following characteristic properties:
- in water they form solutions with a *pH* below 7
- they change the colours of *acid–base indicators*
- they react with *metals* such as magnesium to produce hydrogen gas:

$$Mg(s) + 2HCl(aq) \rightarrow MgCl_2(aq) + H_2(g)$$

- they react with carbonates such as calcium carbonate to form carbon dioxide gas:

$$CaCO_3(s) + 2HCl(aq) \rightarrow CaCl_2(aq) + CO_2(g) + H_2O(l)$$

- they react with *basic oxides* to form salts and water:

$$CuO(s) + H_2SO_4(aq) \rightarrow CuSO_4(aq) + H_2O(l)$$

Pure acids may be solids (such as citric acid and tartaric acid), liquids (such as *sulfuric acid, nitric acid* and ethanoic acid) or gases (such as hydrogen chloride which becomes *hydrochloric acid* when it dissolves in water).

Definitions of acids are based on the theories used to explain why they have similar properties. Chemists use different theories in different contexts. One of the first was the *Arrhenius theory of acids*. The definition of an acid generally used today is based on the *Brønsted–Lowry theory*. Another definition is based on the *Lewis acid–base theory*.

acid anhydrides are related to *carboxylic acids*. The *functional group* in an anhydride is formed by eliminating a molecule of water from two carboxylic acid groups. Sometimes this happens simply on heating the acid but generally anhydrides are made in other ways.

With water, alcohols and amines, acid anhydrides react in a similar way to *acyl chlorides*. Chemists call this type of reaction *acylation*.

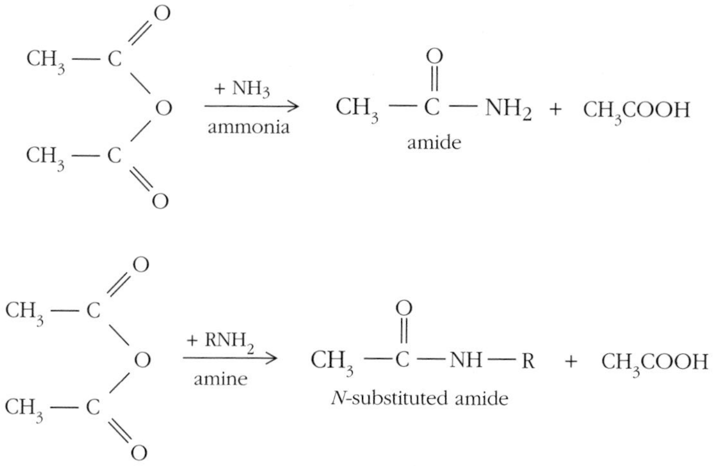

Reactions of ethanoic anhydride

Anhydrides are less reactive than acyl chlorides and so are often preferred for laboratory and industrial syntheses. Ethanoic anhydride, for example, is used to convert 2-hydroxybenzoic acid to *aspirin*.

acid–base equilibria are equilibrium systems involving *acid–base reactions*. Acid–base reactions are *reversible*. This is illustrated by a solution of an ammonium salt in water.

$$NH_4^+(aq) + H_2O(l) \rightleftharpoons NH_3(aq) + H_3O^+(aq)$$
$$\text{acid 1} \quad\quad \text{base 2} \quad\quad \text{base 1} \quad\quad \text{acid 2}$$

There is competition for *protons* (hydrogen ions) between ammonia molecules and water molecules. On the left-hand side of the equation the protons are held by *lone pairs of electrons* on the ammonia molecules. On the right-hand side they are held by lone pairs on water molecules.

The *equilibrium law* applies. The position of equilibrium is determined by the value of the equilibrium of constant for the reaction. For the example in the equilibrium expression, the relevant equilibrium constant is the acid dissociation constant, K_a, for the ammonium ion.

The equilibrium involves two conjugate acid–base pairs:
- NH_4^+ and NH_3
- H_3O^+ and H_2O

An acid turns into its conjugate base when it loses a proton. A base turns into its conjugate acid when it gains a proton.

acid–base indicators are used to show up changes in *pH* of solutions. Indicators are *weak acids* or *weak bases* which change colour when they lose or gain hydrogen ions. When added to a solution, an indicator gains or loses protons depending on the pH of the solution. It is conventional to represent a weak acid indicator as HIn where In is a short-hand for the rest of the molecule. In water:

$$HIn(aq) + H_2O(l) \rightleftharpoons H_3O^+(aq) + In^-(aq)$$

un-ionised
indicator
colour 1

indicator after
losing a proton
colour 2

3

Indicators such as phenolphthalein, methyl orange and bromothymol blue change colour over a range of pH values and are used to detect the *end-point* in *acid–base titrations*.

The pH range over which an indicator changes colour is determined by its strength as an acid (or base). Typically the range is given roughly by $pK_a \pm 1$ (see the *Henderson–Hasselbalch equation* for an explanation).

Indicator	pK_a	Colour change HIn/In⁻	pH range over which colour change occurs
methyl orange	3.6	red/yellow	3.2–4.2
methyl red	5.0	yellow/red	4.2–6.3
bromothymol blue	7.1	yellow/blue	6.0–7.6
phenolphthalein	9.4	colourless/red	8.2–10.0

The structures of methyl orange in acid and alkaline solutions

In acid solution the added hydrogen ion (proton) localises two electrons to form a covalent bond. In alkaline solution the removal of the hydrogen ion allows the two electrons to join the other delocalised electrons. The change in the number of delocalised electrons causes a shift in the peak of the wavelengths of light absorbed, so the colour changes and the molecule acts as an indicator (see *coloured compounds*).

acid–base reactions, according to the *Brønsted–Lowry theory*, are reactions involving the transfer of protons from an acid to a base.

acid–base titration: a practical technique used to determine the *concentration* of an *acid* or an *alkali*. A *titration* measures the volume of a *standard solution* of alkali or acid needed to react exactly with a measured volume of the unknown solution. The procedure for an acid–base titration is the same as for any other titration but the method for finding the end-point is distinctive: the analyst either adds an *acid–base indicator* or uses a pH meter to follow the *pH changes during the acid–base titrations*.

Worked example

Limewater is a saturated solution of calcium hydroxide in water. 25.0 cm³ of 0.04 mol dm⁻³ hydrochloric acid from a burette neutralised 20.0 cm³ of limewater. What was the concentration of the limewater?

Notes on the method

Always start by writing the equation for the reaction. See *titration* for a general method for the calculations.

Remember to convert volumes in cm^3 to volumes in dm^3 by dividing by 1000.

In any titration there is one unknown — in this case the concentration of the limewater, c_B.

Answer

The equation for the reaction is:

$$Ca(OH)_2(aq) + 2HCl(aq) \rightarrow CaCl_2(aq) + 2H_2O(l)$$

The volume of calcium hydroxide in the flask, $V_B = \dfrac{20.0}{1000}$ dm^3

Let the concentration of calcium hydroxide be c_B.

The volume of hydrochloric acid added from the burette, $V_A = \dfrac{25.0}{1000}$ dm^3

The concentration of hydrochloric acid, $c_A = 0.04$ mol dm^{-3}

$$\frac{V_A \times c_A}{V_B \times c_B} = \frac{n_A}{n_B}$$

$$\frac{\dfrac{25.0}{1000} \times 0.04}{\dfrac{20.0}{1000} \times c_B} = \frac{2}{1}$$

Therefore $\qquad c_B = \dfrac{25.0 \times 0.04}{2 \times 20.0} = 0.025$ mol dm^{-3}

The concentration of the limewater was 0.025 mol dm^{-3}.

acid catalysis: any reaction speeded up by an acid *catalyst*. One example is the acid-catalysed *hydrolysis* of an *ester* to give a *carboxylic acid* and an *alcohol*. All the chemicals are in the same solution so this is an example of a reaction using a *homogeneous catalysts*.

acid chlorides: see *acyl chlorides*.

acid deposition is the term used to describe the processes that remove acidic gases, liquids and particles from the atmosphere. *Atmospheric pollutants* form during combustion in engines, furnaces, incinerators and power stations. During combustion, any sulfur in a *fossil fuel* is oxidised to sulfur dioxide (SO_2). Nitrogen in a fuel may end up as the element after combustion or, if the temperature is high enough, combined with oxygen to form a mixture of *nitrogen oxides* (NO_x).

The primary pollutants produced during combustion are converted to secondary pollutants by chemical reactions in the air. Among the secondary pollutants are sulfuric acid, nitric acid and ammonium sulfate. The pollutants cause acidification by being deposited in the environment as gases or particles (dry deposition) or in the form of *acid rain* or mist (wet deposition).

acid dissociation constants measure the strength of acids (see K_a).

acidic oxide: an oxide of a *non-metal* which reacts with water to form an *acid*. Some acidic oxides are insoluble but they can be recognised because they react directly with *basic oxides* to form *salts*. Note that acidic oxides are not themselves *acids* as defined by the *Brønsted–Lowry theory* because they do not contain ionisable hydrogen atoms, so they cannot act as proton donors.

Oxide	Acid formed with water
carbon dioxide, CO_2	carbonic acid, H_2CO_3
sulfur dioxide, SO_2	sulfurous acid, H_2SO_3 [sulfuric(IV) acid]
sulfur trioxide, SO_3	sulfuric acid, H_2SO_4 [sulfuric(VI) acid]
phosphorus(V) oxide, P_2O_5	phosphoric acid, H_3PO_4

Silicon(IV) oxide (SiO_2) is an acidic oxide which is insoluble in water. Silica reacts with basic metal oxides at high temperatures in furnaces during glassmaking and in steelmaking. Calcium oxide (quicklime) is added to a *blast furnace* to remove silicon(IV) oxide (silica) and other impurities. The calcium silicate is a liquid at the temperature of the furnace; it runs towards the bottom where it floats on top of the molten iron as a slag which can be tapped off separately.

$$CaO(s) + SiO_2(s) \rightarrow CaSiO_3(l)$$

acid rain: see *acid deposition*. Rain is naturally acidic with a pH around 5.6 because of dissolved carbon dioxide. In acid rain the pH is lower than this value because of dissolved pollutants.

acid salt: a compound formed by partly neutralising *acids* such as *sulfuric acid*, H_2SO_4, or *phosphoric(V) acid*, H_3PO_4. These are acids with two or three ionisable hydrogen atoms. Sodium hydrogensulfate, $NaHSO_4$, and sodium dihydrogenphosphate, NaH_2PO_4, are examples of acid salts.

The negative ion in an acid salt can act either as an acid by giving away a further proton, or as a *base* by accepting a proton. The pH of an *aqueous solution* of an acid salt depends on the *acid strength* of the parent acid.

A solution of sodium hydrogensulfate, a salt of a *strong acid*, is acidic with a pH below 7 because the hydrogensulfate ion is also an acid, giving protons to water molecules:

$$HSO_4^- + H_2O \rightleftharpoons SO_4^{2-} + H_3O^+$$

A solution of sodium hydrogencarbonate, a salt of the *weak acid* called carbonic acid, is alkaline with a pH above 7 because the hydrogencarbonate ion is a base, taking protons (H^+) from water molecules:

$$HCO_3^- + H_2O \rightleftharpoons H_2CO_3 + OH^-$$

acid strength: see *strong acid*, *weak acid* and *acid dissociation constant*.

acid strength of organic hydroxy compounds: the order of acid strength for organic compounds with —OH groups is:

carboxylic acids > phenols > water > alcohols.

Carboxylic acids are *weak acids* which ionise to a significant extent by giving protons to water forming a solution with pH 3–4. *Phenol* is acidic enough to lower the pH of a solution in water to 5–6 but it ionises significantly only when mixed with a stronger base such as the hydroxide ions in a solution of sodium hydroxide. Ethanol and other *alcohols* are such weak acids that they do not ionise to a significant extent in water or in a solution of a strong alkali.

Introducing halogen atoms into the structure of carboxylic acids can have a marked effect on their acid strength.

Substituting an electronegative atom on the carbon atom next to the —COOH group makes the acid stronger. The more electronegative the atom, the greater the effect. The *inductive effect* of the electronegative atom pulls electrons away from the carboxylic acid group so that it holds onto to its proton less strongly. This is illustrated by the chlorinated ethanoic acids. The order of acid strength is:

trichloroethanoic acid > dichloroethanoic acid > chloroethanoic acid > ethanoic acid.

acrylics are a group of *addition polymers* based on derivatives of propenoic acid (acrylic acid), $H_2C = CH—COOH$. The most widely used acrylic is the polymer of propenonitrile (acrylonitrile), $H_2C = CH—CN$. Often a small amount of chloroethene is added to the monomer. *Copolymerisation* produces a polymer which, when spun into *fibres*, feels similar to wool. Another common acrylic is the polymer sold under trade names such as Perspex, which is the polymer of an ester, methyl-2-methylpropenoate.

actinide elements (or actinoids) are the *f-block elements* in period 7 of the *periodic table*. They are the 14 elements from thorium to lawrencium which lie between actinium (element 89) and rutherfordium (element 104).

activated complex: a combination of reacting atoms, molecules or ions at that point during a chemical reaction when they are at the top of the *activation energy* barrier between reactants and products. An activated complex is a combination of reacting particles at a higher energy because chemical bonds are stretched. The activated complex is the *transition state* for the reaction step.

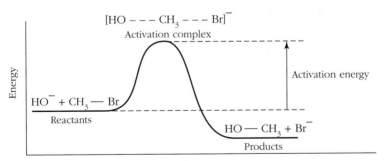

Reaction profile to show the activated complex formed in the course of a nucleophilic substitution in a halogenoalkane

activation energy: the minimum energy needed in a collision between molecules if they are to react. The activation energy is the height of the energy barrier separating the reactants and products during a chemical reaction.

7

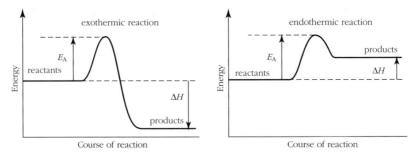

*Energy profile diagrams showing the activation energies for both **exothermic** and **endothermic** reactions*

Activation energy is an important idea in the **collision theory** of reaction rates. It accounts for the fact that reactions go much more slowly than would be expected if every collision between atoms and molecules led to a reaction. Only a very small proportion of collisions bring about chemical change. Molecules react only if they collide with enough energy between them to overcome the energy barrier. At around room temperature only a small proportion of molecules have enough energy to react.

Activation energies account for the way in which reaction rates vary with temperature, *T*. At a higher temperature there are more molecules with enough energy to react when they collide.

The activation energies for many reactions in biochemical systems are around 50 kJ mol^{-1} and for these reactions the rate approximately doubles for each 10 K rise in temperature.

It is possible to determine the activation energy for a reaction by measuring its rate over a range of temperatures and then using the **Arrhenius equation**.

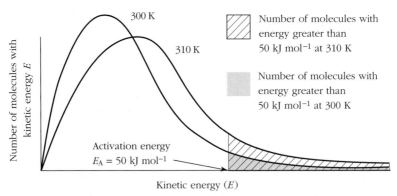

Maxwell–Boltzmann distribution curve for the kinetic energies of molecules at about room temperature, 300 K and 310 K. The shaded area shows the proportion of molecules having at least the activation energy for a reaction. This area is bigger at a higher temperature.

active site: the part of a **catalyst** which is responsible for its catalytic activity. **Zeolites** are examples of inorganic catalysts, for example, which owe their catalytic activity to active sites in the crystal structure which can adsorb specific molecules so that they can react to produce the required products. **Enzymes** are biological catalysts with an active site in the coiled **protein** chain.

acyl chlorides are made from *carboxylic acids* by replacing the hydroxyl (—OH) group with a chlorine atom.

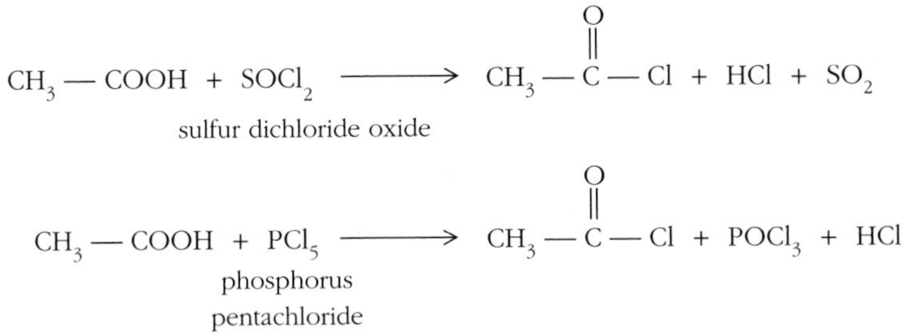

$$CH_3 - COOH + SOCl_2 \longrightarrow CH_3 - \overset{\overset{\displaystyle O}{\|}}{C} - Cl + HCl + SO_2$$

sulfur dichloride oxide

$$CH_3 - COOH + PCl_5 \longrightarrow CH_3 - \overset{\overset{\displaystyle O}{\|}}{C} - Cl + POCl_3 + HCl$$

phosphorus
pentachloride

Formation of ethanoyl chloride from ethanoic acid

Acyl chlorides are very reactive; they are powerful acylating agents. Most are quickly *hydrolysed* back to the acid by cold water. Acyl chlorides react rapidly with *alcohols* and *phenols* to form *esters*. These reactions are all examples of *acylation*.

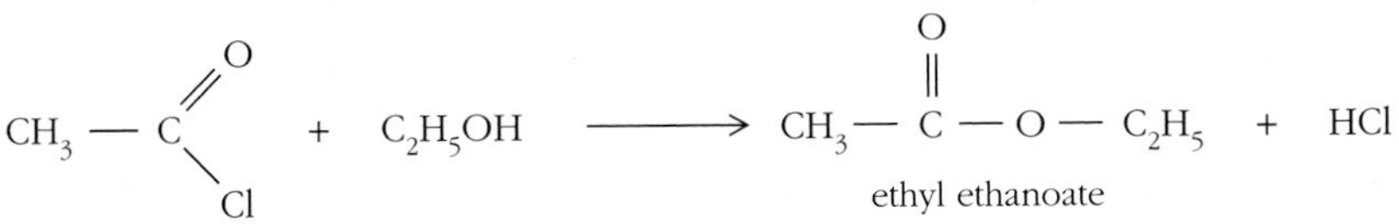

$$CH_3 - \overset{\displaystyle O}{\underset{\displaystyle Cl}{C}} + C_2H_5OH \longrightarrow CH_3 - \overset{\overset{\displaystyle O}{\|}}{C} - O - C_2H_5 + HCl$$

ethyl ethanoate

Formation of an ester

Acyl chlorides react with *ammonia* and *amines* to form *amides*.

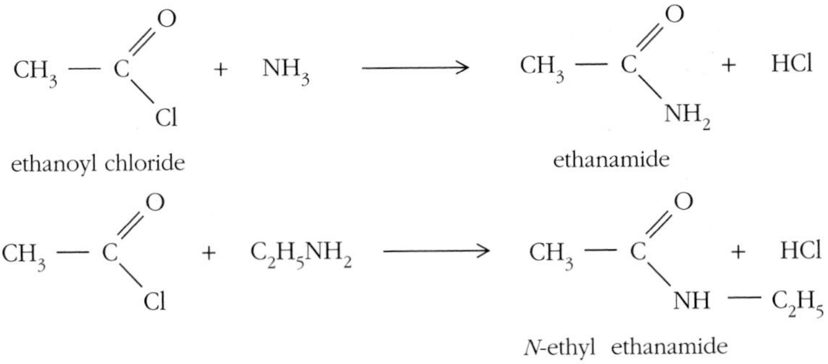

$$CH_3 - \overset{\displaystyle O}{\underset{\displaystyle Cl}{C}} + NH_3 \longrightarrow CH_3 - \overset{\displaystyle O}{\underset{\displaystyle NH_2}{C}} + HCl$$

ethanoyl chloride   ethanamide

$$CH_3 - \overset{\displaystyle O}{\underset{\displaystyle Cl}{C}} + C_2H_5NH_2 \longrightarrow CH_3 - \overset{\displaystyle O}{\underset{\displaystyle NH - C_2H_5}{C}} + HCl$$

N-ethyl ethanamide

Formation of an amide and an N-substituted amide

Acyl chlorides can also acylate *benzene* by the *Friedel–Crafts reaction*.

acylation is a reaction which substitutes an acyl group for a hydrogen atom. The H atom may be part of an —OH group, an —NH$_2$ group or a benzene ring. Acylating agents are either *acyl chlorides* or *acid anhydrides*.

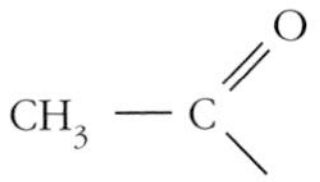

$$CH_3 - \overset{\displaystyle O}{\underset{\displaystyle }{C}}$$

The ethanoyl group – an acyl group

Acylation is the reaction which converts 2-hydroxybenzoic acid to *aspirin*. Acylation of benzene is an example of the *Friedel–Crafts reaction*.

addition–elimination reactions take place when two molecules first add together and then immediately split off a small molecule such as water or hydrogen chloride. They are sometimes called *condensation reactions*.

The reactions take place in three stages:
- nucleophilic addition
- gain and loss of hydrogen ions (protons)
- elimination of water

acyl halides – the reactions of *acyl chlorides* with the *nucleophiles* water, alcohols, ammonia and amines are all addition–elimination reactions.

Mechanism of the reaction of an alcohol (ethanol) with an acyl halide (ethanoyl chloride) to form an ester (ethyl ethanoate)

carbonyl compounds – the addition–elimination reactions of *carbonyl compounds* take place with compounds of the form X—NH$_2$.

Generalised reaction of a carbonyl compound with X—NH$_2$. R and R′ are both alkyl groups in ketones. In an aldehyde R = H while R′ is an alkyl group.

2,4-dinitrophenylhydrazine reacts with carbonyl compounds in this way to form 2,4-dinitrophenylhydrazone *derivatives*.

addition polymerisation is a process for making *polymers* from compounds containing *double bonds*. Many molecules of the *monomer* add together to form a long-chain polymer. Ethene, for example, polymerises to form poly(ethene):

$$n\text{CH}_2\text{—CH}_2 \rightarrow [\text{—CH}_2\text{—CH}_2\text{—}]_n$$

Monomer	Polymer	Notes
ethene 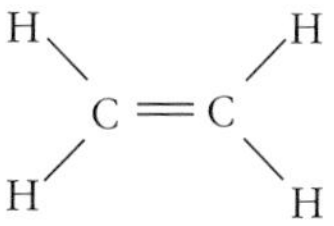	poly(ethene) or polythene 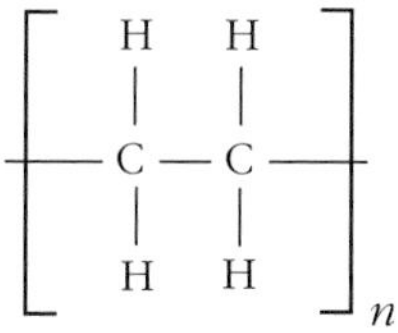	A high-pressure, high-temperature process in the presence of a peroxide initiator produces low-density poly(ethene) with branched chains. The mechanism involves a *free-radical chain reaction.* A low-pressure, low-temperature process with a *Ziegler–Natta catalyst* produces high-density poly(ethene) in which the polymer chains pack closer because they have no side branches.
propene 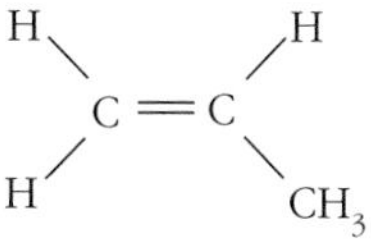	poly(propene) or polypropylene 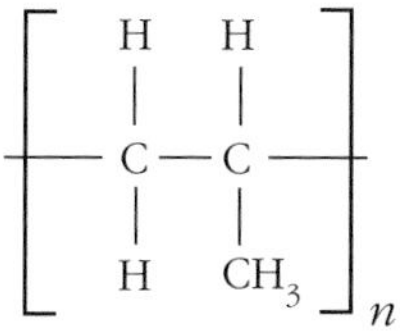	*Isotactic* poly(propene) is made using a Ziegler–Natta catalyst. This strong polymer is used to make pipes, wrapping films, carpet fibres and ropes.
phenylethene 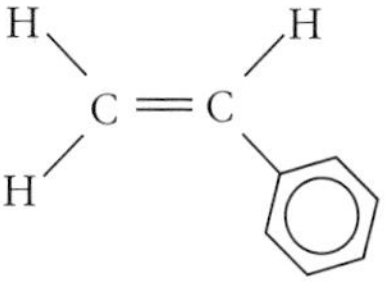	poly(phenylethene) or polystyrene 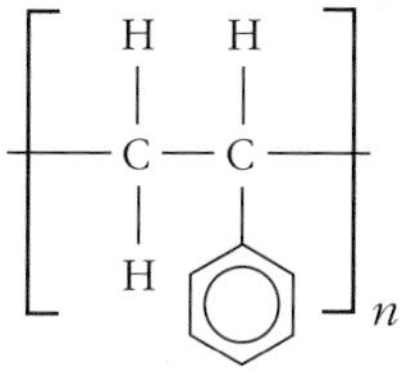	Expanded polystyrene has low density and is an excellent thermal insulator. It is used for packaging because it absorbs shocks.
chloroethene 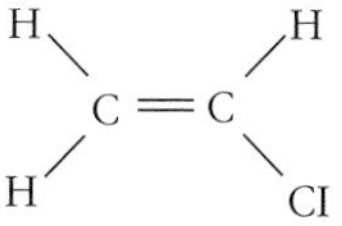	poly(chloroethene) or polyvinylchloride (PVC) 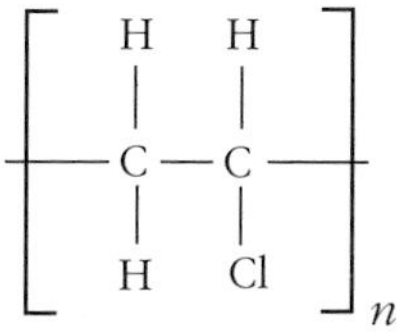	Unplasticised uPVC is a rigid polymer suitable for guttering and window frames. PVC with a *plasticiser* is flexible and used for packaging, flooring and cable insulation.
tetrafluoroethene 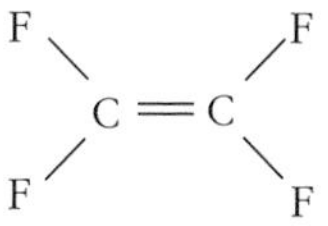	poly(tetrafluoroethene) or PTFE 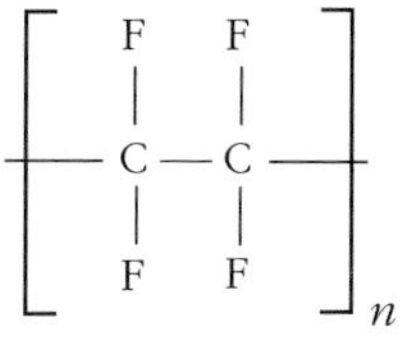	Engineers use PTFE to provide low-friction surfaces to allow bridges and other engineering structures to move slightly as the metals expand and contract with temperature changes. It is the polymer used to coat non-stick pans.

addition reactions are reactions in which two molecules combine to form a single product. Bromine, for example, adds to ethene to form the addition product 1,2-dibromoethane.

1,2-dibromoethane

Structure diagram to show addition of bromine to ethene

Addition reactions are characteristic of **unsaturated compounds** such as the **alkenes** and **carbonyl compounds**.

adsorption is a process in which atoms, molecules or ions are held on the surface of a solid. Adsorption processes are important in reactions using a **heterogeneous catalyst** and in some types of **chromatography**. Adsorption should be carefully distinguished from **absorption**.

aerobic respiration: see **respiration**.

aerosols are suspensions of tiny particles of a solid, or droplets of a liquid, in a gas. They are **colloids**. Smoke is an aerosol in which the finely dispersed particles are solid. Examples of aerosols with liquid droplets are mists, clouds and paint sprays. Aerosols come from many sources including aerosol cans and volcanoes.

agrochemicals include **pesticides** to destroy the organisms that damage crops, such as insects (insecticides), fungi (fungicides) and weeds (**herbicides**). Other agrochemicals are the growth regulators which can stimulate or inhibit plant growth. The chemical industry distinguishes agrochemicals from **fertilisers**, which provide the chemicals needed for healthy plant growth.

air is a mixture of gases as shown in the table below. The proportion of water **vapour** in the air varies widely according to the weather conditions. Power stations, industrial processes and motor vehicles add **atmospheric pollutants** to the air.

Gas	Percentage by volume in dry air	Boiling point/K
nitrogen	78.0	77
oxygen	21.0	90
argon	0.9	87
carbon dioxide (about 0.04%) and small traces of neon, helium and krypton	0.1	

air condenser: a glass tube without a water jacket used for condensing **vapours** during **distillation** or refluxing (see **reflux condenser**). An air condenser is fitted when the temperature of the vapour is so high that the surrounding air is cool enough to condense the liquid. Normally this is when the boiling point of the liquid being distilled is above about 150°C.

alcohols are compounds with the formula R—OH where R represents an *alkyl group*. The hydroxy group —OH is the *functional group* which gives the compounds their characteristic reactions.

The chemical industry makes alcohols by the *hydration* of *alkenes* in the presence of an acid *catalyst*.

Equation for the industrial production of ethanol. Ethanol is also produced by fermentation.

The alcohols are sometimes called alkanols. Alcohols are named by changing the ending of the corresponding *alkane* to -ol. So ethane becomes ethanol.

methanol   ethanol

Names and structures of two alcohols: methanol and ethanol

$$CH_3 — CH_2 — CH_2 — CH_2 — OH$$ butan-1-ol, a primary alcohol

$$CH_3 — CH_2 — CH — CH_3$$
                          |
                          OH   butan-2-ol, a secondary alcohol

$$CH_3 — C — CH_3$$ 2-methylpropan-2-ol, a tertiary alcohol

Primary, secondary and tertiary alcohols

Even the simplest alcohols, methanol and ethanol, are liquids at room temperature because of *hydrogen bonding* between the hydroxy groups. For the same reason alcohols with relatively short hydrocarbon chains mix freely with water.

Alcohols burn in air with a clean, colourless flame. This makes them attractive as fuels and explains why methanol and ethanol are both fuels and fuel additives. *Bioethanol* is one of the main *biofuels*.

$$C_2H_5OH(l) + 3O_2(g) \rightarrow 2CO_2(g) + 3H_2O(l)$$

The reactions of ethanol are typical of the alcohols in general.

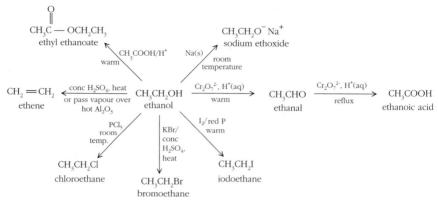

Reactions of ethanol

Oxidation reactions distinguish primary alcohols, secondary alcohols and tertiary alcohols.

primary alcohol $\xrightarrow[\text{warm}]{Cr_2O_7^{2-},\ H^+(aq)}$ aldehyde $\xrightarrow[\text{reflux}]{\text{excess } Cr_2O_7^{2-},\ H^+(aq)}$ carboxylic acid

secondary alcohol $\xrightarrow[\text{reflux}]{Cr_2O_7^{2-},\ H^+(aq)}$ ketone

tertiary alcohol   no reaction

Use of oxidation with an acidified solution of dichromate(VI) ions to distinguish primary, secondary and tertiary alcohols

aldehydes are *carbonyl compounds* in which the carbonyl group is attached to two hydrogen atoms or an *alkyl group* and a hydrogen atom. So the carbonyl group is at the end of a carbon chain. The —CHO group is the *functional group* which gives aldehydes their characteristic reactions.

The aldehydes are sometimes called alkanals. They are named after the alkane with the same carbon skeleton by changing the ending 'e' to 'al' (so, for example, ethane becomes ethanal).

methanal   ethanal   propanal

Structures and names of aldehydes

Methanal is a gas at room temperature. Ethanal boils at 21°C so it may be a liquid or gas at room temperature depending on the conditions.

Aldehydes are formed by **oxidation** of primary **alcohols** on heating with a mixture of dilute sulfuric acid and potassium dichromate(VI) under conditions which allow the aldehyde to distil off as it forms. Unlike **ketones**, aldehydes can easily be oxidised further to **carboxylic acids** by longer heating with an excess of the reagent.

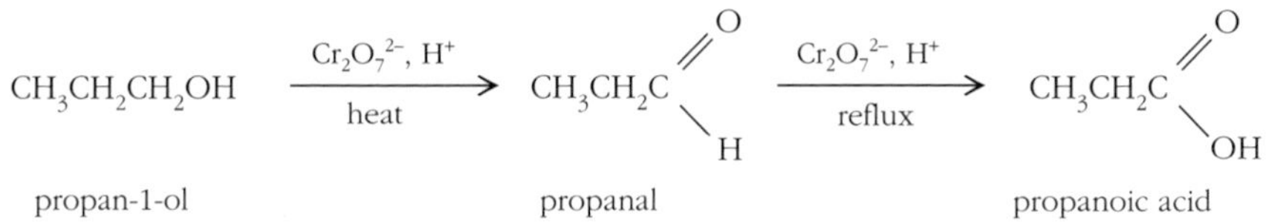

$$CH_3CH_2CH_2OH \xrightarrow[\text{heat}]{Cr_2O_7^{2-},\ H^+} CH_3CH_2C\!\!\begin{array}{c}O\\//\\\backslash\\H\end{array} \xrightarrow[\text{reflux}]{Cr_2O_7^{2-},\ H^+} CH_3CH_2C\!\!\begin{array}{c}O\\//\\\backslash\\OH\end{array}$$

propan-1-ol   propanal   propanoic acid

Two-stage oxidation of propan-1-ol

Fehling's solution and **Tollens' reagent** are mild oxidising agents, used to distinguish aldehydes from ketones. **Benedict's solution** can also be used but it is not so reliable with aldehydes. Preparing a crystalline derivative with **2,4-dinitrophenylhydrazine** makes it possible to identify aldehydes. Ethanal is the only aldehyde to undergo the **triiodomethane reaction**.

Sodium tetrahydridoborate(III) ($NaBH_4$) reduces aldehydes to primary alcohols. An alternative reducing agent is **lithium tetrahydridoaluminate(III)** ($LiAlH_4$) but it is not so easy to use because it has to be kept dry and dissolved in ether, which is a highly flammable solvent.

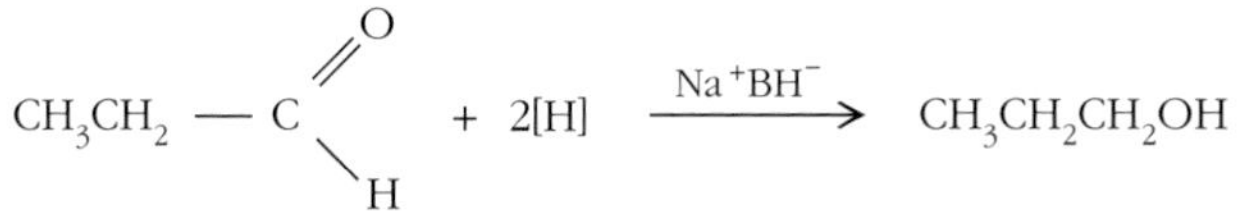

$$CH_3CH_2 - C\!\!\begin{array}{c}O\\//\\\backslash\\H\end{array} + 2[H] \xrightarrow{Na^+BH^-} CH_3CH_2CH_2OH$$

Reduction of propanal to propan-1-ol. The 2[H] comes from the reducing agent. This is a shorthand way of balancing a complex equation involving reduction.

Because of the double bond in the carbonyl group, aldehydes (like ketones) undergo **addition reactions**. These are **nucleophilic addition** reactions.

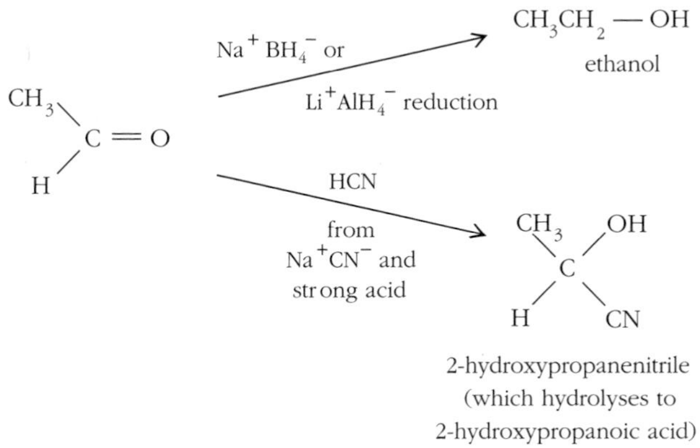

Addition reactions of ethanal

In some reactions addition is immediately followed by elimination of water. These are *addition–elimination reactions*.

aliphatic hydrocarbons are *hydrocarbons* with no benzene rings. The chains of carbon atoms may be branched, unbranched or in rings. *Alkanes*, *alkenes* and *alkynes* are all aliphatic compounds.

aliquot: a measured volume taken from a liquid sample for analysis. Typically aliquots are taken by a pipette for *titration*.

alkalis are *bases* which dissolve in water. The common laboratory alkalis are the hydroxides of sodium and potassium, calcium hydroxide (in limewater) and *ammonia*. Alkalis form solutions with a *pH* above 7 so they change the colours of *acid–base indicators*. What alkalis have in common is that they dissolve in water to produce hydroxide (OH^-) ions. Sodium hydroxide (Na^+OH^-) and potassium hydroxide (K^+OH^-) contain hydroxide ions in the solid as well as in solution. Ammonia produces hydroxide ions by reacting with water. Ammonia acting as a base takes *protons* from water molecules:

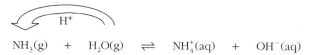

$$NH_3(g) \quad + \quad H_2O(g) \quad \rightleftharpoons \quad NH_4^+(aq) \quad + \quad OH^-(aq)$$

alkali metals: the elements in *group 1* of the periodic table. All the elements react with water to form alkaline solutions of the metal hydroxide. Sodium, for example, reacts to form *sodium hydroxide*. It is important to remember that the hydroxides are alkaline because of the presence of OH^- ions. Alkali metal ions themselves, such as sodium ions, Na^+, do not make solutions alkaline.

alkaline cell: an *electrochemical cell* in which the electrolyte is the alkali, potassium hydroxide. Alkaline cells are suitable for devices with a motor that draws a heavy or continuous current.

alkaline earth metals: the elements in *group 2* of the periodic table. The alkaline earths are the *oxides* of these metals which are much less soluble in water than the corresponding group 1 oxides.

alkaloids are organic nitrogen compounds extracted from plants. Alkaloids can have a powerful effect on the human nervous system and can be very poisonous. Caffeine, the stimulant in tea and coffee, is an alkaloid. Other alkaloids include the *drugs* quinine, morphine and codeine.

alkanes are the *hydrocarbons* which make up most of crude oil and natural gas. They are present in *fuels* such as *petrol* and *diesel fuel*.

Alkanes are *saturated compounds* with the general formula C_nH_{2n+2}. The carbon atoms in alkane molecules may be in straight chains or branched chains but all the bonds are single bonds.

Name	Molecular formula
methane	CH_4
ethane	C_2H_6
propane	C_3H_8
butane	C_4H_{10}
pentane	C_5H_{12}
hexane	C_6H_{14}
heptane	C_7H_{16}
octane	C_8H_{18}
nonane	C_9H_{20}
decane	$C_{10}H_{22}$

There are many structural *isomers* of alkanes. Alkane molecules can be represented by *structural formulae*, *displayed formulae* and *skeletal formulae*. The names of branched alkanes are based on the longest straight chain in the molecule with the positions of the side-chain *alkyl groups* identified by numbering the carbon atoms.

Alkane molecules are non-polar so they do not mix with, or dissolve in, *polar solvents* such as water. The molecules are only held together by weak *intermolecular forces*. The longer the molecules, the greater the attraction between them. The boiling points rise as the number of carbon atoms per molecule increases. Alkanes in the range C_1 to C_4 are gases at room temperature and pressure. Under the same conditions, alkanes in the range C_5 to C_{17} are liquids while those with more than 17 carbon atoms per molecule are solids. Liquid alkanes with longer chain lengths are *viscous liquids* used as lubricants.

The *bond enthalpies* for C—C and C—H bonds are high so the bonds are relatively hard to break. Also the bonds are not polar. This means that alkanes are very unreactive towards reagents in water such as acids and alkalis, as well as oxidising and reducing reagents. Three important reactions of alkanes are *combustion*, *halogenation* and cracking (see *thermal cracking* and *catalytic cracking*). Cracking and the halogenation of alkanes are examples of *free-radical chain reactions*.

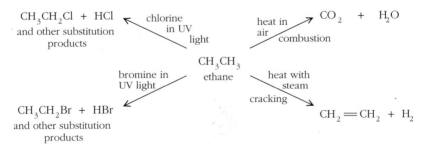

Reactions of alkanes, with ethane as the example

17

alkenes, such as ethene and propene, are products of the ***thermal cracking*** of hydrocarbons in natural gas or ***crude oil***. They are important starting points for making other chemicals because of the reactivity of the double bonds in their molecules.

Alkenes are ***unsaturated compounds***. They are hydrocarbons with the general formula C_nH_{2n}. The characteristic ***functional group*** of the alkenes is a carbon–carbon double bond. The presence of the double bond makes alkenes more reactive than ***alkanes***.

The name of an alkene is based on the name of the corresponding alkane with the ending change to 'ene'. Where necessary, a number in the name shows the position of the double bond in the structure, as in the two structural ***isomers*** but-1-ene and but-2-ene. Counting starts from the end of the chain that will give the lowest possible number in the name. The number in the name shows the first of the two atoms connected by the double bond. In but-1-ene, for example, the double bond is between the first and the second atoms in the carbon chain.

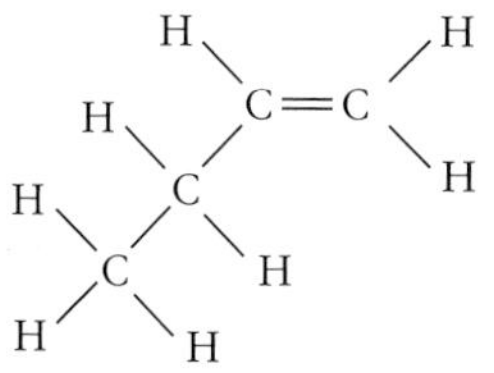

but-1-ene, C_4H_8

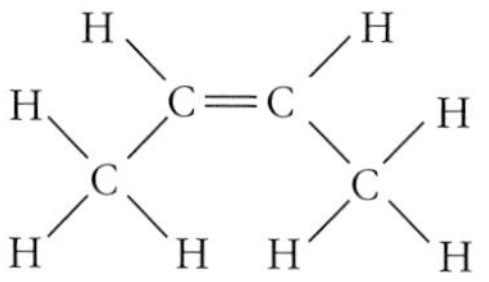

but-2-ene, C_4H_8

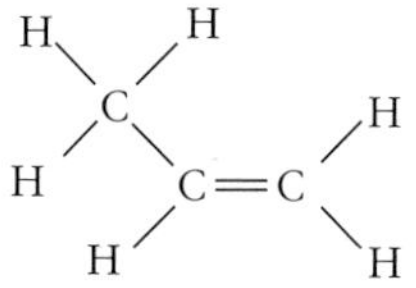

propene, C_3H_6

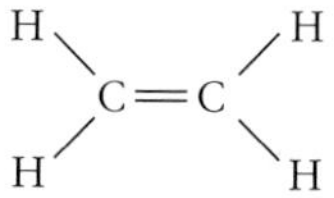

ethene, C_2H_4

Names and structures of alkenes. Note that in an ethene molecule all the atoms lie in the same plane.

The ***pi bonds*** in alkenes prevent free rotation about the double bonds. This means that some alkenes can show ***cis–trans isomerism (E–Z isomerism)***.

The boiling points of alkenes increase as the number of carbon atoms in the molecules increase. Ethene, propene and the butenes are gases at room temperature. Alkenes with more than four carbon atoms are liquids or even solids. Alkenes, like other hydrocarbons, do not mix with or dissolve in water.

The characteristic reactions of alkenes are ***addition reactions***. The reactions with reagents such as bromine and hydrogen bromide are ***electrophilic addition*** reactions. The high-pressure process for making poly(ethene) is a process of ***addition polymerisation*** that involves a ***free-radical chain reaction***.

Addition reactions of ethene

Markovnikov's rule helps to predict the main product of an addition reaction when an unsymmetrical molecule adds to an unsymmetrical alkene such as propene.

Potassium manganate(VII) oxidises alkenes. The products depend on the conditions. A dilute, acidified solution of potassium manganate(VII) converts an alkene to a diol at room temperature.

The reaction of ethene with dilute, acidified manganate(VII) ions producing ethane-1,2-diol

A solution of manganate(VII) ions is purple. The colour disappears as it reacts with an alkene. So the reaction with cold $MnO_4^-(aq)$ ions can be used to distinguish unsaturated and saturated hydrocarbons. Another test for unsaturated hydrocarbons uses a solution of bromine (see *organic analysis*).

A hot, concentrated solution of acidic potassium manganate(VII) breaks apart the double bond in an alkene.

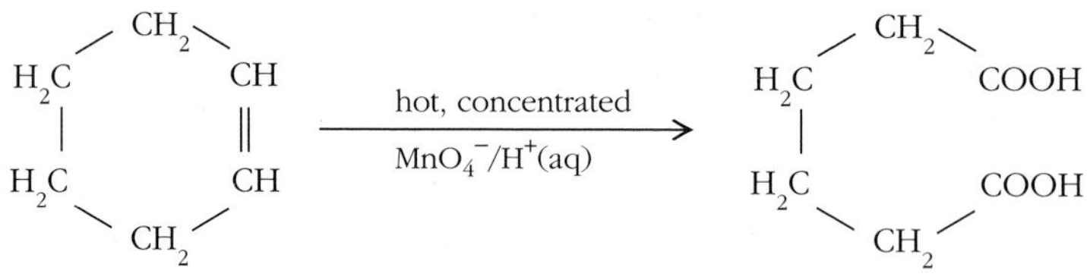

cyclohexene → hexane-1,6-dioic acid

Using hot, acidic potassium manganate(VII) to break a carbon chain at a double bond. In this example the reaction converts cyclohexene to hexane-1,6-dioic acid.

alkoxides are metal *salts* of alcohols. Ethanol, for example, reacts with *sodium* to produce sodium ethoxide.

$$C_2H_5OH(l) + Na(s) \rightarrow C_2H_5O^-Na^+(s) + H_2(g)$$

The ethoxide ion is a very *strong base*. It rapidly gains a proton if water is added, to turn back to ethanol.

Sodium methoxide is used as the catalyst for the production of *biodiesel* from vegetable oils.

alkyl group: a group of carbon and hydrogen atoms which forms part of the structure of a molecule. The simplest example is the methyl group, CH_3-, which is methane with one hydrogen atom removed. In general, alkyl groups are *alkane* molecules minus one hydrogen atom.

Alkyl group	Formula
methyl	CH_3-
ethyl	CH_3CH_2-
propyl	$CH_3CH_2CH_2-$
butyl	$CH_3CH_2CH_2CH_2-$

A useful shorthand for any alkyl group is the capital letter R. Further alkyl groups are then represented by R′ or R″. So, for example a tertiary *amine* with three different alkyl groups attached to the nitrogen atom can be written as:

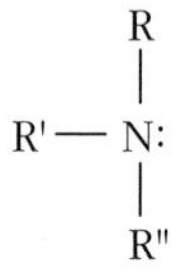

$$R' - \underset{\underset{R''}{|}}{\overset{\overset{R}{|}}{N}}:$$

General formula of a tertiary **amine**

alkylation is a reaction which introduces an *alkyl group* into a molecule by an *addition reaction* or a *substitution reaction*. Industrially the alkylation of alkenes produces branched alkanes needed to raise the octane number of *petrol*.

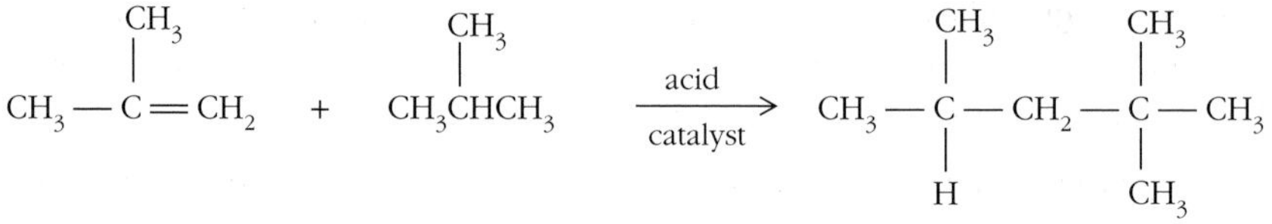

methylpropene + methylpropane → 2,2,4-trimethylpentane

Production of a branched alkane by alkylation of an alkene

The Friedel–Crafts reaction adds alkyl groups to benzene rings in arenes.

alkynes are hydrocarbons with a triple bond between two carbon atoms. They are *unsaturated compounds* with the general formula C_nH_{2n-2}. The best known example is *ethyne* (historically known as acetylene) which is the first member of the series.

allotropes are different forms of the same element in the same physical state. *Oxygen* has gaseous allotropes with different molecular structures – normal diatomic oxygen (O_2) and *ozone* (O_3). *Sulfur* has solid allotropes with the same molecular structure (S_8) but different crystal structures – rhombic and monoclinic sulfur.

Carbon has allotropes consisting of different giant structures of atoms – diamond and graphite. Carbon also exists in molecular form as the *fullerenes*.

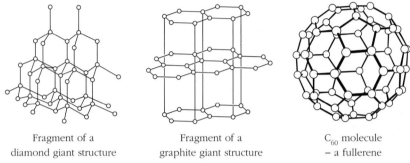

| Fragment of a diamond giant structure | Fragment of a graphite giant structure | C_{60} molecule – a fullerene |

Allotropes of carbon

Other elements with allotropes are *phosphorus* and *tin*.

alloys are usually mixtures of metals, but *steel* is an alloy of iron with the non-metal carbon. Alloys are often more useful than pure metals. Controlling the composition of an alloy makes it possible to vary its properties. Examples of alloys are:

- mild steel – an alloy of iron with 0.12–0.25% carbon; it is relatively cheap, easily rolled into sheets and can be pressed into shape
- *brass* – an alloy of copper with less than 20% zinc; it is easily worked, has a gold colour and does not corrode
- *bronze* – an alloy of copper with up to 12% tin; it is strong, hardwearing and resistant to *corrosion*
- solder – an alloy of lead and tin which melts at a low temperature and can be used to join metals
- aluminium alloys – alloys of aluminium, with elements such as copper, zinc, manganese, silicon or magnesium; they are much lighter and more resistant to corrosion than mild steel, but not quite as corrosion resistant as pure aluminium.

alpha decay is the emission of an alpha particle from the nucleus of an *atom* during *radioactive decay*. Alpha (α) particles consist of two protons and two neutrons and are identical with the nuclei of helium atoms. An alpha particle emitted from a radioactive nucleus is usually represented as $_2^4He$ but chemically it is $_2^4He^{2+}$.

Alpha decay produces an atom of an element two places to the left in the periodic table because the *atomic number* of the atom decreases by two. At the same time, the *mass number* decreases by four. When an atom of uranium-238 decays by loss of an alpha particle it turns into an atom of thorium-234.

$$^{238}_{92}U \rightarrow {}^{234}_{90}Th + {}^{4}_{2}He$$

aluminium (Al) is a silvery metal. It is very useful because it has a relatively low density while being resistant to *corrosion* and is a good conductor of thermal energy and electricity. Aluminium is commonly mixed with other elements to make *alloys*.

Aluminium is the second element in group 3, coming below boron in the periodic table. The *electron configuration* of aluminium is $[Ne]3s^23p^1$. Aluminium has a cubic *close-packed giant structure* with the atoms held together by *metallic bonding*.

Aluminium is a reactive metal with a *standard electrode potential* $E^{\ominus} = -1.66$ V for the $Al^{3+}(aq)|Al(s)$ half-cell. This puts aluminium above zinc in the *electrochemical series*. A thin, tough and transparent layer of aluminium oxide on the surface of the metal stops most reactions. As a result aluminium does not corrode in air or dissolve in acids as might be expected. The oxide layer is so important that it is often thickened by *anodising*.

The oxide layer protects the metal from dilute acids but aluminium is attacked by fairly concentrated hydrochloric acid.

$$2Al(s) + 6HCl(aq) \rightarrow 2AlCl_3(aq) + 3H_2(g)$$

The oxide layer and the metal itself dissolve in concentrated, strong alkalis.

$$2Al(s) + 2OH^-(aq) + 6H_2O(l) \rightarrow 2[Al(OH)_4]^-(aq) + 3H_2(g)$$

On heating, aluminium reacts directly with non-metals, including oxygen, sulfur, nitrogen and the halogens. Aluminium combines very strongly with oxygen and is used to reduce the oxides of other metals in the *thermite reaction.*

Important aluminium compounds include *aluminium oxide*, the hydroxide and *aluminium chloride*.

In all its compounds aluminium is in the *oxidation state* +3. The relatively high charge and small size of the Al^{3+} ion helps to account for the differences between aluminium compounds and the compounds of metals in group 1 and group 2. In some ways aluminium compounds have properties more characteristic of non-metals than metals. Compared with the larger Na^+ ions, Al^{3+} ions have a strong *polarising power*, giving rise to *polar covalent bonds* rather than *ionic bonding*. Only the fluoride and oxide are ionic.

Aluminium(III) ions in aluminium salts are hydrated in aqueous solution and are present as a *complex ion* $[Al(H_2O)_6]^{3+}$. Six water molecules form *coordinate bonds* with each metal ion. The hydrated aluminium ion is a proton donor. It is as strong an *acid* as ethanoic acid. (See the *hydrolysis of metal aqua ions.*)

aluminium extraction: *aluminium* is obtained by *electrolysis* of a solution of aluminium oxide in molten cryolite, Na_3AlF_6. Pure aluminium oxide for the process is obtained by purifying *bauxite*.

The discovery that aluminium oxide dissolves in molten cryolite was essential to the development of the process because the pure oxide melts at 2015°C – much too high for economic industrial processing.

Electrolysis takes place in carbon-lined steel tanks. The carbon lining is the *cathode* of the cell. The *anodes* are blocks of carbon. The currents used are high, of the order of 200 000 A, so the process is generally carried out where electricity is relatively cheap, often close to a source of hydroelectric power.

    Reduction at the cathode: $Al^{3+}(l) + 3e^- \rightarrow Al(l)$

The aluminium is liquid at the temperature of the molten electrolyte (970°C) and it collects at the bottom of the pot. The molten metal is tapped off from time to time.

    Oxidation at the anode: $2O^{2-}(l) \rightarrow O_2(g) + 4e^-$

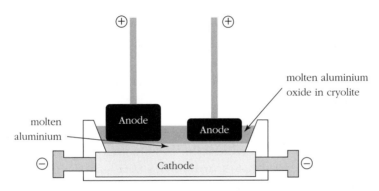

Cross-sectional diagram of an electrolysis cell for extracting aluminium

Much of the oxygen reacts with the carbon of the anodes, forming carbon dioxide. The anodes burn away and have to be replaced regularly. The energy from the burning anodes helps to heat the electrolyte, keeping it at the working temperature around 960°C.

Waste gas from the cell contains fluorides and has to be thoroughly cleaned to avoid pollution of the region surrounding the plant.

aluminium chloride ($AlCl_3$) is an off-white solid which *sublimes* on heating and fumes in moist air because of *hydrolysis* to hydrogen chloride and the hydroxide. Aluminium chloride is manufactured on a large scale by injecting a stream of chlorine gas into molten aluminium at 800°C. At this temperature the product vaporises and is collected as a solid in condensers.

Aluminium atoms in $AlCl_3$ have only six electrons and have a strong tendency to accept two more. As a result, aluminium chloride vapour contains Al_2Cl_6 dimers.

A dot-and-cross diagram for $AlCl_3$ and the bonding in an Al_2Cl_6 dimer. Lone pairs on chlorine atoms form dative covalent bonds with aluminium atoms.

Solid $AlCl_3$ has a layer lattice with bonding intermediate between ionic and covalent.

$AlCl_3$ is a strong acceptor of electron pairs, making it a powerful Lewis acid.

Aluminium chloride is the catalyst for the *Friedel–Crafts reaction* and other related reactions. In industry this means that it is an important catalyst for the manufacture of a wide range of useful products including *polymeric materials*, pigments, pharmaceuticals, *dyes* and *detergents*.

aluminium oxide (Al_2O_3) is a white solid with a very high melting point. Aluminium oxide exists naturally as a group of minerals called corundum. Corundum is an important abrasive; its hardness is 9 on the *Mohs scale of hardness*. Emery is a greyish-black variety of corundum made of aluminium oxide mixed with iron oxide minerals. Some gemstones are varieties of corundum, including rubies which are red because of the presence of some chromium(III) ions in place of aluminium(III) in the crystal structure. Sapphires are blue because some of the aluminium ions are replaced by a mixture of titanium(IV) and cobalt(II) or iron(II) ions.

Aluminium oxide is an *amphoteric compound* but the pure solid is insoluble and this property is more easily demonstrated by adding alkali to a solution of aluminium(III) ions to precipitate aluminium hydroxide. The white hydroxide dissolves in excess alkali or in acid. The amphoteric behaviour is used in the production of the pure oxide from *bauxite*.

alums are *double salts* which crystallise easily from solutions in water forming attractive octahedral crystals. Alums have the general formula $M(I)M(III)(SO_4)_2.12H_2O$. $M(I)$ is an ion with a 1+ charge such as Na^+, K^+ or NH_4^+ and $M(III)$ is a metal ion with a 3+ charge such as Al^{3+}, Cr^{3+} or Fe^{3+}.

amides are organic nitrogen compounds derived from *carboxylic acids* by replacing the —OH group with an —NH_2 group.

Amides are white crystalline solids at room temperature, apart from methanamide which is a liquid. The simple amides are soluble in water.

ethanamide   benzamide

Names and structures of amides

Amides form rapidly at room temperature when *acyl chlorides* or *acid anhydrides* react with ammonia. Amines react in a similar way to form *N*-substituted amides.

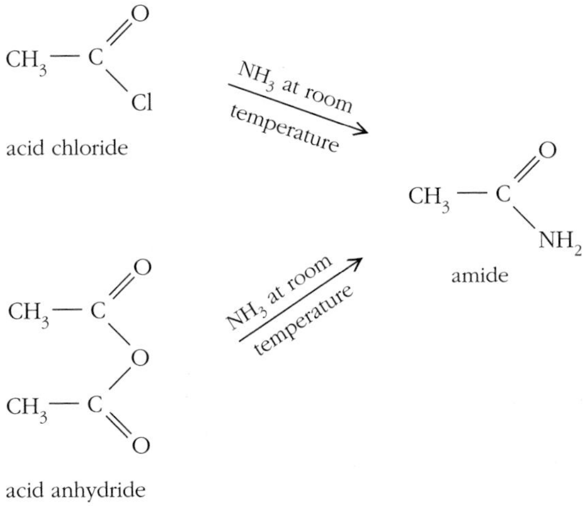

CH$_3$—C(=O)Cl

acid chloride

→ NH$_3$ at room temperature →

CH$_3$—C(=O)NH$_2$

amide

CH$_3$—C(=O)—O—C(=O)—CH$_3$ (acid anhydride)

→ NH$_3$ at room temperature →

Reactions which produce amides

Another way of making amides is to convert a carboxylic acid to an ammonium salt, and then to dehydrate the salt by heating it under reflux for several hours.

$$CH_3COOH + NH_3 \rightarrow CH_3COO^-NH_4^+ \rightarrow CH_3CONH_2 + H_2O$$

carboxylic   ammonia   ammonium salt   amide
acid

Unlike *amines*, amides are very weak bases. They give neutral solutions in water and do not form salts with acids such as hydrochloric acid. The lone pair of electrons on the nitrogen atom is delocalised with the carbonyl group making it unavailable to form a dative bond to a proton.

CH$_3$ —— C(=O)NH$_2$

Delocalisation of electrons in the amide group

Heating with phosphorus(v) oxide dehydrates an amide to a *nitrile*.

Heating with dilute acid or dilute alkali hydrolyses amides to the corresponding acid.

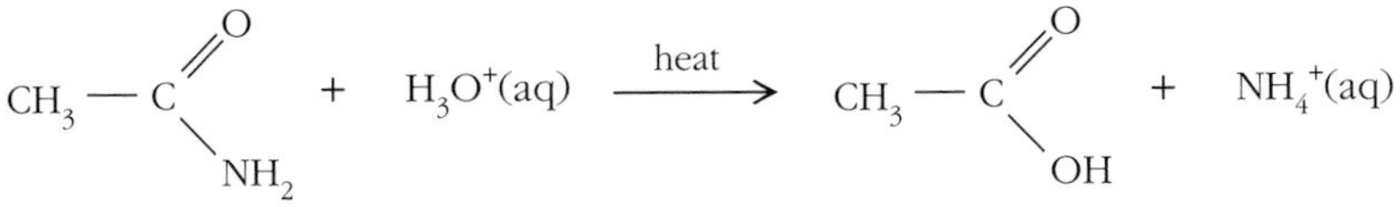

CH$_3$—C(=O)NH$_2$ + H$_3$O$^+$(aq) —heat→ CH$_3$—C(=O)OH + NH$_4^+$(aq)

Hydrolysis of ethanamide

amines are nitrogen compounds in which one or more of the hydrogen atoms in ammonia (NH$_3$) is replaced by an *alkyl group* or *aryl group*. The number of alkyl groups determines whether the compound is a primary amine, a secondary amine or a tertiary amine.

The names of amines are based on the nature and number of alkyl groups attached to the nitrogen atom.

$$CH_3 — \overset{\displaystyle H}{\underset{\displaystyle H}{N:}} \qquad CH_3 — \overset{\displaystyle CH_3}{\underset{\displaystyle H}{N:}} \qquad CH_3 — \overset{\displaystyle CH_3}{\underset{\displaystyle CH_3}{N:}}$$

methylamine	dimethylamine	trimethylamine
(primary)	(secondary)	(tertiary)

Names and structures of primary, secondary and tertiary amines.

Note the meaning of primary, secondary and tertiary here – it is not the same as for alcohols and halogenoalkanes.

Amines can be very smelly. Ethylamine has a fishy smell. Animal flesh smells putrescent because it gives off diamines.

Two methods of making primary amines are:
- the reduction of a *nitrile*
- the substitution reaction of ammonia with a *halogenoalkane*.

There is *a lone pair of electrons* on the nitrogen atom in primary amines. This means that, like ammonia, they can act as *bases*, form *complex ions* with metal ions and react as *nucleophiles*. Primary alkyl amines are stronger bases than ammonia (see *base strength of amines*).

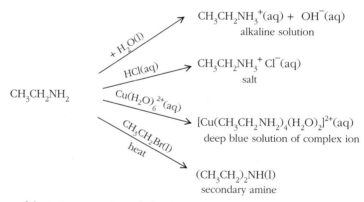

Reactions of the primary amine ethylamine

Acid anhydrides and *acyl chlorides* both acylate primary amines. For the properties of aryl (aromatic) amines see *phenylamine*.

amino acids are the compounds which join together in long chains to make *proteins*. They are carbon compounds with two *functional groups* – a basic amino group and a carboxylic acid group. Proteins consist of long chains of amino acids. There are about 20 different amino acids which link together to make proteins.

Chemists sometimes call them α-amino acids because the amino group is attached to the alpha carbon atom – the one next to the carboxylic acid group.

Amino acids are crystalline solids. They are soluble in water and crystallise as *zwitterions*. A mixture of amino acids in solution can be separated by *paper chromatography*.

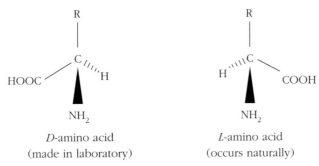

$$H_2N - \underset{\underset{H}{|}}{\overset{\overset{H}{|}}{C}} - COOH \qquad H_2N - \underset{\underset{H}{|}}{\overset{\overset{CH_3}{|}}{C}} - COOH \qquad H_2N - \underset{\underset{H}{|}}{\overset{\overset{CH_2OH}{|}}{C}} - COOH$$

glycine   alanine   serine

Examples of amino acids in proteins. Note that the amino group is attached to the carbon atom next to the carboxylic acid group. Alanine is 2-aminopropanoic acid.

Each of the amino acids in a protein has a central carbon atom attached to four other groups so, except for glycine which has two hydrogen atoms, these are *chiral compounds* which can exist in mirror image forms. The two *optical isomers* of an amino acid are traditionally designated L or D according to their configuration in three dimensions. The amino acids that occur naturally all take the L-form.

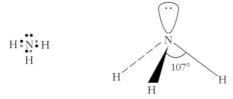

D-amino acid   *L*-amino acid
(made in laboratory)   (occurs naturally)

Mirror image forms of an amino acid

amino group: the —NH$_2$ functional group found in primary *amines* and *amino acids*. As in *ammonia*, there is a *lone pair of electrons* on the nitrogen atom. The amino group, like ammonia, can act as a *base*, form *complex ions* with metal ions and react as a *nucleophile*.

ammonia (NH$_3$) is the simplest compound of nitrogen and hydrogen. It is a colourless gas with a pungent smell. It is the only common alkaline gas. The molecule has a pyramidal shape.

Dot-and-cross diagram for ammonia and the shape of an ammonia molecule.

There are three bonding pairs and one lone pair in the outer shell the nitrogen atom. The four electron pairs point to the vertices of a distorted tetrahedron. The bond angle is less than the regular tetrahedral angle (109.5°) because repulsion between the lone pair and the bonding pairs is greater than the repulsion between bonding pairs (see *shapes of molecules*).

The properties of ammonia are affected by *hydrogen bonding*. Ammonia molecules can form hydrogen bonds with each other and with water molecules. As expected for a compound with small molecules, ammonia is a gas at room temperature but because of hydrogen bonding it is quite easily liquefied by cooling or increasing the pressure. Ammonia is also very soluble in water because of hydrogen bonding.

Many of the important properties of ammonia involve the lone pair of electrons on the nitrogen atom. The lone pair means that ammonia is:

- a *weak base*: $NH_3(g) + HCl(g) \rightarrow NH_4^+Cl^-(s)$
- a *ligand* in complex ions:

  $[Cu(H_2O)_6]^{2+}(aq) + 4NH_3(aq) \rightarrow Cu(NH_3)_4(H_2O)_2]^{2+}(aq) + 4H_2O(aq)$

- a *nucleophile*: $CH_3CH_2CH_2Br + NH_3(aq) \rightarrow CH_3CH_2CH_2NH_3^+Br^-(aq)$

Other reactions of ammonia involve the N—H bonds. In particular, the reactions in which ammonia acts as a *reducing agent*. Ammonia burns in oxygen to form nitrogen and steam. It reacts with chlorine to form nitrogen and ammonium chloride. It reduces metal oxides, such as copper(II) oxide:

$$3CuO(s) + 2NH_3(g) \rightarrow 3Cu(s) + N_2(g) + H_2O(l)$$

The catalytic oxidation of ammonia is important in *nitric acid manufacture*.

ammonia manufacture: the Haber process for the synthesis of ammonia combines nitrogen with hydrogen in the presence of an iron catalyst.

$$N_2(g) + 3H_2(g) \rightleftharpoons 2NH_3(g) \qquad \Delta H = -92.4 \text{ kJ mol}^{-1}$$

Hydrogen for the process comes from natural gas and steam (see *steam reforming*). The nitrogen comes from the air.

The reaction is very slow at room temperature. Raising the temperature increases the rate of reaction but the *reversible* reaction is *exothermic* so, according to *Le Chatelier's principle*, the higher the temperature the lower the yield of ammonia at equilibrium. A catalyst makes it possible for the reaction to go fast enough without the temperature being so high that the yield is too low. Also, according to Le Chatelier's principle, increasing the pressure raises the percentage of ammonia at equilibrium.

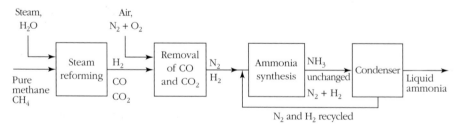

Flow diagram for the synthesis of ammonia

The process typically operates at a pressure of about 150 times atmospheric pressure and with temperatures in the range 650–720 K. Under these conditions around 20–30% of the mixture of nitrogen and hydrogen is converted to ammonia as the gases pass through the catalyst beds in the reactor.

The main uses of ammonia are in making:
- *fertilisers* (85%)
- *nitric acid*
- the *polyamide*, nylon, and processing other fibres such as cotton

The manufacture of ammonia is a major user of energy resources. About 2% of all natural gas supplies are used to make ammonia. Artificial *nitrogen fixation* by the Haber process is having a major impact on the *nitrogen cycle*.

ammoniacal silver nitrate: see *Tollens' reagent*.

ammonium hydroxide is the traditional name for 'ammonia solution' or 'aqueous ammonia'. The older name was used because a solution of *ammonia* in water contains some ammonium ions and hydroxide ions.

$$NH_3(aq) + H_2O(l) \rightleftharpoons NH_4^+(aq) + OH^-(aq)$$

However ammonia is a *weak base*, so most of the ammonia in the solution is not ionised. Aqueous ammonia is a better name, because almost all the ammonia is present as dissolved molecules linked to water molecules by *hydrogen bonds*.

ammonium ion: the ion present in ammonium salts. The ammonium ion (NH_4^+), forms when *ammonia* reacts with acids. Examples are ammonium chloride, ammonium sulfate and ammonium nitrate.

$$2NH_3(aq) + H_2SO_4(aq) \rightarrow (NH_4)_2SO_4(aq)$$

$$NH_3(aq) + HNO_3(aq) \rightarrow NH_4NO_3(aq)$$

An ammonium ion forms when the *lone pair of electrons* on the nitrogen atom in ammonia forms a dative bond with a hydrogen ion (proton) from an acid. The ammonium ion has four bonding electron pairs around the nitrogen atom, so it has a tetrahedral shape.

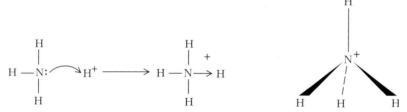

Formation of an ammonium ion and the shape of the ammonium ion. Once formed, all four bonds are the same.

Ammonium salts such as ammonium sulfate and ammonium nitrate are soluble nitrogen compounds that are manufactured and used on a large scale as *fertilisers*. Ammonium nitrate is also used as an *explosive*.

amorphous solids have no regular crystal structure; their particles are in a random jumble. *Glasses* are amorphous solids.

amount of substance: the measurement which allows chemists to find formulae, write equations and make fair comparisons between equal numbers of atoms, molecules or ions.

Amount of substance is a physical quantity (symbol n) which is measured in the unit *mole* (symbol mol). In chemistry the word 'amount' has this precise meaning. There is no measuring instrument for determining amounts directly, unlike balances for determining *masses* in kilograms or graduated glassware for measuring *volumes* in dm³ (litres). Instead, chemists first measure masses or volumes and then calculate the amount in moles given the *molar mass*, *concentration* or *molar volume* of the specified substance or *entity*.

For any pure substance:

$$\text{amount of substance (mol)} = \frac{\text{mass of substance (g)}}{\text{molar mass (g mol}^{-1})}$$

For a solution:

$$\text{amount of substance (mol)} = \frac{\text{volume of}}{\text{solution (dm}^3)} \times \frac{\text{concentration of}}{\text{solution (mol dm}^{-3})}$$

For a gas:

$$\text{amount of substance (mol)} =$$

$$\frac{\text{volume of gas at a specified temperature and pressure (m}^3)}{\text{molar volume of any gas under the same conditions (m}^3 \text{ mol}^{-1})}$$

(See also the *Avogadro constant*.)

ampere: the SI unit of current. A current of one ampere is a flow of one *coulomb* of charge per second. $1 \text{ A} = 1 \text{ C s}^{-1}$.

amphetamines are *drugs* which stimulate the central nervous system and make people feel more alert. For a while amphetamines, such as benzedrine, were used medicinally until it was found that they cause strong psychological dependence.

amphoteric compounds are substances which can behave both as *acids* and as *bases*. Water is an example. Water can both accept *protons* from acids forming *oxonium ions* and donate protons to stronger *bases* forming hydroxide ions.

Water acting as a base (taking protons from an acid):

$$H_2O(l) + \underset{\text{from acid}}{H^+} \rightleftharpoons H_3O^+(aq)$$

Water acting as an acid (giving protons to a base):

$$H_2O(l) \rightleftharpoons OH^-(aq) + \underset{\text{to base}}{H^+}$$

Alternatively, these substances are described as being amphiprotic. Other examples are the hydrogencarbonate ion (HCO_3^-) and *amino acids*.

amphoteric oxides are oxides which react both like *acidic oxides* and *basic oxides*. Examples are: ZnO, Al_2O_3, Cr_2O_3, SnO, SnO_2, PbO, PbO_2. The anhydrous oxides are often relatively *inert* to aqueous reagents so that it is easier to demonstrate their amphoteric behaviour on a test-tube scale with freshly formed samples of the metal hydroxides. Zinc hydroxide, for example, dissolves in acids to form *salts* (so acting as a *base*); it also dissolves in a solution of a strong alkali, forming zincate ions (thus acting as an *acid*).

$$Zn(OH)_2(s) + 2H^+(aq) \rightleftharpoons Zn^{2+}(aq) + 2H_2O(l)$$

$$Zn(OH)_2(s) + 2OH^-(aq) \rightleftharpoons ZnO_2^{2-}(aq) + 2H_2O(l)$$

The reaction of an amphoteric hydroxide with acids and bases can also be described in terms of the addition or removal of protons from a complex of the metal ion, such as $Al(H_2O)_3(OH)_3$. In a hydrated aluminium(III) ion the electrons in the water molecules are pulled towards the highly polarising aluminium ion, making it easier for the water molecules linked to the aluminium ion to give away protons. The hydrated ion is partially ionised in solution. The solution is acidic enough to release carbon dioxide when added to sodium carbonate.

$$[Al(H_2O)_6]^{3+} \rightleftharpoons [Al(H_2O)_5OH]^{2+} + H^+$$

$$[Al(H_2O)_5OH]^{2+} \rightleftharpoons [Al(H_2O)_4(OH)_2]^+ + H^+$$

Adding a base such as hydroxide ions to a solution of aluminium ions removes a third proton, producing an uncharged complex. The uncharged complex is much less soluble in water and precipitates as a white jelly-like precipitate – hydrated aluminium hydroxide.

$$[Al(H_2O)_4(OH)_2]^+(aq) + OH^-(aq) \rightleftharpoons [Al(H_2O)_3(OH)_3](s) + H_2O(l)$$

Adding excess alkali removes yet another proton, producing a negatively charged ion which is soluble in water so that the precipitate redissolves. This demonstrates the amphoteric properties of aluminium hydroxide.

$$[Al(H_2O)_3(OH)_3](s) + OH^-(aq) \rightleftharpoons [Al(H_2O)_2(OH)_4]^-(aq) + H_2O(l)$$

All these changes are reversed by adding a solution of a strong acid such as hydrochloric acid which turns the hydroxide ions in the complex back to water molecules.

anaerobic respiration: see *respiration*.

anaesthetics: total anaesthetics are used in medicine to induce pain-free sleep during surgery. Early anaesthetics were trichloromethane (chloroform), ethoxyethane (ether) and dinitrogen oxide (laughing gas). From 1956 the most commonly used anaesthetic was halothane ($CF_3CHBrCl$). Unfortunately the liver metabolises halothane producing toxic products which can cause hepatitis. To avoid these side effects new anaesthetics were developed based on *ethers*. Ethers are highly flammable but can be converted to safe and effective anaesthetics by replacing hydrogen atoms with fluorine atoms. The latest anaesthetics were introduced in the mid 1990s. They include Sevoflurane, FCH_2—O—$CH(CF_3)_2$.

Local anaesthetics prevent pain just in the region of the body where they are applied or injected. They work by stopping the transmission of nerve impulses without causing unconsciousness.

analgesics: drugs which relieve pain. There are mild analgesics such as *aspirin*, *paracetamol* and *ibuprofen*. Other analgesics are powerful, like the *narcotic* morphine from the opium poppy.

analytical chemistry is concerned with determining the *qualitative* and *quantitative* composition of substances.

anhydrides: see *acid anhydrides*.

anhydrous salts are the compounds left after removing the water from a *hydrated* salt. Hydrated copper(II) sulfate, for example, is blue. Heating drives off the *water of crystallisation* as steam leaving a white solid, anhydrous copper(II) sulfate.

$$CuSO_4 \cdot 5H_2O(s) \rightarrow CuSO_4(s) + 5H_2O(l)$$

aniline is the traditional name for *phenylamine*. Aniline and related compounds are used to make *azo dyes* which are therefore also called aniline dyes.

anions are negative ions attracted to the *anode* during *electrolysis*.

Charge	Anion	Symbol
1–	bromide	Br^-
	chloride	Cl^-
	ethanoate	CH_3COO^-
	hydrogencarbonate	HCO_3^-
	hydroxide	OH^-
	iodide	I^-
	nitrate [nitrate(v)]	NO_3^-
	nitrite [nitrate(III)]	NO_2^-
	manganate(VII)	MnO_4^-
2–	carbonate	CO_3^{2-}
	oxide	O^{2-}
	sulfate [sulfate(VI)]	SO_4^{2-}
	sulfide	S^{2-}
	sulfite [sulfate(IV)]	SO_3^{2-}
	thiosulfate	$S_2O_3^{2-}$
	dichromate(VI)	$Cr_2O_7^{2-}$
3–	nitride	N^{3-}
	phosphate	PO_4^{3-}

anion tests are used in *qualitative analysis* to identify the negative ions in salts. The tests use reagents which can produce colour changes, gases and precipitates. Knowledge of the chemistry involved makes it possible to interpret the changes and identify the ions.

Anion tests

Test	Observations	Inference
Test for carbonate, sulfite and nitrite		
Add dilute hydrochloric acid to the solid salt. Warm gently if there is no reaction at first.	Gas which turns limewater milky white	carbon dioxide from a carbonate
	Gas which is acidic, has a pungent smell and turns acid–dichromate paper from orange to green.	sulfur dioxide from a sulfite
	Colourless gas given off which turns brown where it meets the air.	nitrogen oxide (NO) from a nitrite turning to nitrogen dioxide (NO_2)

Test for halide ions

Make a solution of the salt. Acidify with nitric acid, then add silver nitrate solution. Test the solubility of the precipitate in ammonia solution.	White precipitate soluble in dilute ammonia solution.	chloride
	Cream precipitate soluble in concentrated ammonia solution.	bromide
	Yellow precipitate insoluble in excess ammonia.	iodide

Test for sulfate and sulfite ions

Make a solution of the salt. Add a solution of barium nitrate or chloride. If a precipitate forms add dilute nitric acid.	White precipitate which redissolves on adding acid.	sulfite
	White precipitate which does not redissolve in acid.	sulfate

Test for nitrates

Make a solution of the salt, add sodium hydroxide solution and then a piece of aluminium foil or a little Devarda alloy. Heat.	Alkaline gas evolved which turns red litmus blue. Pungent smell.	ammonia from a nitrate (or nitrite)

anisotropic solids have properties that depend on the direction in which they are measured. Examples of highly anisotropic substances are fibrous materials such as asbestos, and layer structures such as mica. The carbon atoms in graphite are arranged in layers (see *allotropes*). The properties of large single crystals of graphite show the differences between measurements parallel to the planes of atoms and at right-angles to the planes of atoms.

annealing is a process of controlled heating and cooling used to modify the properties of a material, for which purpose it is kept for a time in a furnace at a temperature below its melting point. Annealing is used to remove internal stresses from *glass* objects after blowing or casting. Annealing is an important part of the heat treatment used to modify *crystal structures of metals* and hence control their properties.

anode: the *electrode* at which *oxidation* takes place. During *electrolysis* the external power supply removes electrons from the anode so that it becomes the positive electrode, attracting negative ions; the negative anions lose electrons (oxidation) at the anode, thus turning into atoms or molecules.

The term anode is also sometimes used in *electrochemical cells*, where the oxidation process at this electrode in one of the *half-cells* forces electrons onto the electrode, which becomes the negative terminal of the cell.

Note that electrons flow out of the anode to the external circuit both during electrolysis and when a current is drawn from a chemical cell.

anodising is an electrolytic process for thickening the oxide layer on the surface of *aluminium*. It is an example of anodic *oxidation*. In the process, an aluminium sheet or component is the *anode* of an *electrolysis* cell containing sulfuric acid. The newly formed oxide layer can absorb dyes so that the surface of anodised aluminium is easily coloured. The thicker oxide film helps to protect the metal, so anodised aluminium is more resistant to *corrosion*.

antacids are ingredients of indigestion tablets taken to neutralise acid in the stomach. After a meal the cells lining the stomach produce gastric juice which contains, among other things, a mixture of hydrochloric acid and enzymes to digest food. Antacids are **bases** which neutralise acids. Examples are sodium hydrogencarbonate, calcium carbonate, aluminium hydroxide and magnesium hydroxide.

anthropogenic changes are brought about by human activities. The term is often used in connection with activities that affect the environment, including those which add **greenhouse gases** to the atmosphere.

antibiotics are chemicals which prevent **bacteria** growing or kill them. Thus antibiotics can cure bacterial diseases. Most antibiotics are the products of micro-organisms. Some of the antibiotics used in medicine are natural antibiotics which have been modified chemically, such as the semi-synthetic penicillins. Other antibiotics are now made entirely by chemical synthesis. The misuse of antibiotics in medicine and agriculture has led to the emergence of antibiotic-resistant bacteria.

The **antibiotic** penicillin destroys bacteria by disrupting their cell walls. Penicillin molecules resemble the shapes of molecules used to make the membrane around bacterial cells. The resemblance is enough for enzymes to build penicillin molecules into the structure. But penicillin lacks the part of the natural molecules which **cross-links** the structure. The walls are fatally weakened so that the bacterial cells break open and die.

antifreeze is a chemical added to the cooling system of engines to prevent the water freezing and damaging the engine during frosty weather. Antifreeze lowers the **freezing point** of water well below 0°C. Ethane-1,2-diol is used as antifreeze because it mixes freely with water but has a sufficiently high boiling point (198°C) so that it does not evaporate from the coolant when the engine is hot. Ethane-1,2-diol has a high boiling point and mixes freely with water because it has two —OH groups allowing extensive **hydrogen bonding** both between molecules of the diol and between the diol and water molecules. Another advantage of adding ethane-1,2-diol to the water in cooling systems is that it helps to prevent **corrosion** of iron and steel.

anti-knock additives prevent knocking in engines by raising the octane number of **petrol** and preventing the fuel from igniting too early. From the 1920s the main anti-knock additive was a volatile lead compound tetraethyl lead, $Pb(C_2H_5)_4$. Lead compounds in exhaust gases are toxic and so poison the metal catalyst in **catalytic converters**. Petrol companies now produce fuels with high **octane numbers** by refining the fuel further and adding oxygenates.

antioxidants are **food additives** which prevent, or slow down, oxidation reactions which can reduce the nutritional value or taste of food. Antioxidants are particularly important for preventing foods which contain **fats** and **vegetable oils** becoming rancid. There are three main types of antioxidants:
- BHA (E320) and tocopherols (E306) which interrupt the propagation of the **free-radical chain reactions** that lead to rancidity
- reducing agents such as vitamin C (E300) that remove or reduce the concentration of oxygen
- chemicals such as salts of **EDTA** (E386) which form **chelates** and reduce the concentration of free metal ions in solution

antiseptics are chemicals which kill micro-organisms, but, unlike **disinfectants**, can safely be used on the skin. **Phenol** was the first antiseptic used in surgery, by Joseph Lister in 1857. Lister found that a spray of phenol controlled the growth of bacteria in open wounds but it also made the wounds difficult to heal so he moved on to other methods. Dettol and TCP are improved antiseptics related to phenol.

aqua ions of metals are *complex ions* formed between metal ions and water molecules in aqueous solution. Dative covalent bonds hold the water molecules to the metal ions.

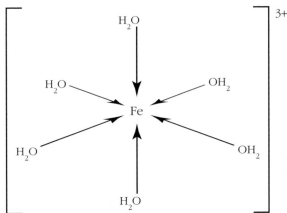

Structure of a hydrated iron(III) ion showing the octahedral arrangement of water molecules

aqua regia was well known to alchemists for its ability to dissolve noble metals such as gold. Hence the name, which is Latin, meaning 'royal water'. It is a mixture of three parts concentrated **hydrochloric acid** and one part concentrated **nitric acid**. The nitric acid oxidises gold atoms to Au^{3+} ions which then form complex ions, $[AuCl_4]^-$, because of the high concentration of chloride ions.

aqueous solution: a solution of one or more *solutes* dissolved in water. The state symbol (aq) in *chemical equations* shows that a *species* is in aqueous solution.

arenes are *hydrocarbons*, such as *benzene* and *methylbenzene*, with rings of carbon atoms stabilised by *delocalised electrons*. Traditionally, chemists have called the arenes 'aromatic hydrocarbons', ever since Kekulé was struck by the fragrant smell of oils such as benzene. In the modern name, the 'ar-' comes from 'aromatic' and the ending '-ene' means that these hydrocarbons are unsaturated compounds like the *alkenes*.

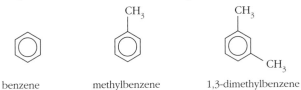

benzene   methylbenzene   1,3-dimethylbenzene

Structures of arenes

The circle in the benzene ring represents six delocalised electrons. This representation explains the shape and **stability of benzene**. The **Kekulé structure** can be more helpful when describing the mechanism of the reactions of benzene.

As well as arenes related to benzene, there are others with fused ring systems such as naphthalene and anthracene.

Structure of naphthalene with fused rings, showing both the Kekulé structure and delocalisation. The delocalised electrons are shared across both the rings.

Arenes are non-polar. The forces between the molecules are weak *intermolecular forces*. The boiling points of arenes depend on the sizes of the molecules. The bigger the molecules, the higher the boiling points. Benzene and methylbenzene are liquids at room temperature. Naphthalene is a solid.

The important reactions of benzene and other arenes are very different from *alkenes* because of the delocalised electrons. Instead of *electrophilic addition* reactions, the useful changes are *electrophilic substitution* reactions.

argon (Ar) is the third member of the family of *noble gases* coming below neon in the periodic table with the *electron configuration* $[Ne]3s^2 3p^6$.

The element is a colourless, odourless gas consisting of single atoms. Argon makes up about 0.9% of dry *air* but it was not identified until the late nineteenth century because of its inertness.

aromatic hydrocarbons: the traditional term for *arenes*.

aromatic nitro compounds: see *benzene* and *phenylamine*.

aromatic substitution: another name for *electrophilic substitution* in arenes.

Arrhenius equation: an equation which describes how the *rate constant* for a reaction varies with temperature and makes it possible to determine the *activation energy* for the reaction. The *collision theory* of reaction rates helps to account for the form of the equation.

$$k = Ae^{-E_A/RT}$$

k is the rate constant for the reaction and E_A is the activation energy, R is the gas constant and T the temperature in kelvin. A is another constant.

Putting the equation into logarithmic form (using *logarithms* to base e) makes it easier to use:

$$\ln k = \ln A - \frac{E_A}{RT} = \text{constant} - \frac{E_A}{RT}$$

This equation gives a straight line if $\ln k$ is plotted against $1/T$. The gradient of the line is $-E_A/R$.

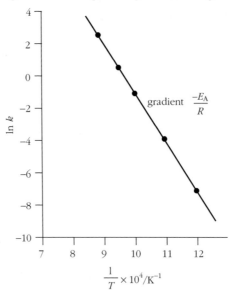

Plot of ln k against 1/T for a reaction

Arrhenius theory of acids: this theory explains the behaviour of acids in terms of hydrogen ions. What *acids* have in common, according to this theory, is that they produce hydrogen ions when they dissolve in water. The theory is illustrated by the ionisation of hydrogen chloride in water.

$$HCl(g) \rightarrow H^+(aq) + Cl^-(aq)$$

This theory was first proposed by the Swedish chemist Svante Arrhenius (1859–1927).

arsenic (As) is a greyish-white *metalloid* which forms toxic compounds. The element comes below phosphorus in group 5 of the *periodic table* and has the *electron configuration* $[Ar]3d^{10}4s^24p^3$.

aryl group: a group of carbon and hydrogen atoms which forms part of the structure of a molecule. The simplest example is the *phenyl group*, C_6H_5— which is *benzene* with one hydrogen atom removed. In general aryl groups are arene molecules minus one hydrogen atom.

ascorbic acid (vitamin C) is a white crystalline compound found dissolved in the juices of fresh fruit and vegetables. Lack of ascorbic acid in the diet leads to scurvy. Ascorbic acid is relatively soluble in water and may be washed out of foods on cooking. It is thermally unstable and is progressively destroyed during cooking.

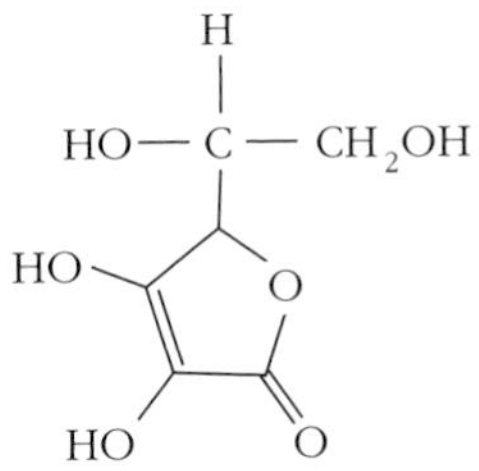

Structure of ascorbic acid

The ascorbic acid molecule does not contain a free *carboxylic acid* group because its carboxyl group reacts with its hydroxyl group, eliminating water to form a ring compound. Ascorbic acid is a good *reducing agent*. It is added to some processed foods as preservative; as an *antioxidant*, it prevents oxidation of other food components.

aspirin is the common name for the most widely used medical drug. Chemically it is the ethanoyl (acetyl) *ester* of 2-hydroxybenzoic acid (salicylic acid):

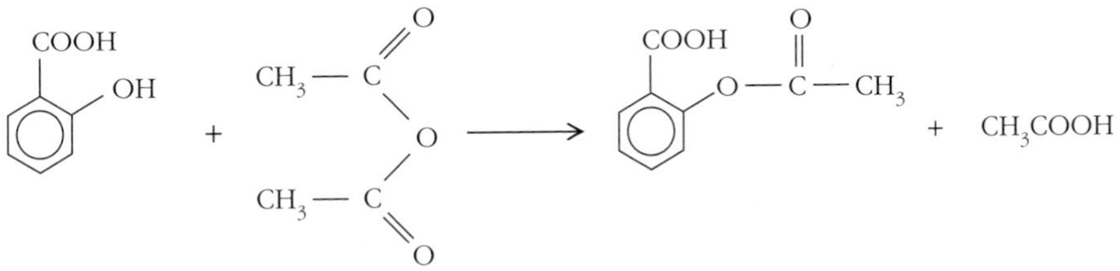

Formation of aspirin by acylation of 2-hydroxybenzoic acid with ethanoic anhydride

Aspirin is a mild painkiller (*analgesic*); it also helps to lower fevers and reduce inflammation of joints. Large-scale studies have shown that people who have had a heart attack are less likely to have a second attack if they take half an aspirin tablet a day. The main side effect of taking aspirin is stomach bleeding.

Children under 12 years old should not take aspirin because of the slight risk of a rare illness called Reye's syndrome, which damages the liver and brain. The risk is greater if children have a viral infection.

assaying is a process of quantitative analysis which determines how much of a given sample is the substance indicated by its name. So a piece of gold is assayed to see what proportion of the metal sample is gold. Gold and silver articles are assayed before being hallmarked. A range of assaying techniques has been developed to determine the purity or biological activity of drugs, enzymes and other substances which affect living things.

astatine (At) is the rarest element in the halogen family. It is the element below iodine in group 7 of the periodic table. It has the electron configuration [Xe]$4f^{14}5d^{10}6s^{2}6p^{5}$. There are twenty known isotopes of astatine, all of which are highly radioactive. The longest lived is astatine-210 with a half-life of 8.3 hours. The element was discovered in 1940 when it was prepared artificially by bombarding bismuth with alpha particles in a cyclotron.

asymmetric molecules are molecules with no centres, axes or planes of symmetry. Asymmetric molecules are *chiral compounds* and can exist in distinct mirror image forms giving rise to *optical isomerism*. Any molecule containing a carbon atom with four different groups or atoms attached to it is asymmetric.

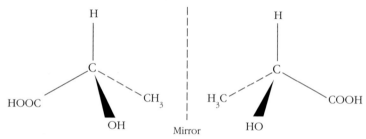

The mirror image forms of the asymmetric molecule, 2-hydroxypropanoic acid (lactic acid)

atactic polymer: a form of addition polymer, such as poly(propene), in which the side groups along the polymer chain are randomly orientated. Atactic poly(propene) is an amorphous, rubbery polymer of little value, unlike poly(propene) in the form of an *isotactic polymer.*

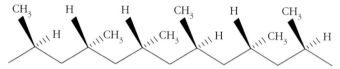

Part of a chain of atactic poly(propene)

atmosphere: the zone around the Earth which contains all the gases in the *air.* By convention the upper limit of the atmosphere is taken to be 1000 km above the Earth's surface. However, because of gravity most of the air is concentrated in the lower regions. About 50% by mass lies within 5.6 km of the surface and 99% within 40 km. Most of the climate and weather processes take place in the troposphere, which extends up to about 17 km.

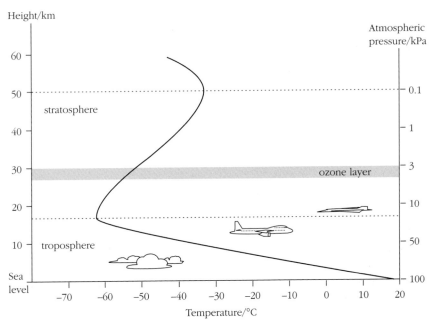

Vertical structure of the Earth's atmosphere showing how temperature and pressure vary with altitude

atmospheric pollutants are chemicals from natural and ***anthropogenic changes*** which are released into the ***air*** where they have harmful effects both locally and globally. Primary pollutants are converted to secondary pollutants by chemical reactions in the air.

Primary pollutant	Origin of the pollutant
carbon dioxide, CO_2	complete combustion of hydrocarbons in fuels
carbon monoxide, CO	incomplete combustion of fuels
nitrogen oxides, NO_x	reaction of nitrogen and oxygen from the air in furnaces and hot engines
sulfur dioxide, SO_2	from combustion of fuels, such as coal, which contain sulfur, also from the roasting of sulfide ores
volatile organic compounds, VOCs	from unburnt fuels, solvents and chemical processes
CFCs	from old refrigeration and air conditioning systems

Air pollutants can lead to ***acid deposition, photochemical smog*** and ***global warming*** as well as depletion of ***ozone in the stratosphere***.

atmospheric pressure usually means the ***pressure*** of the atmosphere at ground level. Standard atmospheric pressure (1 atm) = 101 325 N m⁻² = 101.325 kPa. The choice of this value dates back to the days when atmospheric pressure was normally measured by a column of mercury in a barometer. A pressure of one atmosphere was defined as the pressure needed to support a column of mercury 760 mm high.

atom: the smallest particle of an element. An atom consists of a tiny nucleus surrounded by a cloud of electrons. The nucleus consists of protons and neutrons. All the atoms of the same element have the same *atomic number*. The number of neutrons may vary and this accounts for the existence of isotopes. (See also *fundamental particles*.)

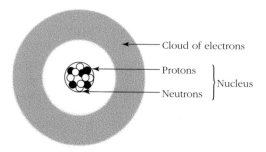

Atomic structure

Chemical reactions involve the electrons in the outer shells of atoms. As a result, the electron configuration of an element helps to account for its chemical behaviour.

Radioactive decay and the changes in nuclear reactors involve the nuclei of atoms. When describing nuclear changes it helps to use symbols which show the mass number and atomic number:

Mass number ———— $^A_Z\mathbf{X}$ ———— Element symbol
Atomic number ————

Symbols for atoms.

The atomic number gives the number of protons in the nucleus. In a neutral atom the number of electrons is the same. The difference between mass number and atomic number gives the number of neutrons in the nucleus.

Atom economy is a measure of how efficiently a chemical reaction converts the atoms in its reactants to atoms in the product. The atom economy for a reaction is calculated from the balanced equation for the reaction.

$$\text{atom economy} = \frac{\text{molar mass of the atoms in the product}}{\text{molar mass of the atoms in all the reactants}} \times 100\%$$

The *yield* of a reaction is not the only indicator of its efficiency. Even with a high yield a chemical process can produce much unwanted waste products if the atom economy is low. One of the aims of **green chemistry** is to develop processes with high atom economies.

Worked example

Calculate the atom economy for the reaction of benzene with oxygen to form *cis*-butenedioic anhydride (maleic anhydride).

Notes on the method

Start by writing the equation for the reaction, then look up (or work out) the molar masses of the product and the reactants. Finally, substitute in the formula for atom economy.

Answer

Molar mass	Molar mass	Molar mass
$= 78 \text{ g mol}^{-1}$	$= 4.5 \times 32 \text{ g mol}^{-1}$	$= 98 \text{ g mol}^{-1}$
	$= 144 \text{ g mol}^{-1}$	

$$\text{atom economy} = \frac{98 \text{ g mol}^{-1}}{78 \text{ g mol}^{-1} + 144 \text{ g mol}^{-1}} \times 100\% = 44\%$$

Even if the reaction gives 100% yield, only 44% of the mass of reactants ends up in the desired product. The rest goes to waste. Note that some wastes are more harmful than others. Carbon dioxide, for example, is a greenhouse gas.

Some types of organic reaction give better atom economies than others:
- *addition reactions* have an atom economy of 100% because all the reactant atoms join together
- *substitution reactions* and *condensation reactions* have atom economies less than 100% because there are other products.

atomic absorption spectroscopy is a technique of quantitative analysis for measuring the amounts of elements in samples such as blood, soils and food. The procedure depends on the fact that different atoms absorb radiation at particular frequencies.

The sample is vaporised and split into gaseous atoms by a very hot flame at around 2000°C. Light from a special lamp produces radiation with a wavelength absorbed by atoms of the element to be measured.

By measuring the absorbance at a particular wavelength it is possible to determine the amounts of certain elements in the sample. The procedure is *calibrated* using samples of known concentration.

Atomic absorption spectroscopy provides a sensitive and reliable method for determining more than 60 elements. Examples include the determination of traces of mercury in the environment, lead in blood, or heavy metals in the effluent from a factory. The technique is widely used in commercial laboratories, for example in the water industry.

atomic emission spectroscopy is a very sensitive method of analysis used to identify and measure the elements in a sample. Atomic emission spectra of pure elements also provide the evidence needed to determine the *energy levels* and *electron configurations* in atoms (see *atomic spectrum* and *hydrogen emission spectrum*).

Carrying out a *flame test* and examining the coloured light from the flame with a hand-held spectroscope is a simple demonstration of this type of spectroscopy.

The first step in a more sophisticated analysis is to vaporise the sample using a very hot flame or an electric spark to atomise the sample and produce a gas consisting of free atoms. Energy from the flame or spark also excites some of the electrons in the atoms so that they jump to higher energy levels. As the electrons drop back to lower levels they emit radiation.

41

The radiation passes through a prism or diffraction grating to produce the *atomic spectra* of the elements in the sample.

Important applications of atomic emission spectroscopy include the measurement of sodium, potassium, lithium and calcium ions in biological fluids such as blood and urine.

atomic number is the number of protons in the nucleus of an *atom*. The alternative term is proton number.

atomic orbitals are the subdivisions of the electron shells in atoms. The main shells divide into subshells labelled s, p, d and f. The subshells are further divided into atomic orbitals.

Each orbital is defined by its:
● energy
● shape
● direction in space.

There is one orbital in the first shell, four in the second shell, nine in the third shell and 16 in the fourth shell. Each orbital can contain up to two electrons.

The study of the *atomic spectrum* of an element makes it possible to determine the energies of its atomic orbitals. The terms 'energy level' and 'orbital' are often used interchangeably.

The energies of atomic orbitals in atoms

The labels s, p, d and f are left over from early studies of atomic spectra which used the words sharp, principal, diffuse and fundamental to describe different series of lines. These terms have no significance in an advanced level course. The pattern of energy levels accounts for the electron configurations of atoms and the arrangement of elements in the periodic table.

The shapes of orbitals are derived from theory. The shapes are determined by solving a mathematical equation (the Schrödinger wave equation) which makes it possible to calculate the probability of finding an electron at any point in an atom.

The probable location of the electrons in the first shell is spherical. It is an example of an *s orbital* (1s).

The four orbitals in the second shell are made up of one s orbital (2s) and three dumbbell shaped *p orbitals*. The three p orbitals ($2p_x$, $2p_y$, $2p_z$) are arranged at right angles to each other:

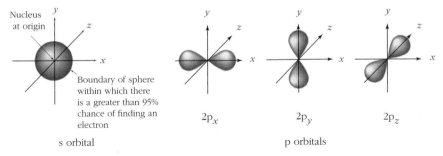

Electron density plots derived from theory showing the shapes of s and p orbitals

atomic radius: this is a measure of the size of an atom in a crystal or molecule. Chemists use *X-ray crystallography* and other techniques to measure the distance between the nuclei of atoms. The atomic radius of an atom depends on the type of bonding and on the number of bonds. Usually atomic radii for metals are calculated from the distances between atoms in metal crystals (metallic radii). Atomic radii for non-metals are calculated from the lengths of covalent bonds in crystals or molecules (*covalent radii*).

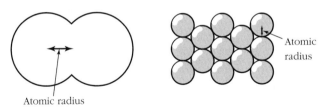

Atomic radii in a molecule and in a crystal

Atomic radii increase down any group in the periodic table as the number of electron shells increases. In *group 1* the metallic radii rise from 0.157 nm for lithium to 0.272 nm for caesium.

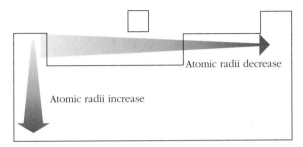

Trends in atomic radii in the periodic table

Atomic radii decrease from left to right across a period. Across the period Na to Ar, atomic radii decrease from 0.191 nm for sodium to 0.099 nm for chlorine. From one element to the next

43

across a period, the charge on the nucleus increases by one as the number of electrons in the same outer shell increases by one. **Shielding** by electrons in the same shell is limited so the 'effective nuclear charge' increases and the electrons are drawn more tightly to the nucleus.

atomic spectrum: the pattern of lines seen with a spectroscope when the atoms of an element emit radiation. Atoms emit radiation when excited by heat or electricity. This type of spectrum is the basis of **atomic emission spectroscopy**.

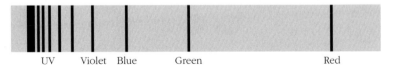

| UV | Violet | Blue | Green | | Red |

*The line **hydrogen emission spectrum** in the visible and ultraviolet regions*

Some elements emit radiation in the visible region of the spectrum and these are the elements which give coloured flames in **flame tests**.

Quantum theory explains why atomic emission spectra consist of a series of sharp lines. Each line in an emission spectrum corresponds to an energy jump of a definite size as electrons drop back from a higher to a lower energy level.

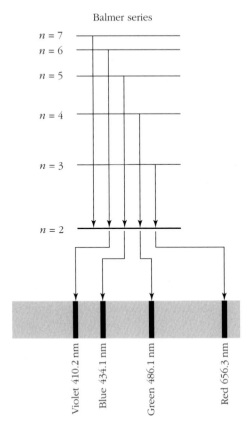

Electron jumps between energy levels in the hydrogen atom giving rise to the visible lines in the emission spectrum

Atoms also absorb radiation. If radiation passes through the vapour of an element and is viewed through a spectroscope, the *atomic absorption spectrum* appears as dark lines where particular frequencies are absorbed from the continuous spectrum.

atomic theory: the theory that all matter is made up of very tiny particles called atoms. Modern atomic theory was foreshadowed in ancient Greek philosophy but chemists now trace the development of the theory back to the work of John Dalton. It is the ideas developed between 1800 and 1930 that are significant for the model of atoms used in advanced chemistry courses. Studies of the structure of atomic nuclei continue with the help of huge particle accelerators such as those of CERN in Switzerland.

1804 John Dalton's theory of atoms as solid spheres explains the differences between elements and compounds. In this theory all the atoms of the same element weigh the same.

1896 Henri Becquerel discovers radioactivity leading to the work of Marie and Pierre Curie who separate and identify more radioactive elements.

1897 J J Thompson discovers the electron showing that atoms are themselves made up of even smaller particles.

1911 Hans Geiger and E Marsden's alpha particle scattering experiment leads to the Rutherford model with a nucleus surrounded by orbiting electrons.

1913 Niels Bohr model explains the *atomic spectrum* of hydrogen by proposing that the electrons orbiting the nucleus can only have definite energies. According to *quantum theory* atoms absorb and emit radiation at particular frequencies depending on the size of the energy jumps between *energy levels*.

1919 Francis Aston uses his mass spectrograph to demonstrate the existence of *isotopes* which had earlier been proposed by Soddy to describe atoms of the same element which have different masses.

1923 Louis-Victor de Broglie puts forward the theory that electrons have wave-like properties as well as particle properties. This theory is confirmed by experiment in 1927 when scientists in the USA and UK obtained diffraction effects with beams of electrons.

1926 Erwin Schrödinger publishes his mathematical theory of wave mechanics. The Schrödinger wave equation explains the pattern of energy levels in hydrogen atoms and gives rise to the picture of an electron in *atomic orbitals* with particular energies and shapes.

1932 James Chadwick discovers the neutron, making it possible to explain the existence of isotopes and radioactivity.

atomisation is a process in which an element or compound is converted into gaseous atoms. (See *enthalpy change of atomisation*.)

ATP (adenosine triphosphate) plays a crucial role in the energy transfer processes of living organisms. *Hydrolysis* of ATP to ADP and a phosphate ion, under the control of *enzymes*, splits off one of the three phosphate groups, releasing 31 kJ mol^{-1} to drive biochemical processes and for muscle movement.

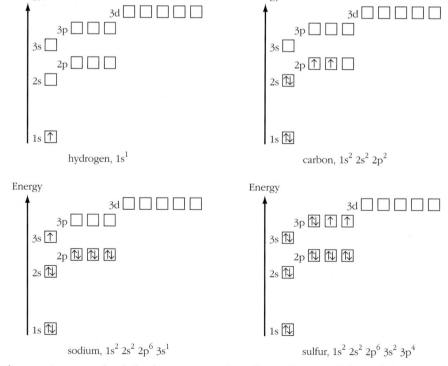

Structure of ATP

attacking group is a term used in describing the mechanism of chemical reactions. *Nucleophiles* and *electrophiles* are examples of attacking groups.

aufbau principle: the principle that the *electron configurations* of atoms build up according to a set of rules. The three rules are that:

● electrons go into an *orbital* at the lowest available *energy level*
● each orbital can contain at most two electrons (with opposite *spins*)
● where there are two or more orbitals at the same energy, they fill singly before the electrons pair up.

hydrogen, $1s^1$

carbon, $1s^2\,2s^2\,2p^2$

sodium, $1s^2\,2s^2\,2p^6\,3s^1$

sulfur, $1s^2\,2s^2\,2p^6\,3s^2\,3p^4$

Electrons in energy levels for four atoms to show the application of the aufbau principle

autocatalysis is catalysis of a reaction by one of its products. An autocatalytic reaction starts slowly but then speeds up as the catalytic product starts to form. The oxidation of ethanedioate ions by manganate(VII) ions in acid solution is catalysed by manganese(II) ions, thus:

$$2MnO_4^-(aq) + 5C_2O_4^{2-}(aq) + 16H^+(aq) \rightarrow 10CO_2(g) + 2Mn^{2+}(aq) + 8H_2O(l)$$

At first bubbles of carbon dioxide appear very slowly but once the reaction produces some manganese(II) ions it speeds up and gas is evolved more rapidly. This is an example of a reaction using a *homogeneous catalysts*.

auxochrome: see *chromophore*.

Avogadro constant: the number of atoms, molecules, ions or other chemical *entities* in one mole of a substance. The Avogadro constant (L or N_A) = 6.02×10^{23} mol^{-1}.

amount of substance (mol) $\times$ Avogadro constant (mol^{-1}) = no. of specified entities

So 0.25 mol carbon atoms contains 1.50×10^{23} atoms while 0.5 mol oxygen molecules contains 3.10×10^{23} molecules.

Avogadro's law states that equal *volumes* of gases under the same conditions of temperature and *pressure* contain equal numbers of molecules. This means that one *mole* of any gas occupies the same volume under the same conditions. The volume occupied by one mole of a gas is the *molar volume* which varies with the conditions.

Amedeo Avogadro (1776–1856) was the Italian scientist who proposed this law to account for *Gay-Lussac's law of combining volumes*.

azeotropic mixture: a mixture of liquids which on *boiling* produces a *vapour* with the same composition as the liquid. This means both that the liquids cannot be separated by fractional distillation and that azeotropes are constant boiling mixtures. A mixture containing 95.6% ethanol with 4.4% water is an azeotrope which boils at 78.2°C. As a result it is not possible to produce pure ethanol (anhydrous or absolute ethanol) by distillation alone.

azo compound: a compound with the group —N=N— in its structure.

azo dyes are made by coupling a *diazonium salt* with one of a variety of coupling agents. Benzene diazonium chloride, for example, couples with naphthalene-2-ol to make a yellow azo compound.

benzene diazonium chloride naphthalene-2-ol an azo dye

A coupling reaction to make an azo dye

Many azo *dyes* are manufactured by linking combinations of 50 diazonium salts with 52 coupling agents. Most of the dyes are red, orange or yellow. The *acid–base indicator* methyl orange is an azo dye.

back titration is an analytical technique used when the reaction between the standard solution and the substance to be analysed is slow. The procedure is to add a measured excess of the standard solution, allow time for the reaction to finish and then to use a titration with a second standard solution to measure how much of the first standard solution remains unused.

bacteria are a group of micro-organisms. Some bacteria are harmful and responsible for diseases such as cholera, tuberculosis and food poisoning. Most bacteria play a vital part in ecological processes. Soil bacteria, for example, help to break down the remains of living things, releasing the nutrient elements such as nitrogen in a form which plants can use for growth.

Bacteria are the basis of much traditional and modern *biotechnology*. In a traditional process, for example, bacteria are used to make yoghurt. In new methods they are altered by genetic engineering to produce substances such as human insulin for diabetics.

bactericide: a chemical which kills bacteria. See *chlorine water treatment* and *water treatment*.

Bakelite® is the phenol–methanal plastic discovered by the Flemish-born chemist Leo Baekland in 1905 who was then living in the USA. This was the start of the modern plastics industry. Bakelite® is a cross-linked, *thermosetting polymer* formed when phenol and *methanal* (formaldehyde) are mixed and heated with an acid *catalyst*.

balanced equations show the amounts (in moles) of reactants and products involved in chemical reactions. There is no change in the total number of atoms of each element as reactants turn into products during a chemical reaction. Four steps lead to the balanced equation for a reaction. Take for example the reaction of sodium with water:

Step 1 Identify the reactants and products by name:
sodium reacts with water to form sodium hydroxide and hydrogen.

Step 2 Write down the correct formulae for reactants and products:

$Na + H_2O \rightarrow NaOH + H_2$

Step 3 Balance the numbers of atoms of each element by inspection and by writing numbers in front of the formulae as necessary. In this example there has to be an even number of hydrogen atoms on the left (in H_2O molecules) so the number on the right must be even too:

$2Na + 2H_2O \rightarrow 2NaOH + H_2$

Never change any of the formulae to balance an equation. Do not add an extra formula to balance an equation without carefully checking that it is right to do so.

Step 4 Add *state symbols*:

$$2Na(s) + 2H_2O(l) \rightarrow 2NaOH(aq) + H_2(g)$$

The balanced equation shows the *amounts* (in moles) of reactants and products. In the example, 2 mol sodium atoms react with 2 mol water molecules to give 2 mol sodium hydroxide and 1 mol hydrogen. This is the *stoichiometry* of the reaction. (See also *redox reactions*.)

ball-and-stick models are used to show in three dimensions the structure and bonding of crystals, molecules and complex ions. They are better than *space-filling models* for showing bond angles, but they do not show the relative size of atoms and ions.

Balmer series: a series of lines which appear in the visible region of the *hydrogen emission spectrum*.

barium (Ba) is a silvery-white, soft metal which is one of the alkaline earth metals in *group 2* of the periodic table. Barium occurs naturally as barytes ($BaSO_4$) and also as witherite ($BaCO_3$).

Barium sulfate absorbs X-rays strongly so it is the main ingredient in 'barium meals' used to diagnose disorders of the stomach or intestines. Soluble barium compounds are toxic but barium sulfate is very insoluble and so cannot be absorbed into the bloodstream from the gut. X-rays cannot pass through the 'barium meal' which therefore creates a shadow on the X-ray film.

base: a molecule or ion which can accept a hydrogen ion (proton) from an acid. This is the usual definition based on the *Brønsted–Lowry theory*. A base has a *lone pair of electrons* which can form a *dative covalent bond* with a proton. (See also the *Lewis acid–base theory*.)

$$H \overset{\bullet\bullet}{\underset{\bullet\bullet}{\times\text{O}}}^{-} \quad \searrow H^+ \longrightarrow \quad H - \overset{\bullet\bullet}{\underset{\bullet\bullet}{\text{O}}} \rightarrow H$$

A hydroxide ion has a lone pair of electrons which can form a dative bond with a hydrogen ion

Common bases are the oxide and hydroxide ions, *ammonia* and *amines* as well as the carbonate and hydrogencarbonate ions. There is a lone pair on the nitrogen atom of ammonia which allows it to act as a base:

$$NH_3(aq) \quad + \quad HNO_3(aq) \rightarrow NH_4^+NO_3^-(aq)$$
$$\text{base} \qquad\qquad \text{acid}$$

In *biochemistry* the term base often refers to one of the five nitrogenous bases which make up nucleotides and *nucleic acids DNA* and *RNA*. These compounds (adenine, guanine, cytosine, uracil and thymine) are bases in the chemical sense because they have lone pairs on nitrogen atoms which can accept hydrogen ions.

base catalysis: any reaction speeded up by a base *catalyst*. One example is the base catalysed *hydrolysis* of a *nitrile* to the salt of a carboxylic acid.

$$CH_3CH_2CH_2 - C \equiv N \xrightarrow[\text{heat}]{H_2O/NaOH} CH_3CH_2CH_2 - C \overset{O}{\underset{O^- Na^+}{\big\backslash}}$$

Hydrolysis of butanenitrile to sodium butanoate with a base catalyst

base dissociation (ionisation) constants measure the strength of bases (see K_b)

base strength of amines: the order of base strength for organic compounds with $-NH_2$ groups is primary alkyl *amine > ammonia > phenylamine*.

Primary alkyl *amines* are stronger bases than ammonia because of the *inductive effect*. The push of electrons from the alkyl group to the nitrogen atom helps to stabilise the ion formed when the amine accepts a proton.

| methylamine | methylammonium ion |

Phenylamine is a weaker base than ammonia because the lone pair on the nitrogen atom interacts with the *delocalised electrons* of the benzene ring. This extended delocalisation stabilises the free amine and lessens the tendency of the molecule to form a dative covalent bond with a proton.

basic oxide: an oxide of a *metal* which reacts with acids to form salts and water. Copper(II) oxide, for example reacts with acids to produce copper(II) salts

$$CuO(s) + H_2SO_4(aq) \rightarrow CuSO_4(aq) + H_2O(l)$$

Note that it is the oxide ion in the basic oxide which acts as a base by taking a hydrogen ion from the acid:

$$O^{2-} + 2H^+ \rightarrow H_2O$$

Basic oxides which dissolve in water are *alkalis*. As the compound dissolves, the oxide ion, acting as a base, takes a hydrogen ion from water and forms a hydroxide ion

$$CaO(s) + H_2O(l) \rightarrow Ca(OH)_2(aq)$$

batch process: a process which produces a specified amount of a product in a single operation. In industry a batch process is used to manufacture chemicals needed in relatively small amounts so that a *continuous process* is not worthwhile. Batch processing is typically used to make fine chemicals on a scale of up to 100 tonnes per year. An industrial batch process is essentially a large-scale version of a synthesis carried out in laboratory glassware. After producing a batch of a chemical the apparatus is cleaned and used again.

Some metals are extracted by batch processes (see *titanium extraction*).

battery: two or more *electrochemical cells* connected in series make up a battery of cells. This is the strictly correct meaning of the term battery. It is a term which scientists borrowed from the army where a line of guns is still called a battery. A car battery consists of six rechargeable *lead–acid cells* connected in series to give a 12 volt supply. In everyday life, however, it is common to use the word 'battery' when referring to a single cell.

bauxite: an ore of aluminium consisting of impure aluminium oxide. Impurities are removed by heating powdered bauxite with sodium hydroxide solution. Aluminium oxide, which is an *amphoteric oxide*, dissolves but other oxides such as iron(III) oxide and titanium(IV) oxide do not. After filtering, *seed crystals* are added and hydrated aluminium oxide crystallises as the solution cools. Heating the hydrate crystals at 1000°C produces anhydrous aluminium oxide ready for *aluminium extraction*.

becquerel (Bq) is the *SI unit* for measuring radioactivity. If one atomic nucleus decays per second the activity is one becquerel.

Benedict's solution is a reagent for detecting *reducing sugars*. The deep-blue reagent contains copper(II) ions complexed with citrate ions in alkaline solution. A reducing sugar reduces the reagent to copper(I) oxide. The blue colour goes and a reddish-brown precipitate forms. The reagent is similar to Fehling's solution, but safer to use because it is less *corrosive*. Benedict's solution is also more stable and does not have to be stored as two reagents.

benzene (C_6H_6) is the simplest *arene* hydrocarbon. It is a non-polar liquid at room temperature and was once widely used as a solvent until it was realised to be a *carcinogen*.

Three ways of representing the structure of benzene. Note that all the atoms in a benzene molecule are in the same plane. The molecule is planar.

At first sight benzene is an unsaturated compound with three double bonds which should be highly reactive like the *alkenes*. Benzene is far more stable and less reactive than the *Kekulé structure* suggests (see *stability of benzene*).

The 'three double bonds' in the benzene ring are not isolated. The electron clouds in these double bonds overlap and merge to form a double doughnut-shaped electron cloud of *delocalised electrons* that sandwiches the six-sided carbon skeleton.

Like other arenes, benzene burns with a very smoky, yellow flame. The characteristic reactions of benzene are *electrophilic substitution* reactions.

Summary of the substitution reactions of benzene (see also sulfonation of benzene)

51

Functional groups attached to the benzene ring can affect its reactivity. Electrophilic substitution happens more readily if the ring is activated by attached —R, —OR or —NR$_2$ groups (where R represent and alkyl group). Reaction is slower if the ring is deactivated by —NO$_2$ or —COR groups.

Although the characteristic reactions of benzene are electrophilic substitution reactions, the compound does take part in some addition reactions. For example, an addition reaction produces cyclohexane when a mixture of benzene vapour and hydrogen gas pass over a finely divided nickel catalyst at 150°C.

$$C_6H_6(g) + 3H_2(g) \rightarrow C_6H_{12}(g)$$

benzene derivatives consist of a *benzene* ring with one or more side chains or functional groups substituted for the hydrogen atoms of the ring. Chemists continue to use a mixture of systematic and traditional names for these derivatives.

The properties of functional groups such as —OH in *phenol* or —NH$_2$ in *phenylamine* are modified when bonded directly to a benzene ring. These functional groups show their normal behaviour, however, when substituted into a side chain of compounds such as phenylmethanol (benzyl alcohol), $C_6H_5CH_2OH$.

beryllium is a strong metal with a high melting point and is much less dense than iron. The element makes useful alloys with other metals. Beryllium is the first element in *group 2* of the periodic table. Its electron configuration is [He]2s^2. The element occurs naturally as beryl, an aluminosilicate mineral, which is colourless if pure but a brilliant green in emerald, where chromium(III) ions replace some of the aluminium(III) ions.

Beryllium is not a typical group 2 element. It has similarities to aluminium in much of its chemistry. For example:
- beryllium metal, like aluminium, is coated with an inert layer of oxide which prevents further reaction with oxygen unless the metal is heated to a high temperature
- the bonding in beryllium compounds is mainly covalent
- the oxide and hydroxide are amphoteric
- the anhydrous chloride, BeCl$_2$, vaporises easily and forms dimers in the vapour (Be$_2$Cl$_4$).

BeCl$_2$(g)
molecules

BeCl$_2$(s)

The structure of beryllium chloride.

Beryllium chloride vapour consists of BeCl$_2$ molecules which are linear (see *shapes of molecules*). The beryllium atom in BeCl$_2$ can accept two electron pairs. Solid beryllium chloride has an extended chain-like structure, with chlorine atoms forming *dative covalent bonds* with beryllium atoms.

The chemistries of beryllium and aluminium show these similarities because the *polarising power* of the small Be^{2+} ion is about the same as that of the larger Al^{3+} ion which has a bigger charge. The polarising power of an element determines to a large extent the type of bonding between the element with non-metals such as oxygen and chlorine, and hence the chemical characteristics of the compounds

beta decay is the emission of a beta particle (β) from the nucleus of an *atom* during *radioactive decay*. Beta particles are electrons given off from the nuclei of unstable atoms. During beta decay a neutron turns into a proton so the *atomic number* of the atom increases by one while the *mass number* is unchanged. An atom of a new element forms. The new element is one place to the right in the *periodic table*.

$$^{228}_{88}\text{Ra} \quad \rightarrow \quad ^{228}_{89}\text{Ac} \quad + \quad ^{0}_{-1}\text{e}$$

radium   actinium   beta
                                  particle

The symbol $^{0}_{-1}\text{e}$ for a beta particle means that the mass numbers and atomic numbers balance on the two sides of the nuclear equation.

bicarbonate is the traditional name for hydrogencarbonate. Sodium bicarbonate (or 'bicarb') can be used as an ingredient in baking powder and as an alkali in *antacids*.

bidentate ligands form two *dative covalent bonds* with metal ions in *complexes*. Examples of bidentate (or 'two-toothed') *ligands* are 1,2-diaminoethane, *amino acids* and the ethanedioate ion, $C_2O_4^{2-}$. The complexes formed by bidentate ligands are examples of *chelates* which are more stable than complexes with *monodentate ligands*.

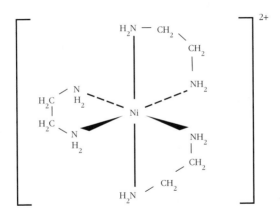

Structure of a complex formed by the bidentate ligand, 1,2-diaminoethane (en)

bifunctional compounds are compounds with two *functional groups*. Important examples are the *amino acids*.

biochemistry is the study of chemical changes in living organisms. Metabolism describes the many chemical processes in all living thing. Many of these *enzyme*-catalysed reactions are involved in the hydrolysis and oxidation of food. Others build up small molecules into the complex molecules which cells need as they grow. The techniques of molecular biology helped to transform biochemistry because they allowed scientists to work out the structures and shapes of large molecules such as *proteins* and *nucleic acids*, and hence to begin to discover the way in which genetic material in cells controls metabolism.

biodegradable materials break down in the environment due to the action of micro-organisms. Most synthetic polymers, especially polymers made by *addition polymerisation*, are not biodegradable so polymer waste does not break down in *landfill* sites.

Condensation polymers are more likely to be biodegradable by *hydrolysis* of ester or amide bonds. One example is polylactic acid (PLA) made from lactic acid (2-hydroxypropanoic acid) formed when bacteria break down cane sugar or corn starch (both *renewable resources*). In the presence of a catalyst, lactic acid molecules join up in pairs to form a cyclic compound which polymerises in the presence of a second catalyst. PLA can be recycled but it is also biodegradable. It has a range of medical applications including stitches for wounds. It is also used for packaging, waste sacks and disposable eating utensils.

$$
\left[\begin{array}{c} \underset{\displaystyle \text{O} - \text{CH} - \text{C}}{\overset{\displaystyle CH_3 \quad\; O}{\displaystyle |\quad\quad\; \|}} \end{array}\right]_n \overset{\displaystyle CH_3 \quad\; O}{\underset{\displaystyle O - \text{CH} - \text{C} -}{\displaystyle |\quad\quad\; \|}}
$$

The structure of polylactic acid. The polymer breaks down by hydrolysis of the ester links if composted under suitable conditions.

Biopol was the first totally biodegradable plastic to be manufactured for use as a plastic. Biopol is a trade name for a *polyester* made by bacteria. The bacteria make the natural polyester as an energy store. Manufacturers have developed techniques for growing the bacteria on a large scale, extracting the polymer and converting it to a useful plastic. The plastic is more expensive than poly(ethene) but micro-organisms in the soil can break it down by hydrolysing the ester links.

biodiesel is a fuel made from *renewable resources* such as the *triglycerides* in *vegetable oils* or animal *fats*. The triglyceride is heated with methanol or ethanol in the presence of a base to act as a catalyst. If the base catalyst is sodium hydroxide it can be dissolved in the alcohol before being added to the triglyceride. When the reaction is complete, the *fatty acids* are converted to methyl or ethyl esters and glycerol (propane-1,2,3-triol) separates as a by-product. Chemists call this transesterification.

$$
\begin{array}{l}
CH_2 - O - \overset{\displaystyle O}{\overset{\displaystyle \|}{C}} - R_1 \\[4pt]
CH - O - \overset{\displaystyle O}{\overset{\displaystyle \|}{C}} - R_2 \quad + \; 3CH_3OH \;\longrightarrow \\[4pt]
CH_2 - O - \overset{\displaystyle O}{\overset{\displaystyle \|}{C}} - R_3
\end{array}
\qquad
\begin{array}{l}
R_1 - \overset{\displaystyle O}{\overset{\displaystyle \|}{C}} - O - CH_3 \quad CH_2 - OH \\[4pt]
R_2 - \overset{\displaystyle O}{\overset{\displaystyle \|}{C}} - O - CH_3 \; + \; CH_2 - OH \\[4pt]
R_3 - \overset{\displaystyle O}{\overset{\displaystyle \|}{C}} - O - CH_3 \quad CH_2 - OH
\end{array}
$$

Transesterification of fatty acids in a triglyceride with methanol. R_1, R_2 and R_3 represent the hydrocarbon chains of three fatty acids.

The use of biodiesel is increasing because of the urgency of the need to reduce carbon dioxide emissions to the atmosphere (see *greenhouse gases*). The carbon dioxide released when biodiesel burns is equivalent to the carbon dioxide taken in by *photosynthesis* during plant growth. However, across the whole life cycle of these fuels they are not *carbon neutral* because of the energy resources used during the production of the fuel, including the manufacture of fertilisers, plant cultivation, harvesting, extraction of the oil and processing of the oil into fuel.

bioethanol is a fuel made by *fermentation* of sugars from *carbohydrates* obtained from plants crops or biomass. The apparent advantage of bioethanol is that it is made from *renewable resources* and is *carbon neutral*. The crops grown to make the ethanol take in carbon dioxide during photosynthesis as they grow. The same quantity of carbon dioxide is then given out as the bioethanol burns.

The problem with bioethanol in the USA is that its production from maize uses a lot of energy from *crude oil*. This includes fuel to make and run farm machinery, energy and chemicals to manufacture pesticides and fertilisers, as well as energy to process the corn crop and produce ethanol.

Increasing the use of agricultural crops to make biofuels pushes up the price of food. Instead of making bioethanol from maize, a better solution is to produce ethanol from non-food sources such as woody plants and agricultural wastes including straw. Using micro-organisms to break down cellulose to sugars does not compete with food supplies and makes it possible to process large amounts of waste. However this approach to making second generation biofuels is still at the stage of research and development.

The conditions for producing bioethanol in Brazil are more favourable. There the ethanol is manufactured by fermenting sugars from sugar cane. The refineries that make bioethanol in Brazil have the advantage that they can meet all their energy needs for heating and electricity by burning the sugar-cane waste. However there are several negative aspects of this industry in Brazil where, among other things, large-scale deforestation has been carried out to make way for sugar-cane plantations.

biofuels are fuels produced from vegetable matter and organic wastes. Biofuels include *bioethanol*, *biodiesel*, *biogas* and wood.

biogas is methane gas produced by the action of bacteria on animal and plant wastes under anaerobic conditions. Many sewage works in the UK produce biogas. Biogas also forms in *landfill* sites as the waste rots down. The gas can be hazardous unless collected and burnt. Small scale biogas digesters are used in many parts of the world to supply nearby homes with gas for cooking.

biological oxygen demand (BOD) is a measure of water *pollution* by organic matter from sources such as sewage. BOD is the quantity of dissolved oxygen (measured in mg dm^{-3}) removed from a sample of water by micro-organisms when incubated for five days at 20°C. A sample of water containing much organic matter will support large numbers of bacteria which use dissolved oxygen as they break down the pollutants.

biopolymers are naturally occurring polymers in living things. They include *carbohydrates* such as starch and cellulose, *proteins* and *nucleic acids*.

biotechnology makes use of micro-organisms or *enzymes* to produce useful products. Traditional biotechnology uses yeast to make beer, bread and wine as well as bacteria to make yoghurt. Modern biotechnology, based on genetic engineering, makes it possible to use genetically modified bacteria to produce medically important substances such as insulin, growth hormone and the blood clotting agent called factor VIII.

Biuret test: a test to detect *proteins*. The procedure is to add sodium hydroxide to a test sample followed by a few drops of copper(II) sulfate solution. The solution turns mauve if protein is present. The test actually detects *peptide* bonds.

blank determinations are used in *quantitative analysis* to eliminate *errors of measurement*. The analyst works through the whole procedure with a blank solution containing the solvent and all the reagents but none of the sample. In a *titration* a blank determination makes it possible to allow for the volume of reagent needed to change the colour of the indicator at the *end-point*. Blank determinations in *colorimetry* and *spectroscopy* help to detect, and allow for, errors caused by contamination of reagents or the effects of the apparatus used.

blast furnace: a furnace for extracting metals which depends on a blast of pre-heated air entering near the bottom of the furnace. *Iron extraction* is a large-scale, *continuous process* carried out in blast furnaces.

bleaching is a process for destroying unwanted colours by *oxidation* or *reduction*. Domestic liquid bleach is a solution of sodium chlorate(I) formed by the reaction of *chlorine* with cold sodium hydroxide solution. Bleaching powder is calcium chlorate(I), $Ca(OCl)_2$ made by absorbing chlorine gas in calcium hydroxide. These are oxidising bleaches. Another oxidising bleach is *hydrogen peroxide*. Oxidising bleaches not only remove colour but also kill micro-organisms.

Sulfur dioxide and sodium sulfite are reducing bleaches. Paper bleached white by a reducing bleach may turn yellow with age as oxygen in the air reverses the process.

Delocalised electrons in sequences of alternating double and single bonds (*conjugated systems*) are responsible for organic *coloured compounds*. Oxidising bleaches break some of the double bonds. Reducing bleaches convert double bonds to single bonds.

blood buffers closely control the pH of blood within the narrow range 7.38 to 7.42. Chemical reactions in cells tend to upset the normal pH. Respiration, for example, produces *carbon dioxide* all the time. The carbon dioxide diffuses into the blood where it is mainly in the form of carbonic acid. However the blood pH stays constant because it is stabilised by *buffer solutions*, in particular by the buffer system based on the equilibrium between carbon dioxide, water and hydrogencarbonate ions. This is the carbonic acid–hydrogencarbonate buffer.

$$CO_2(g) + H_2O(l) \rightleftharpoons H^+(aq) + HCO_3^-(aq)$$

Proteins in blood, including *haemoglobin*, can also contribute to the buffering of blood pH. This is because the molecules contain both acidic and basic functional groups.

blowing agents are used to create expanded or foam *plastics*. Blowing agents have to be affordable and chemically stable while having low flammability and low toxicity. In many cases it is important that the blowing agent improves the effectiveness of the foamed polymer as a thermal insulator. *CFCs* proved to be excellent blowing agents but they have been phased out since the discovery that they destroy *ozone in the stratosphere* and act as *greenhouse gases*. Carbon dioxide is often used as the blowing agent for *polyurethane* foams.

body-centred cubic structure: a crystal structure with an atom at each corner of a cube surrounding one atom at the centre of the cube. This structure is more open than the two *close-packed structures*.

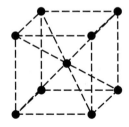

Body-centred cubic structure. Each metal atom is surrounded by eight nearest neighbours.

Metals with the body-centred cubic structure are the **group 1** metals lithium, sodium and potassium, also the **d-block elements** chromium, vanadium and tungsten.

boiling: a liquid boils when it is hot enough for bubbles of vapour to form within the body of the liquid. This happens when the **vapour pressure** of the liquid equals the external **pressure**.

boiling point: the boiling point (or boiling temperature) of a liquid varies with pressure. Raising the external pressure raises the boiling point. Boiling points are usually measured at atmospheric pressure. The normal boiling point is the temperature at which the vapour pressure of the liquid equals the standard **atmospheric pressure**.

Boltzmann distribution: see **Maxwell–Boltzmann distribution**.

bond angle: the angle between two **covalent bonds** in a molecule or **giant structure**. Instrumental analysis shows that **covalent bonds** have a definite direction as well as length.

Electron-pair repulsion theory makes it possible to predict the **shapes of molecules** and ions and bond angles with surprising accuracy.

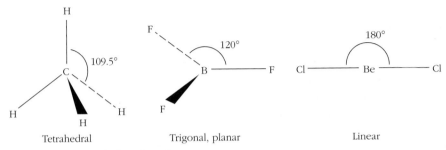

| | | |
| Tetrahedral | Trigonal, planar | Linear |

Molecules showing the bond angles

bond breaking: see **heterolytic bond breaking** and **homolytic bond breaking**. Note that it takes energy to break bonds. Bond breaking is an **endothermic** change.

bond dissociation enthalpy is the **enthalpy change** on breaking one mole of a covalent bond in a gaseous molecule. Energy is needed to break covalent bonds. Bond breaking is **endothermic** so the enthalpy change is positive.

Chemists use spectroscopy to measure bond dissociation enthalpies. The bond dissociation enthalpies for the hydrogen halides are as follows:

$$HX(g) \rightarrow H(g) + X(g)$$

$$\Delta H = +562 \text{ kJ mol}^{-1} \text{ for HF}$$
$$\Delta H = +432 \text{ kJ mol}^{-1} \text{ for HCl}$$
$$\Delta H = +366 \text{ kJ mol}^{-1} \text{ for HBr}$$
$$\Delta H = +299 \text{ kJ mol}^{-1} \text{ for HI}$$

For these compounds the bonds get easier to break as the halogen atoms get larger and the bonds get longer down **group 7**.

In molecules with two or more bonds between similar atoms, the successive bond dissociation enthalpies are not the same. In water, for example the energy needed to break the first O—H bond in H—O—H(g) is 498 kJ mol^{-1} but the energy needed to break the second O—H bond in O—H(g) is only 428 kJ mol^{-1}.

bond enthalpies (or bond energies) are the average (mean) values of *bond dissociation enthalpies* used in approximate calculations to estimate enthalpy changes for reactions. The average values of bond enthalpies take into account the facts that:

- the successive bond dissociation enthalpies are not the same in compounds such as water and methane
- the bond dissociation enthalpy for a specific covalent bond varies slightly from one molecule to another.

Worked example

Use mean bond enthalpies to estimate the enthalpy of formation of hydrazine.

Notes on the method

Write out the equation showing all the atoms and bonds in the molecules to make it easier to count the numbers of bonds broken and formed.

Look up the mean bond energies in a book of data. The symbol $E(N—H)$ stands for the bond energy of a covalent bond between a nitrogen atom and a hydrogen atom.

Answer

The equation for the reaction:

$$N \equiv N(g) + 2H — H(g) \longrightarrow \begin{array}{c} H \\ \diagdown \\ N — N \\ \diagup \\ H \end{array} \begin{array}{c} H \\ \diagup \\ \diagdown \\ H \end{array} (g)$$

The energy needed to break the bonds in the reactants

$$= E(N \equiv N) \text{ kJ mol}^{-1} + 2E(H — H) \text{ kJ mol}^{-1}$$
$$= 945 \text{ kJ mol}^{-1} + 2 \times 436 \text{ kJ mol}^{-1}$$
$$= 1817 \text{ kJ mol}^{-1}$$

The energy given out when the product forms

$$= E(N — N) \text{ kJ mol}^{-1} + 4E(N — H) \text{ kJ mol}^{-1}$$
$$= 158 \text{ kJ mol}^{-1} + 4 \times 391 \text{ kJ mol}^{-1}$$
$$= 1722 \text{ kJ mol}^{-1}$$

More energy is needed to break bonds than is given out when bonds are formed so the reaction is endothermic and the enthalpy change is positive.

$$\Delta H = +1817 \text{ kJ mol}^{-1} - 1722 \text{ kJ mol}^{-1} = +95 \text{ kJ mol}^{-1}$$

bond length: the distance between the nuclei of two atoms linked by one or more *covalent bonds*. For the same two atoms, *triple bonds* are shorter than *double bonds*, which in turn are shorter than single bonds.

Bond	Bond length/nm
C—C	0.154
C=C	0.134
C≡C	0.120
C⋯C in benzene	0.140

Bond lengths that are intermediate between single and double bonds, as in *benzene*, are a sign of *delocalised electrons*.

bond order is the number of bonds between a pair of atoms. In nitrogen N≡N the bond order is 3. In ethene $H_2C=CH_2$ the bond order between the two carbon atoms is 2 and the bond order between the carbon and hydrogen atoms is 1. The higher the bond order the shorter the *bond length*.

The bond order need not be a whole number. In benzene where there are six *delocalised electrons* over the six carbon atoms, in addition to the sigma bonds, the bond order between each pair of carbon atoms is 1.5.

bond polarity is the extent to which the shared pair of electrons in a *covalent bond* is attracted towards one of the atoms joined by the bond. Bonding electrons are pulled towards the more *electronegative* atom. The more electronegative atom carries a partial negative charge (δ−) and the other atom carries a partial positive charge (δ+). (See also *polar covalent bonds*.)

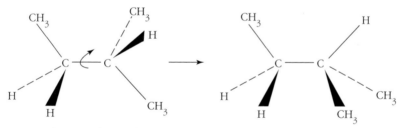

Examples of molecules with polar covalent bonds

bond rotation is possible about single covalent bonds but is prevented under normal conditions by *double bonds* or *triple bonds*. The lack of rotation about double bonds gives rise to *geometric isomerism*.

The groups at the end of a single C—C bond can rotate freely relative to each other giving rise to different conformations of molecules

bonding: see *metallic bonding*, *covalent bonds*, *ionic bonding*, *hydrogen bonding* and *intermolecular forces*.

bonding molecular orbital – see *molecular orbitals*.

Born–Haber cycle: a thermochemical cycle which can be used to calculate the *lattice energy* for a compound of a metal with a non-metal.

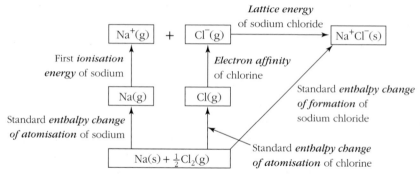

Born–Haber cycle for sodium chloride (see the separate entries in this book for definitions of the terms)

All but one of the terms in the cycle can be measured experimentally. Applying *Hess's law* makes it possible to calculate the one unknown term, which is the lattice energy (enthalpy).

A Born–Haber cycle can help reveal the factors that determine the stability of compounds. For a *stable compound*, such as magnesium chloride, the lattice energy must be greater than the total energy needed to produce gaseous ions from the elements. Removing two electrons from a magnesium atom is highly endothermic. What accounts for the stability of magnesium chloride ($MgCl_2$) is the large, negative lattice energy when the gaseous ions assemble themselves into a crystal lattice. $MgCl_3$ is not stable because of the very large third ionisation energy of magnesium. MgCl might be stable but not as stable as $MgCl_2$. $MgCl_2$ has a much larger lattice energy because an Mg^{2+} ion not only has a larger charge but is also smaller than a Mg^+ ion because it has lost both electrons in the outer shell. (In this way Born–Haber cycles can help to account for the *octet rule*.)

A Born–Haber cycle can also help to determine whether the bonding in a compound is truly ionic. The experimental lattice energy calculated from a Born–Haber cycle can be compared with the theoretical value calculated using the laws of electrostatics and assuming that the only bonding in the crystal is ionic.

Compound	Experimental lattice energy from a Born–Haber cycle/kJ mol^{-1}	Theoretical lattice energy calculated assuming that the only bonding is ionic/kJ mol^{-1}
NaCl	−780	−770
KCl	−711	−702
AgCl	−905	−833

The values in the table show that ionic bonding can account for the lattice energies of sodium and potassium chlorides. The lattice energy of silver chloride is greater than can be explained by ionic bonding. There must be a contribution from covalent bonding.

(Note that chemists also use a second type of Born–Haber cycle to analyse the energy changes when an ionic crystal dissolves in water – see *enthalpy change of solution*.)

boron (B) is a hard, black solid with a high melting point and is the only non-metal in *group 3* of the periodic table. Its *electron configuration* is $[He]2s^22p^1$. Boron is in oxidation state +3 in all its compounds. The bonding between boron and other elements is covalent.

Boron has become important as an element since the development of the nuclear industry. This is because the *isotope* boron-10 absorbs neutrons strongly.

Boron trichloride (BCl_3) is a colourless liquid, and like *aluminium chloride* is a strong electron pair acceptor (Lewis acid).

Sodium tetrahydridoborate(III) ($NaBH_4$) is a useful reducing agent.

boron nitride is *isoelectronic* with pure carbon. Like carbon, boron nitride exists in forms with different crystal structures. One form has the structure of carbon's *diamond allotrope*. This diamond-like form is one of the hardest materials known. In this form boron nitride is used to make industrial tools. It has the advantage over diamond that it does not react with metals such as iron so that it can be used to machine steel.

The other form of boron nitride has a structure like *graphite* which makes it a useful lubricant. It is a good insulator and so can be used where graphite would be unsuitable because it is an electrical conductor. It is also very inert. Unlike graphite, it does not burn.

It is also been possible to make boron nitride with structures that resemble *fullerenes* including forms that are similar to carbon *nanotubes*. These can be used as *semiconductors*.

borosilicate glass is used to make oven glassware and laboratory equipment because it has a higher melting point than cheaper, soda lime glass and is much less likely to crack as it heats up or cools down. Borosilicate glass has a high refractive index, making it useful for the production of lenses.

Boyle's law states that the volume of a fixed amount of gas is inversely proportional to its pressure at constant temperature.

$$p \; \alpha \frac{1}{V} \text{ at constant temperature for a fixed amount of a gas}$$

Hence $p_1V_1 = p_2V_2 = \text{constant}$

Brady's reagent: see *2,4-dinitrophenylhydrazine*.

brass is an *alloy* of copper (60–80%) and zinc (20–40%). Brass is easily worked, has an attractive gold colour and does not corrode.

breathalyser: an instrument for estimating the concentration of alcohol in blood. A breathalyser measures the ethanol in a sample of air from the lungs. There is an equilibrium between ethanol dissolved in blood and ethanol vapour in the air in the lungs and the constant ratio at body temperature makes it possible to infer the blood alcohol concentration.

The first successful breathalyser was based on the chemical reduction of orange dichromate(VI) ions to green chromate(III) ions as they oxidise ethanol to ethanal. The driver breathed through a tube containing the orange crystals, and the extent to which they turned yellow was *calibrated* to measure the blood alcohol concentration.

Roadside breathalysers are now *fuel cells*. At one electrode ethanol is oxidised to ethanoic acid, while at the other electrode oxygen is reduced to water. The higher the concentration of ethanol in the driver's breath, the greater the voltage of the cell.

After a roadside test, suspects are taken to a police station where the alcohol concentration in their breath is measured using a instrument which may combine the technologies of a fuel cell and *infrared spectroscopy*. If this measurement shows that drivers are above the limit, they have the right to ask for an analysis of ethanol in a blood sample, determined by *gas chromatography*.

brine is the term used industrially for a solution of sodium chloride in water. Brine is produced by 'solution mining' of underground salt deposits. Water is pumped into the salt down one pipe. The mineral dissolves and flows to the surface through a second pipe. This is the start of the chlor-alkali industry in Cheshire (see *chemical industry*).

Natural brines are solutions of a mixture of salts and can be a useful source of chemicals such as potassium, magnesium and bromine or their compounds.

bromination is a reaction which replaces a hydrogen atom in an organic molecule with a bromine atom. One example is the bromination of *alkanes* – a *free-radical chain reaction* initiated by ultraviolet radiation.

Another example is the bromination of *benzene* by bromine in the presence of iron bromide as the catalyst. This is an *electrophilic substitution* reaction (see also *methylbenzene*).

bromine (Br) is a corrosive, dark red liquid which evaporates easily to give an orange vapour. It is the only liquid non-metal at room temperature. Bromine is a halogen – the third element in *group 7* – with the electron configuration $[Ar]3d^{10}4s^24p^5$.

Bromine consists of diatomic molecules with pairs of atoms held together by single covalent bonds. The molecules are non-polar so the *intermolecular forces* are relatively weak. Bromine mixes freely with hydrocarbon solvents and the orange colour of the solution is the same as that of bromine vapour (see also *bromine water*).

Bromine, like the other halogens, is an oxidising element. It is a less powerful *oxidising agent* than chlorine.

Bromine reacts with *s-block elements* to form ionic bromides in which the bromine atoms gain one electron to fill the 4p energy levels.

$$\overset{\times\times}{\underset{\times\times}{\times}}\overset{-}{Br}{}^{\bullet}_{\times}$$

Outer electron configuration of a bromide ion

Bromine oxidises *non-metals* such as sulfur and hydrogen on heating, forming covalent, molecular bromides.

$$H_2(g) + Br_2(g) \rightarrow 2HBr(g)$$

$$H\overset{\times\times}{\underset{\times\times}{\times}}Br\times$$

Covalent bonding in hydrogen bromide

Hydrogen bromide is a fuming, acidic gas. Like the other *hydrogen halides* it is soluble in water and a strong acid.

Bromine reacts with water to only a limited extent but it reacts with sodium hydroxide rapidly *disproportionating* at room temperature to form bromate(v) ions.

oxidised from 0 to +5

$$2Br_2(aq) + 6OH^-(aq) \longrightarrow 5Br^-(aq) + BrO_3^-(aq) + 3H_2O(l)$$

reduced from 0 to –1

Bromine oxidises a range of ions or molecules in solution:

- iodide ions to iodine
- sulfite ions to sulfate ions
- thiosulfate ions to sulfate ions
- hydrogen sulfide to sulfur.

bromine extraction: a process for obtaining bromine from seawater or natural *brine*. The four-stage process concentrates and separates the bromine from seawater, which contains about 65 ppm of bromide ions. The process produces one tonne of bromine from 20 000 tonnes of seawater.

Stage 1 *Oxidation* of bromide ions to bromine.
Seawater is filtered and acidified to pH 3.5 to prevent chlorine and bromine reacting with water, then chlorine displaces bromine:

$$Cl_2(g) + 2Br^-(aq) \rightarrow 2Cl^-(aq)\ Br_2(aq)$$

Stage 2 Separation of bromine vapour.
A blast of air through the reaction mixture carries away the displaced bromine and helps to concentrate it:

$$Br_2(aq) \rightarrow Br_2(g)$$

Stage 3 Formation of hydrobromic acid.
The air with bromine vapour meets sulfur dioxide gas and a fine mist of water, producing hydrobromic acid, which is ionised in solution. After this stage the concentration of bromine in the solution is 1500 times greater than in seawater:

$$Br_2(g) + SO_2(g) + 2H_2O(l) \rightarrow 4H^+(aq) + 2Br^-(aq) + SO_4^{2-}(aq)$$

Stage 4 Displacement and purification of bromine.
The solution from stage 3 flows down a tower with a flow of chlorine gas and steam passing up it. The chlorine oxidises bromide ions to bromine which evaporates in the steam. The mixture of steam and bromine is cooled and condensed producing a dense lower bromine layer under a layer of water.

Natural brines contain more bromine than seawater so that only stage 4 is needed to separate the bromine.

Bromine is used to make flame retardants, agricultural chemicals, synthetic rubber for the inner lining of tubeless tyres, pharmaceuticals, dyes and a range of chemical intermediates.

bromine water is the traditional name for a solution of bromine in water used as a laboratory reagent. The reagent is orange because it consists largely of a mixture of bromine molecules and water molecules, hence the preferred name 'aqueous bromine'. Bromine *disproportionates* in water to a very limited extent, unlike *chlorine*.

Aqueous bromine is decolorised when shaken with excess of an *alkene* because the *addition reaction* produces colourless products. Aqueous bromine loses its orange colour when mixed with *phenol* and gives a white precipitate of tribromophenol. This is an example of *electrophilic substitution*, which happens more readily with phenol than with benzene.

Brønsted–Lowry theory: the theory that *acids* are hydrogen ion (*proton*) donors and that *bases* are hydrogen ion (proton) acceptors. Johannes Brønsted (1879–1947) was a Danish physical chemist. He published his theory in 1923 at the same time as Thomas Lowry (1874–1936) of the University of Cambridge, but the two worked independently.

The Brønsted–Lowry theory is more general than the *Arrhenius theory of acids* because it also covers reactions that happen without water. According to the theory, hydrogen chloride molecules give hydrogen ions to water molecules when they dissolve in water producing hydrated hydrogen ions called *oxonium ions*. The water acts as a base:

$$H^+$$

$$HCl(g) \quad + \quad H_2O(l) \rightarrow H_3O^+(aq) + Cl^-(aq)$$

An example of an acid–base reaction in the absence of water is the formation of a white smoke of ammonium chloride when *ammonia* gas mixes with hydrogen chloride gas:

$$H^+$$

$$NH_3(g) \quad + \quad HCl(g) \rightarrow NH_4^+Cl^-(s)$$

bronze is an *alloy* of copper with up to 12% tin. Bronze is a strong, hardwearing alloy with good resistance to corrosion. It is used to make gear wheels, bearings, propellers for ships, statues and coins.

Brownian motion is the rapid, random motion of minute particles suspended in a liquid or gas. The phenomenon was first seen by the British botanist Robert Brown looking at pollen grains in water through a microscope. Brownian motion is evidence for the *kinetic theory of gases*. The erratic random motion of pollen, smoke particles or other *colloid*-sized specks of matter is due to continuous bombardment by fast-moving molecules which are too small to be seen through a light microscope.

Buchner flask and funnel: the apparatus used for filtering solutions, especially when isolating and purifying solid products of reactions by *recrystallisation*. A pump (usually a water pump) lowers the pressure inside the flask so that a pressure difference across the filter paper speeds up filtration.

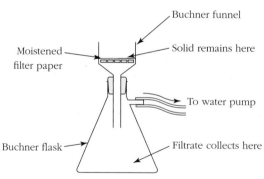

Apparatus for Buchner filtration

buckminsterfullerene: a form of carbon consisting of C_{60} particles (see *allotropes*) and named after the architect Richard Buckminster Fuller who designed the geodesic dome for Expo 67 in Montreal. The popular name for C_{60} molecules is 'buckyballs'.

Buckminsterfullerene was discovered in 1985 by Harry Kroto and his team at the University of Sussex. It was the first of a family of *fullerenes*.

C_{60} molecules are formed as a carbon vapour mixed with helium slowly cools and condenses. C_{60} is special because it is the only structure with pentagons and hexagons which is smooth because every atom has the same curvature as every other atom. Carbon atoms spontaneously assemble themselves into C_{60} and other fullerenes, including *nanotubes*.

buffer solutions are mixtures of molecules and ions in solution which help to keep *pH* more or less constant. A buffer solution cannot prevent pH changes but it evens out the large changes in pH which can happen without a buffer.

Many household products include buffer mixtures, for instance washing powders, shampoos, skin creams and a variety of medicines including eye drops. Advertisers sometimes claim that their products are 'pH balanced' if the formulation includes a buffer solution.

Buffers are important in living organisms (see *blood buffers*). Chemists use buffers when they want to investigate chemical reactions at a fixed pH.

Buffers are equilibrium systems that illustrate the practical importance of *Le Chatelier's principle*. A typical buffer mixture consists of a solution of a *weak acid* and one of its salts. For example a mixture of ethanoic acid and sodium ethanoate. There must be plenty of the acid and its salt.

$$CH_3COOH(aq) + H_2O(aq) \rightleftharpoons H_3O^+(aq) + CH_3COO^-(aq)$$

acid molecules – a reservoir of H^+ ions	stays roughly constant – so the pH hardly changes	base ions – with the capacity to accept H^+ ions

Plenty of the weak acid to supply more H^+ ions if alkali is added

Plenty of the ions from the salt to combine with extra H^+ ions if acid is added

The action of a buffer solution

By choosing the right weak acid, it is possible to prepare buffers at any pH value throughout the pH scale. If the concentrations of the weak acid and its salt are the same, then the pH of the buffer is equal to pK_a for the acid.

More generally, the pH of a buffer mixture can be calculated with the logarithmic form of the equilibrium law (see *Henderson–Hasselbalch equation*).

$$pH = pK_a + \lg\frac{[salt]}{[acid]}$$

Diluting a buffer solution with water does not change the ratio of the concentrations of the salt and acid so the pH does not change (unless the dilution is so great that the assumptions made when deriving the equation no longer apply).

Worked example

What is the pH of a buffer solution containing 0.40 mol dm^{-3} methanoic acid and 1.00 mol dm^{-3} sodium methanoate?

Notes on the answer

Look up the value of pK_a in a book of data. The pK_a of methanoic acid is 3.8.

Answer

$$pH = 3.8 + lg\frac{1}{0.4}$$
$$pH = 3.8 + lg2.5 = 3.8 + 0.4$$
$$pH = 4.2$$

bumping is violent boiling which shakes the apparatus and can throw liquid from the container in which it is being heated. Adding a few fragments of porous pottery or some jagged anti-bumping granules cuts the risk of bumping by helping the bubbles of vapour to form smoothly.

burettes are graduated tubes with taps or valves used to measure the volumes of liquids or solutions during quantitative investigations such as **titrations**. The **accuracy** of a burette, when clean and properly used, depends on the **precision** with which it is manufactured and calibrated. (See **measurement uncertainty**.)

by-products are unwanted products of chemical synthesis or manufacturing (see **ethanoic acid manufacture**). By-products may be formed by the main reaction that creates the product. The formation of by-products in this way reduces the **atom economy** of a process.

By-products may also be produced by side reactions which happen at the same time as the main reaction and lower the **yield**. In the laboratory preparation of 1-bromobutane from butan-1-ol, for example, the reagent is a mixture of sodium bromide and concentrated sulfuric acid. There are two side reactions which cut the yield. In the presence of acid, some of the butan-1-ol dehydrates to an alkene and some reacts to form an ether.

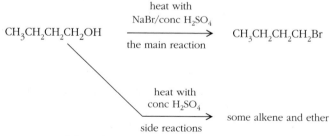

The main reaction and the side reactions.

The main reaction also produces sodium hydrogensulfate and water so that the atom economy is only 50%. Because of the side reactions, the actual yield is typically only about 80% of the theoretical yield.

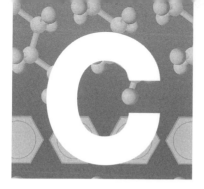

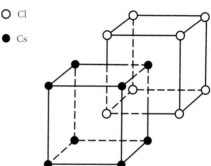

caesium chloride structure is the cubic crystal structure of the ionic compound caesium chloride (CsCl). In general it is a structure for a compound M^+X^- in which each positive ion is surrounded by eight nearest neighbours at the corners of a cube and each negative ion is similarly surrounded by eight positive ions. So the *coordination numbers* for both elements are 8.

○ Cl

● Cs

Structure of caesium chloride showing 8:8 coordination. The structure consists of a simple cubic array of positive ions interpenetrating a cubic array of negative ions.

Other compounds with this structure include CsBr, CsI and NH_4Cl.

This structure is only possible in compounds with a positive ion which is relatively large so it can be in contact with eight neighbouring negative ions. Caesium, in period 6 of the periodic table, forms larger ions than, for example, sodium. (See also *unit cell*.)

caffeine is a *drug* found in tea, coffee, chocolate and cola drinks. Caffeine gives people a 'lift' and helps them to be mentally alert. The drug is used medically to stimulate the nervous, respiratory and cardiovascular systems. Caffeine is also added to medicines to counteract sleepiness caused by other ingredients.

Caffeine can be isolated from tea or coffee as a white crystalline solid by *solvent extraction*.

Cahn–Ingold–Prelog priority rules: an elaborate set of rules for determining the names of *E–Z isomers* and *R–S isomers*. The full rules are designed to deal with complex structures. The following steps provide a simplified way of applying the rules to determine the priority for atoms and groups attached to a particular carbon atom.

1 Look at the atoms bonded to the carbon atom. The atom with the higher *atomic number* has the higher priority: Br > Cl > O > N > C > H.
2 If the atoms attached to the carbon atom are the same, then move along the atoms in the chains in parallel until there is a difference of atomic number. For example propyl, $CH_3CH_2CH_2-$, has priority over ethyl, HCH_2CH_2-, and $ClCH_2CH_2-$ has priority over HCH_2CH_2-.
3 Multiple-bonded atoms count as if they were an equal number of singly bonded atoms.

calcium (Ca) is a reactive, metallic element in group 2 of the periodic table. Its electron configuration is [Ar]4s². Samples of the silvery metal usually look grey because they are covered with a layer of calcium oxide.

Calcium burns brightly in air with a brilliant red flame forming the white solid calcium oxide (CaO). The metal also reacts with cold water producing hydrogen and a white precipitate of calcium hydroxide.

$$Ca(s) + 2H_2O(l) \rightarrow Ca(OH)_2(s) + H_2(g)$$

Calcium forms ionic compounds with non-metals in which the metal is in the +2 oxidation state as Ca^{2+}.

Calcium oxide is a white solid made by heating *calcium carbonate*. It is a *basic oxide*. Its reaction with cold water to make calcium hydroxide is highly *exothermic*. Calcium hydroxide is only sparingly soluble in water forming an alkaline solution often called limewater.

Anhydrous calcium chloride ($CaCl_2$) is a cheap drying agent. The chloride crystallises from solution as a hydrate, $CaCl_2 \cdot 6H_2O$.

calcium carbonate occurs naturally as limestone, chalk and marble. It is also present in the skeletons and shells of some organisms such as molluscs.

Limestone is an important mineral. Some of the rock is quarried for road building and construction. Pure limestone is also used in the chemical industry. Heating limestone in a furnace at 1200 K converts it to calcium oxide (quicklime). The reaction of quicklime with water produces calcium oxide (slaked lime).

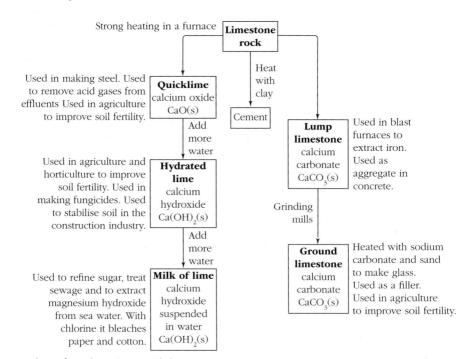

Products from limestone and their uses

calcium phosphate: in the form of hydroxyapatite, $Ca_5(PO_4)_3(OH)$, is found in bones and teeth (see also **phosphoric(v) acid**).

calibration: a procedure used in quantitative analysis. Calibration makes it possible to convert the measurement of temperature, volume or light absorption into an accurate *concentration* in moles per litre. When using colorimeters or spectrometers, for example, it is common to plot a calibration curve. Instrument readings are taken for a series of *standard solutions* of known concentration.

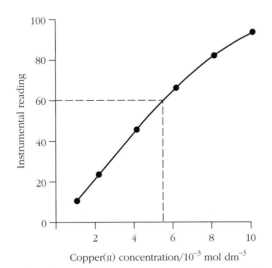

Calibration curve used to determine the concentration of copper(II) ions in aqueous solution obtained by taking instrument readings with a series of standard solutions. The instrument reading for an unknown sample is 60.0. The concentration of the sample can be read from the graph. The concentration of copper(II) ions in the sample is 5.5×10^{-3} mol dm^{-3}.

calorimeter: apparatus for measuring the energy change during a chemical reaction. Typically a calorimeter is insulated from its surroundings and contains water. The energy from the reaction heats up the water and the rest of the apparatus. An accurate thermometer measures the temperature rise. The apparatus is *calibrated* to determine its overall *heat capacity* by measuring the temperature rise for a reaction with a known *enthalpy change*.

Enthalpy changes for reactions in solution can be compared quickly using an expanded polystyrene cup, with a lid, as the calorimeter. Expanded polystyrene is an excellent insulator and has a negligible specific heat capacity. If solutions are dilute it is sufficiently accurate to assume that their density and specific heat capacity are the same as those of water.

Worked example

When 50 cm³ of 2.0 mol dm^{-3} hydrochloric acid mixes with 50 cm³ 2.0 mol dm^{-3} sodium hydroxide in a polystyrene cup the temperature rise is 13.5°C. What is the enthalpy change for the neutralisation reaction?

Notes on the method

Assume that the density of the solutions is the same as water (1 g cm^{-3}), and that for both the specific heat capacity is 4.18 J g^{-1} K^{-1}.

Note that the total volume of solution is 100 cm^3 so the mass of solution heating up can be taken as 100 g.

Use the formula that the energy change $= mc\Delta T$ where m is the mass, c the specific heat capacity and ΔT the temperature change.

The enthalpy change is the energy exchanged with the surroundings when a reaction proceeds as constant temperature and pressure. Here the energy from the exothermic reaction is kept in the system to heat the water. The calculation shows the energy which would otherwise be lost to the surroundings during a constant temperature change.

The enthalpy change, ΔH, for a reaction is the energy change for the amounts (in moles) shown in the chemical equation.

Answer

The amount of hydrochloric acid = the amount of sodium hydroxide

$$= \frac{50}{1000} \text{ dm}^3 \times 2.0 \text{ mol dm}^{-3}$$

$$= 0.1 \text{ mol}$$

Energy given out by the reaction and used to heat the water in the cup

$$= 100 \text{ g} \times 4.18 \text{ J g}^{-1} \text{ K}^{-1} \times 13.5 \text{ K}$$

$$= 5643 \text{ J}$$

Energy given out per mole of acid $= \dfrac{5643 \text{ J}}{0.1 \text{ mol}}$

$$= 56\ 430 \text{ J mol}^{-1} = 56.4 \text{ kJ mol}^{-1}$$

The reaction is exothermic, so the enthalpy change for the system is negative.

$$\text{NaOH(aq)} + \text{HCl(aq)} \rightarrow \text{NaCl(aq)} + \text{H}_2\text{O(l)} \qquad \Delta H = -56.4 \text{ kJ mol}^{-1}$$

capillary rise is an effect seen when a fine capillary tube is dipped into a liquid. The liquid normally rises up the tube. The larger the surface tension of the liquid and the thinner the tube the greater the effect.

More generally, capillary action draws liquids, especially water, through any material which is riddled with fine crevices. Capillary action draws solvent up the paper during *paper chromatography*.

carbocations are *intermediates* formed during organic reactions in which a carbon atom carries a positive charge. Carbocations form, for example, during *nucleophilic substitution* by the S$_N$1 mechanism and during *electrophilic addition* to alkenes.

Three carbocations showing the order of stability: tertiary > secondary > primary. The inductive effect of electron-releasing alkyl groups (shown by arrows on the bonds) helps to spread the positive charge and stabilise the ion. The more alkyl groups the more stable the carbocation.

carbohydrates are compounds of carbon, hydrogen and oxygen. They include sugars, starch and cellulose. The name carbohydrate arises from the formulae, in which hydrogen and oxygen are usually present in the same ratio as in water. The formula for glucose, for example, is $C_6H_{12}O_6$ which could be written $C_6(H_2O)_6$. Neither way of writing the formula gives any idea of how the atoms are arranged in a glucose molecule.

Glucose is a monosaccharide. Sucrose, with two sugar units joined together, is a disaccharide. A condensation reaction between the two sugar molecules joins them together by an *ether* link.

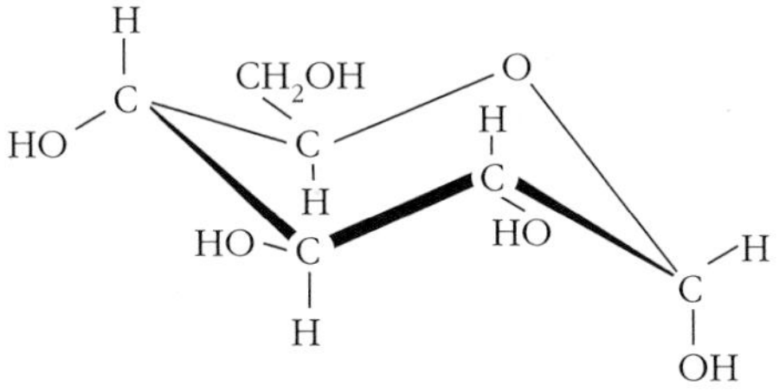

glucose

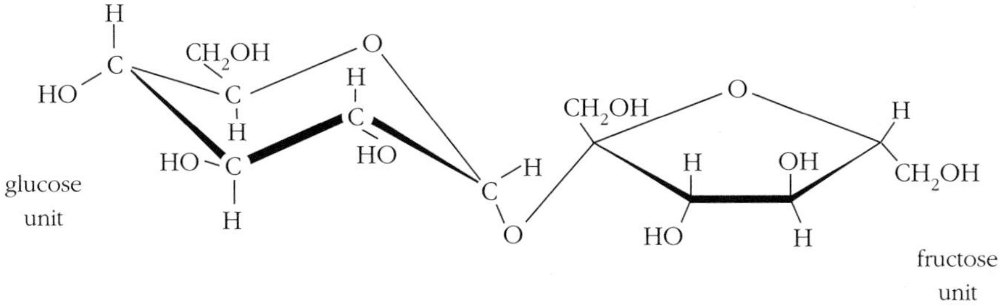

sucrose

Structures of glucose and sucrose (table sugar)

Starch and cellulose are polysaccharides. They both consist of long chains of glucose units. The difference lies in the links between the units.

Plants make glucose by *photosynthesis.* They store some of the glucose as starch, a reserve of energy food. Some of the glucose builds up the cellulose cell walls as the plants grow. When people eat parts of plants, such as potatoes or rice, they can digest starch but not cellulose. Cows feed on grass but rely on bacteria in their guts to break down the cellulose to glucose. Cows chew the cud to give bacteria time to digest the long-chain molecules.

carbon (C) is a non-metal element with three solid *allotropes* – diamond, graphite and *buckminsterfullerene* (see also *fullerenes* and *nanotubes*). *Charcoal* is an impure form of carbon. Carbon is the first element in group 4 of the periodic table. Its electron configuration is $[He]2s^2 2p^2$.

Carbon's remarkable ability to form stable chains and rings of atoms with single, double and triple bonds gives rise to *organic chemistry.*

Carbon forms two oxides both of which are colourless, molecular gases with no smell. *Carbon monoxide* (CO) is a *neutral oxide* while *carbon dioxide* is an acidic oxide.

The simplest chloride of carbon, tetrachloromethane (CCl_4), is a colourless liquid with tetrahedral molecules. Unlike many other non-metal chlorides it is not hydrolysed by water or alkalis (see *hydrolysis of non-metal chlorides*). Carbon does not react directly with chlorine, so tetrachloromethane is made by the reaction of methane with chlorine.

carbon capture and storage (CCS) is one approach being explored for reducing emissions of the *greenhouse gas*, carbon dioxide, on a large scale. This type of approach is only appropriate where there is a concentrated source of carbon dioxide in a flow of waste gas from a power station or industrial site. Also the source has to be fairly close to where the carbon dioxide is to be stored. It is then possible to capture between 85% and 90% of the gas.

Options for CCS include storage in deep geological formations such as disused oil wells, injection deep in the oceans or by reaction with metal oxides to form stable carbonate minerals.

Injection of carbon dioxide into the deep oceans is still in the research phase. So too are the methods being investigated to combine carbon dioxide with metal oxides to form carbonate minerals. Injecting carbon dioxide into oil and gas fields is now feasible and economically viable in some circumstances.

carbon cycle: the cycling of the element carbon in the natural environment as carbon compounds move between the main reservoirs of carbon in the atmosphere, the oceans and on land (see *environmental chemistry*). In the natural environment the balance between *respiration* and *photosynthesis* kept the concentration of carbon dioxide at about 280 ppm by volume from about 2000 years ago up to 1750. Since the start of the Industrial Revolution, burning of fossil fuels and other *anthropogenic changes* have increased the concentration of carbon dioxide in the air to about 387 ppm by volume. Suggested methods of limiting the build-up of carbon dioxide in the atmosphere include the use of *biofuels* and *carbon capture and storage*.

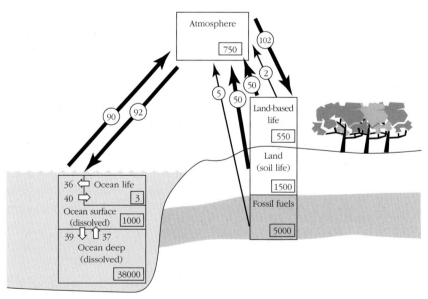

A highly simplified carbon cycle with estimates of the quantities in the main reservoirs where the figures are in gigatonnes of carbon (1 Gt = 1000 million tonnes). Also shown are estimates of the flows of carbon in gigatonnes per year (Gt yr^{-1}). The 2 Gt yr^{-1} from land-based life to the atmosphere is due to deforestation.

carbon dioxide is the acidic oxide of *carbon*. It dissolves in water forming carbonic acid which is a weak acid that ionises to hydrogen ions and hydrocarbonate ions.

$$CO_2(g) + H_2O(l) \rightleftharpoons H^+(aq) + HCO_3^-(aq)$$

The gas is strongly absorbed by alkalis such as potassium hydroxide.

Carbon dioxide exists in the supercritical state above its *critical temperature* of 31.1°C, and its critical pressure which is 73 times atmospheric pressure. Supercritical CO_2 has properties intermediate between those of a liquid and a gas. In this state it can be used as a solvent which is non-toxic and easily separated from a solution by lowering the pressure. Supercritical CO_2 is now being used to decaffeinate coffee, to dry-clean clothes and to extract perfume chemicals from natural sources. The use of supercritical CO_2 in place of other organic solvents, such as hydrocarbons or *halogenoalkanes*, is an example of *green chemistry*.

Dry ice is solid carbon dioxide. Above −79°C it *sublimes* to carbon dioxide gas. This makes it a useful coolant.

Carbon dioxide plays an important part in the *carbon cycle* and is a *greenhouse gas*.

carbon footprint: this measures the total amount of carbon dioxide, and other greenhouse gases, released into the air over the full life cycle of a process or product. This involves calculating the carbon dioxide equivalent, CO_2eq, of the other greenhouse gases in terms of the quantity of carbon dioxide that would have the same *global warming* effect.

On average the carbon footprint for one person in the UK is just under 11 tonnes of carbon dioxide per year from all activities. Of this amount, over 40 per cent comes directly from individual activities such as heating and lighting our homes and driving vehicles. The average driver in the UK produces 4 tonnes of CO_2 per year just from motoring.

carbon monoxide is a colourless gas with no smell, formed by the incomplete *combustion* of fuels containing carbon. The gas is highly poisonous because it is held much more strongly by blood *haemoglobin* than oxygen. Death results when most of the haemoglobin is blocked by carbon monoxide. Lower levels of exposure to the gas reduce the ability of blood to carry oxygen and this puts a strain on the heart.

Carbon monoxide is also an important industrial chemical, both as a cheap reducing agent and as a starting point for the synthesis of organic chemicals. Carbon monoxide is used as a reducing agent in metallurgy, for example during *iron extraction* in a blast furnace.

Steam reforming is the source of synthesis gas, which is a mixture of carbon monoxide and hydrogen.

Carbon monoxide is used on a large scale to manufacture *methanol*. A mixture of carbon monoxide and hydrogen gases combines to make methanol at about 250°C over a catalyst made from copper and zinc oxide. This is an example of an *heterogeneous catalyst*.

$$CO(g) + 3H_2(g) \rightarrow CH_3OH(l)$$

carbon neutral describes any process for which the carbon dioxide (or other greenhouse gases) released at any stage are balanced by actions which remove an equivalent amount of carbon dioxide from the atmosphere. Burning *biofuels* is carbon neutral in theory though not in practice.

carbon-13 NMR spectroscopy is a form of *NMR spectroscopy* that relies on the magnetic properties of the nucleus of carbon-13 atoms. The carbon-13 isotope makes up only about 1 per cent of all naturally occurring carbon atoms but this is enough for a signal to be detected in an NMR machine. Each peak in a carbon-13 NMR spectrum corresponds to one or more carbon atoms in a particular chemical environment.

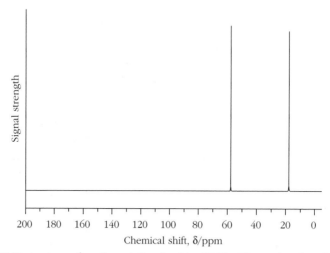

Carbon-13 NMR spectrum for ethanol dissolved in CDCl₃. The molecules of CDCl₃ contain one carbon atom that produces a line in carbon-13 spectra which is easy to recognise. This line has been removed from the spectrum.

There are two peaks in the carbon-13 NMR spectrum for ethanol. This reflects the fact that there are two carbon atoms in an ethanol molecules and they are in different environments. The carbon in the CH₃ group is attached to three hydrogen atoms and a carbon atom. The carbon in the CH₂ group is attached to two hydrogen atoms, one carbon atom and one oxygen atom.

The chemical shifts are measured relative to a reference peak at the zero mark produced by the chemical tetramethylsilane (TMS). The reference peak is not shown.

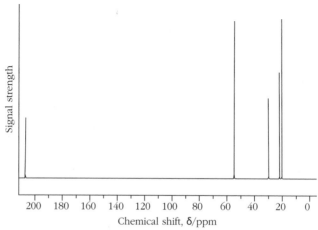

Carbon-13 NMR spectrum for 4-methylpentan-2-one dissolved in CDCl₃. There are six carbon atoms in 4-methylpentan-2-one but only five peaks in the spectrum.

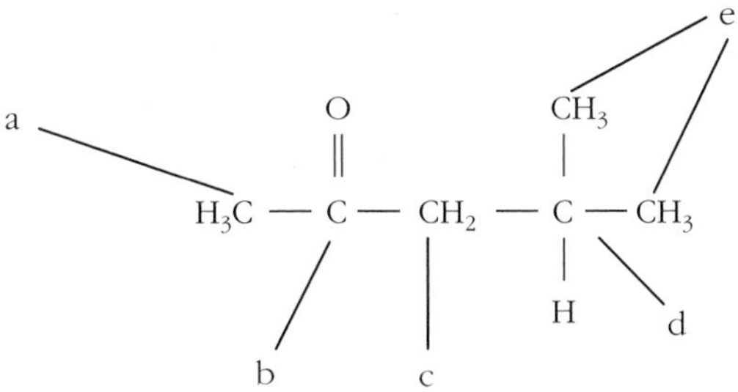

The structure of 4-methylpentan-2-one is labelled to show the five different environments for the six carbon atoms. The two carbon atoms labelled e are in exactly the same chemical environment.

Generally carbon atoms attached to electronegative atoms such as oxygen show the largest chemical shifts. Chemists have found that it is possible to draw up tables to show the likely range of values for the chemical shifts of carbon-13 atoms in different environments. For example carbon-13 atoms in alkanes typically give chemical shifts in the range 10–35 ppm while carbon atoms in the carbonyl groups of aldehydes and ketones give chemical shifts in the range 190–200 ppm.

carbonyl compounds contain the carbonyl group (C=O). The two main classes of carbonyl compounds are the *aldehydes* and the *ketones*. The C=O double bond is polar, with the electrons drawn towards the more electronegative oxygen atom. The characteristic reactions of carbonyl compounds are *nucleophilic addition* reactions and *addition–elimination reactions.*

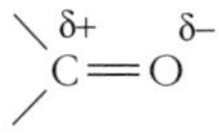

The carbonyl group in aldehydes and ketones

carboxylic acids are compounds with the formula R—COOH where R represents an *alkyl group, aryl group* or a hydrogen atom. The carboxylic acid group (—COOH) is the *functional group* which gives the acids their characteristic properties.

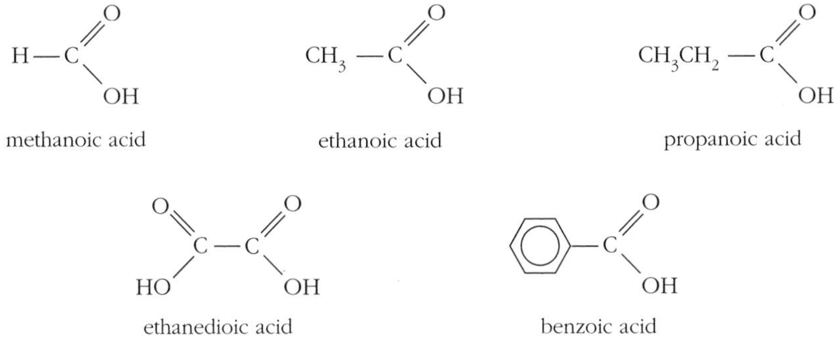

methanoic acid   ethanoic acid   propanoic acid

ethanedioic acid   benzoic acid

Names and structures of carboxylic acids

75

Carboxylic acids are sometimes called alkanoic acids. They are named by changing the ending of the corresponding *alkane* to '-oic' acid. So ethane becomes ethanoic acid.

primary alcohol $CH_3CH_2CH_2OH$ heat with acidic dichromate(VI)   *oxidation*

nitrile $CH_3CH_2C \equiv N$ heat with aqueous acid → hydrolysis $CH_3CH_2 - C \overset{O}{\underset{OH}{}}$

ester $CH_3CH_2 - C \overset{O}{\underset{OCH_3}{}}$ *hydrolysis* heat with aqueous alkali then acidify

*Reactions which produce carboxylic acids. See also **ethanoic acid manufacture**.*

Even the simplest acids, such as methanoic acid and ethanoic acid, are liquids at room temperature because of **hydrogen bonding** between the carboxylic acid groups. Also, because of hydrogen bonding, these acids mix freely with water. Hydrogen bonding leads to **dimerisation** of carboxylic acids in non-polar solvents.

Benzoic acid (benzenecarboxylic acid) is a solid at room temperature. It is only very slightly soluble in cold water but more soluble in hot water.

Carboxylic acids are **weak acids**. The —OH group in the carboxylic acid group is more acidic than the —OH group in **alcohols** because the carboxylate ion formed is stabilised by **delocalised electrons**. Carboxylic acids are sufficiently acidic to produce carbon dioxide when added to a solution of sodium carbonate (or sodium hydrogencarbonate). This reaction distinguishes carboxylic acids from weaker acids such as **phenols** (see also **acid strength of organic hydroxy compounds**).

$$CH_3 - C \overset{O}{\underset{OH}{}} \; + \; H_2O \; \rightleftharpoons \; CH_3 - C \overset{O}{\underset{O}{}} - \; + \; H_3O^+$$

Ionisation of ethanoic acid. Delocalisation of electrons stabilises the carboxylate ion and favours ionisation of the acid.

Carboxylic acids form a number of related compounds (derivatives) – *acyl chlorides, acid anhydrides, esters* and *amides*. The reactions of ethanoic acid are typical of carboxylic acids.

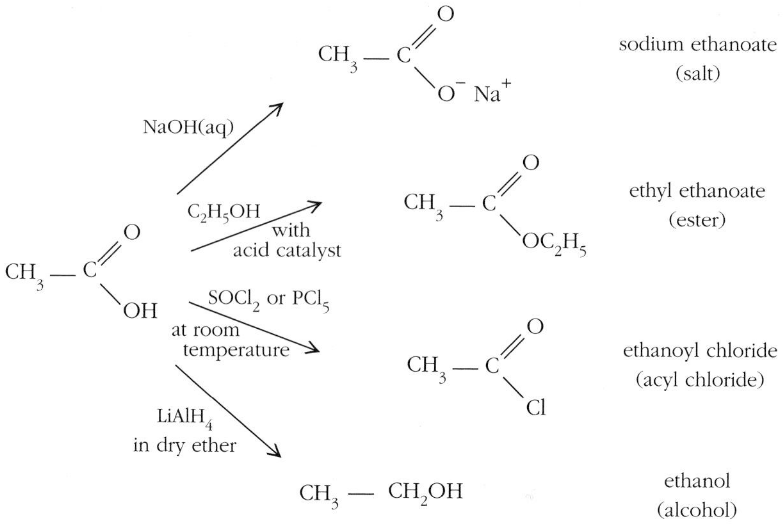

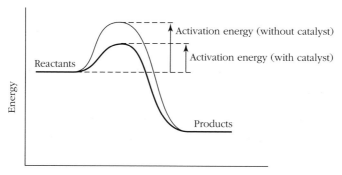

sodium ethanoate
(salt)

ethyl ethanoate
(ester)

ethanoyl chloride
(acyl chloride)

ethanol
(alcohol)

Reactions of ethanoic acid

carcinogens are compounds which can cause cancer. In a healthy body, the rate at which cells divide to produce new cells is strictly controlled. Cancer is a disorder which happens when some cells somehow escape from the normal control and multiply to produce a growth, or tumour. Chemicals that cause cancer probably produce changes (mutations) in the genes that control cell division.

catalysts speed up the rates of chemical reactions without themselves changing permanently. Catalysts can be recovered at the end of the reaction. Often a small amount of catalyst is effective.

Catalysts work by providing an alternative reaction pathway with a lower *activation energy*. A lower activation energy means that there is an increased proportion of molecules with enough energy to react at a particular temperature.

Reaction profile showing the effect of a catalyst on the activation energy of a reaction

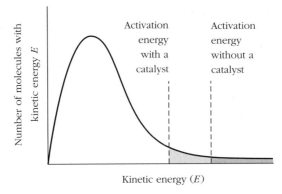

Distribution of molecular energies showing how the proportion of molecules able to react increases when a catalyst lowers the activation energy

Often a catalyst changes the mechanism of the reaction and makes it more productive by increasing the **yield** of the desired product and reducing waste. The alternative routes often involve the formation of **intermediates in the reactions** which are combinations of reacting molecules and the catalyst. These intermediates have a short life. The catalyst regenerates as the product forms. So the catalyst can go on to catalyse the reaction of more molecules. This gives rise to a continuous catalytic cycle.

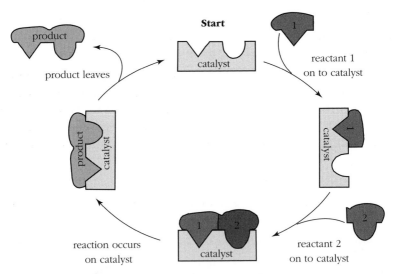

The catalyst is not used up in a catalytic cycle. A long-lasting catalyst can convert a very large amount of reactants into product before it has to be replaced. Ideally the active site on the catalyst is specific to particular reactants.

Many catalysts are specific to a particular reaction. This is especially true of **enzymes**. If the catalyst is in the same phase as the reactants it is a **homogeneous catalyst**. If the catalyst is in a different phase it is a **heterogeneous catalyst**.

Ideally a catalyst should be highly selective. This means that the reaction with the catalyst produces a high yield of the required product with a minimum of by-products. Developing more selective catalysts is one of the aims of **green chemistry**.

catalyst poisons inactivate *catalysts*, especially heterogeneous catalysts. Lead compounds in *petrol*, for example, poison the catalyst in catalytic converters and stop it working. Carbon monoxide and sulfur compounds poison the iron catalyst used for *ammonia manufacture*.

catalytic converter: a device in the exhaust system of a cars with spark-ignition engines which contains a *catalyst* to convert pollutants in the exhaust gases to less harmful substances. Car exhausts produce *atmospheric pollutants* because the engine does not burn all the fuel and because the temperature and pressure in the cylinders are high enough for nitrogen from the air to react with oxygen.

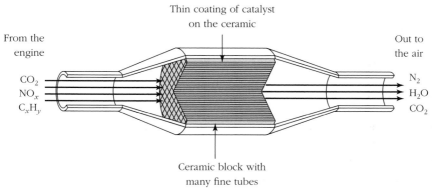

Diagram of a catalytic converter

Unleaded petrol must be used in cars fitted with a catalytic converter because lead is a *catalyst poison*.

The catalyst is a finely divided metal, or alloy, made from platinum, palladium or rhodium supported on an inert ceramic pierced with many fine tubes to give a large surface area. Once the catalyst is hot enough, it converts the pollutants to steam, carbon dioxide and nitrogen. Most vehicles are fitted with three-way catalytic converters in which three reactions take place. There are two stages. Oxygen is produced by the reduction of NO_x, in the first stage:

$$2NO_x \rightarrow xO_2 + N_2$$

The oxygen is used in the second stage to oxidise carbon monoxide and hydrocarbons:

$$2CO + O_2 \rightarrow 2CO_2$$

$$C_nH_m + (n + \frac{m}{4})O_2 \rightarrow nCO_2 + \frac{m}{2}H_2O$$

catalytic cracking: a process in an oil refinery for converting fractions from the *fractional distillation of oil* into more useful products by breaking up larger molecules into smaller ones. The problem in refining is to make the oil products in the amounts needed by customers. Generally crude oil contains too much of the high-boiling fractions (bigger molecules) and not enough of the low-boiling fractions (smaller molecules) needed for fuels such as *petrol*.

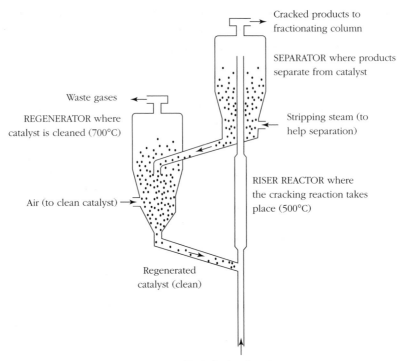

Cracked products to fractionating column

SEPARATOR where products separate from catalyst

Waste gases

REGENERATOR where catalyst is cleaned (700°C)

Stripping steam (to help separation)

RISER REACTOR where the cracking reaction takes place (500°C)

Air (to clean catalyst)

Regenerated catalyst (clean)

Fresh feedstock and steam

The catalyst powder flows to the vertical reactor where cracking takes place. The cracked vapours pass to a fractionating column.

The **heterogeneous catalyst** is a **zeolite**. Cracking takes place on the surface of the catalyst. The mechanism involves **heterolytic bond breaking** and the intermediates are **carbocations**.

Catalytic cracking is a **continuous process**. The finely powdered catalyst made of silicon and aluminium oxides gradually gets coated with carbon so it circulates through a regenerator where the carbon burns away in a stream of air.

catenation: the ability of the atoms of an element to join together in long chains or rings. *Carbon* atoms have an exceptional ability to catenate – hence the wide range of organic compounds. *Silicon*, the second element in group 4, does not have the same ability to catenate. Silicon forms **hydrides**, Si_nH_{2n+2}, analogous to the **alkanes** but the largest known molecule has **n** equal to 8, and only silane (SiH_4) is stable for any length of time.

cathode: the *electrode* at which *reduction* takes place. During *electrolysis* the external power supply adds *electrons* to the cathode so that it becomes the negative electrode attracting positive ions which gain electrons (reduction) turning into atoms or molecules.

The term cathode is also sometimes used in *electrochemical cell* where the reduction process at this electrode in one of the *half-cells* takes electrons from the electrode which becomes the positive terminal of the cell.

Note that electrons flow into the cathode from the external circuit both during electrolysis and when a current is drawn from a chemical cell.

cations are positive ions attracted to the *cathode* during *electrolysis*. (See *cation tests* for examples of common cations.)

cation tests are used in qualitative analysis to identify the positive ions in salts. Adding sodium hydroxide produces a precipitate if the metal hydroxide is insoluble. The precipitate dissolves in excess of the alkali if the hydroxide is *amphoteric*.

Adding *ammonia* solution also precipitates insoluble hydroxides. These redissolve in excess if the metal ion forms stable *complex ions* with ammonia molecules.

These tests can be supplemented with *flame tests* to identify metal ions such as sodium and potassium which do not give precipitates.

Positive ion in solution	Observations on adding sodium hydroxide solution drop-by-drop and then in excess	Observations on adding ammonia solution drop-by-drop and then in excess
calcium, Ca^{2+}	white precipitate but only if the calcium ion concentration is high	no precipitate
magnesium, Mg^{2+}	white precipitate insoluble in excess reagent	white precipitate insoluble in excess reagent
barium, Ba^{2+}	no precipitate	no precipitate
aluminium, Al^{3+}	white precipitate which dissolves in excess reagent	white precipitate insoluble in excess reagent
chromium(III), Cr^{3+}	green precipitate which dissolves in excess reagent to form a dark green solution	green precipitate insoluble in excess reagent
manganese(II), Mn^{2+}	off-white precipitate insoluble in excess reagent, precipitate turns brown on standing	off-white precipitate insoluble in excess reagent
iron(II), Fe^{2+}	green precipitate insoluble in excess reagent, precipitate turns brown m standing	green precipitate insoluble in excess reagent
iron(III), Fe^{3+}	browny-red precipitate insoluble in excess reagent	browny-red precipitate insoluble in excess reagent
cobalt(II), Co^{2+}	blue precipitate from a pink solution, insoluble in excess reagent	blue precipitate from a pink solution dissolving in excess reagent to give a blue solution
nickel(II), Ni^{2+}	green precipitate insoluble in excess reagent	green precipitate which dissolves in excess reagent to give a pinkish solution
copper(II), Cu^{2+}	pale blue precipitate insoluble in excess reagent	pale blue precipitate dissolving in excess reagent to form a dark blue solution
zinc, Zn^{2+}	white precipitate which dissolves in excess reagent	white precipitate which dissolves in excess reagent
lead, Pb^{2+}	white precipitate which dissolves in excess reagent	white precipitate insoluble in excess reagent
ammonium, NH_4^+	alkaline gas (ammonia) given off on heating	no visible change

cell: see *electrochemical cell.*

Celsius temperature scale: a scale in which the melting point of ice is set at 0°C and the boiling point of water at 100°C. The scale is based on the absolute or *Kelvin* temperature scale:

temperature in °C + 273 = absolute temperature in K

cement is manufactured by heating clay with powdered limestone in a rotating kiln and then grinding the lumps of product to a fine powder. The process gives off carbon dioxide both from the decomposition of calcium carbonate in limestone and the use of fossil fuels. Worldwide cement manufacture makes a significant contribution to *greenhouse gas* emissions.

Cement is a complex mixture of calcium silicates and calcium aluminosilicates which becomes hydrated and sets when mixed with water. Stirring cement with sand, gravel and water makes concrete.

ceramics are materials such as pottery, *glasses, cement*, concrete and graphite. Ceramics also include a wide range of crystalline materials such as barium nitride, silicon nitride, magnesium oxide, *aluminium oxide* and magnetic ferrites. What all these materials have in common is that they are non-metallic, inorganic materials which are heated to a high temperature in a furnace at some stage during their manufacture.

Typical advantages of ceramics	Typical disadvantages of many ceramics
very hard	weak in tension
strong in compression	brittle
chemically *inert*	liable to crack if there is a sudden temperature change
refractory	
electrical insulators	

CFCs (chlorofluorocarbons) such as CCl_3F, CCl_2F_2 and CCl_2FCClF_2 are compounds containing carbon, chlorine and fluorine. They have advantages – they are unreactive, do not burn and are not *toxic chemicals*. Also it is possible to make CFCs with different boiling points to suit different applications. These properties made CFCs ideal as the working fluid in refrigerators and air-conditioning units. They were also used as *blowing agents* to make the bubbles in expanded plastics and insulating foams. CFCs were valued as good solvents for dry cleaning and removing grease from electronic equipment.

The problem with CFCs (chlorofluorocarbons) and some other halogen compounds is that they are chemically very unreactive. They escape into the atmosphere where they are so stable that they last for many years, long enough for them to diffuse up to the stratosphere. In the stratosphere, the intense ultraviolet light from the Sun splits up CFCs producing compounds such as hydrogen chloride, HCl.

These chlorine compounds then undergo a series of reactions that only happen in winter in the stratosphere above Antarctica. In the dark winter over Antarctica, the stratosphere is so cold that clouds of ice crystals form. These clouds circle the Pole in a spinning vortex of air that traps the ice crystals. On the surface of the ice crystals, a whole series of reactions take place that cannot happen anywhere else in the atmosphere. The reactions are fast. The end result of the reactions produces chemicals including chlorine, Cl_2, which can destroy ozone in sunlight.

When sunlight returns to the Antarctic in the spring of the southern hemisphere, the chlorine molecules are quickly split into chlorine atoms.

Cl_2 + UV light $\rightarrow$ Cl• + Cl•

Chlorine atoms then react with ozone.

$Cl• + O_3 \rightarrow ClO• + O_2$

$ClO• + O• \rightarrow Cl• + O_2$

Overall: $O• + O_3 \rightarrow 2O_2$

The second reaction involves oxygen atoms which are common in the stratosphere and it reforms the chlorine atom. This means in effect that one chlorine atom can rapidly destroy many molecules of *ozone in the stratosphere*. This effect was noticed in the early 1980s when scientists in the Antarctic detected that the ozone concentration in the stratosphere was much lower than expected. Since then the ozone layer has been monitored by satellites, which have confirmed that there is a 'hole' in the ozone layer not only over the Antarctic but in other regions too.

Now that the damaging effects of CFCs are known they are being phased out wherever possible. The chemical industry has developed *HCFCs*, *HFCs* and other replacement chemicals.

changes of state are changes from one state of matter to another. The following are all examples of changes of state:

- a solid melting to a liquid
- a liquid freezing to a solid
- a liquid evaporating and becoming a gas
- a gas condensing to a liquid.

Energy is taken in from the surroundings during melting and evaporating (they are *endothermic* processes). The energy is needed to break the bonds between atoms, molecules or ions.

Energy is given out to the surroundings during freezing or condensing (they are *exothermic* processes). Energy is given out as the bonds between particles reform.

A few solids turn directly to gas when heated at atmospheric pressure. On cooling, the vapour turns directly back to a solid. This is *sublimation*.

charcoal is a form of *carbon* made by heating wood or bones in the absence of air. Charcoal consists of minute graphite crystals. Heating charcoal in steam at about 1000°C produces activated charcoal by driving out volatile compounds from the pores of the solid. Activated charcoal is an excellent absorbent used to filter out impurities from gases and solutions.

Charles' law states that the volume of a fixed amount of gas at constant pressure is proportional to its absolute temperature:

$$V \propto T \text{ or } \frac{V}{T} = \text{constant}$$

The law follows from the *ideal gas* equation $pV = nRT$ if p and n are constant. Real gases deviate from ideal gas behaviour. Note that the law only applies if there is no chemical change altering the number of gas molecules when the gas gets hotter or colder.

chelates are *complex ions* in which each *ligand* molecule or ion forms more than one *dative covalent bond* with the central metal ion. Chelates are formed by *bidentate ligands* and *polydentate ligands* such as *EDTA*. The term chelate comes from the Greek word for a crab's claw, reflecting the claw-like way in which chelating ligands grab hold of metal ions. Powerful chelating agents trap metal ions and effectively isolate them in solution.

Chelate complexes are generally more stable than complexes formed by *monodentate ligands* (see *stability constants*). The chelate effect can be explained by considering the entropy change for the ligand exchange reactions involved. When a bidentate ligand replaces monodentate ligands in a complex, there is an increase in the number of molecules and ions. The value of the *entropy* change, ΔS_{system}, is positive, increasing the tendency for the reaction to happen. For example:

$$[Ni(NH_3)_6]^{2+}(aq) + 3H_2NCH_2CH_2NH_2(aq) \rightarrow [Ni(H_2NCH_2CH_2NH_2)_3]^{2+}(aq) + 6NH_3(aq)$$

chemical equations describe what happens during reactions by identifying the reactants and products. *State symbols* show the *states of matter* of the chemicals involved. Equations may also show the reaction conditions by including information about temperature, pressure and catalysts above or below the arrow leading from reactants to products. There are various types of chemical equation.

- *Word equations* simply name the reactants and products:
  hydrogen + oxygen → water
- Full, *balanced equations* with formulae are used to calculate the amounts of reactants and products:
  $2H_2(g) + O_2(g) \rightarrow 2H_2O(l)$
- *Ionic equations* are balanced equations leaving out any spectator ions which do not change:
  $H^+(aq) + OH^-(aq) \rightarrow H_2O(l)$
- *Half-equations* are balanced equations which show part of a redox reaction:
  $2H^+(aq) + 2e^- \rightarrow H_2(g)$
- Unbalanced symbol equations show just the main starting chemical and the main product with the conditions for reaction written by the arrow:

$$C_2H_5OH \xrightarrow[\text{heat}]{Cr_2O_7^{2-}/H^+} CH_3COOH$$

- Mechanistic equations show intermediate steps in the *mechanism of a reaction*:

  $H_2 + Br\bullet \rightarrow HBr + H\bullet$

chemical equilibrium: see *dynamic equilibrium* and the *equilibrium law*.

chemical industry: the industry that converts raw materials such as crude oil, natural gas and minerals into useful products such as *pharmaceuticals, fertilisers, detergents*, paints and *dyes*. In recent years the chemical industry has been reinventing many processes in line with the principles of *green chemistry*.

A chemical plant consists not only of the reaction vessels and equipment for separating and purifying products, but also the storage vessels, pumps and pipes, sources of energy and heat exchangers together with the control room.

Bulk chemicals are manufactured on a scale of thousands or even millions of tonnes per year. Examples are ethene, sulfuric acid, ammonia and chlorine. They are mainly used as the starting point for making other substances. Fine chemicals such as pesticides and pharmaceuticals are made on a much smaller scale – a few tonnes or hundreds of tonnes per year.

Speciality chemicals are manufactured for their particular properties as thickeners, stabilisers, flame retardants and so on.

Main raw materials	Large scale processes and products	Uses of the products
Crude oil and *natural gas* for the petrochemical industry	*Fractional distillation, thermal cracking, steam reforming* and isomerisation and many other processes to produce the building blocks for the industry such as *ethene*, methanal, methanol, propene and *benzene*	Manufacture of *polymers*, solvents, pharmaceuticals, *agrochemicals, dyes*, pigments and *detergents*
Salt (sodium chloride) and limestone for the chlor-alkali industry	*Electrolysis of brine* for making chlorine, sodium hydroxide and hydrogen. Solvay process to make sodium carbonate	Manufacture of *bleaches, disinfectants*, solvents, some polymers and in the *glass* and paper industries
Sulfur from underground deposits of the element or from the purification of oil and gas plus oxygen from the air	Contact process for *sulfuric acid manufacture*	Manufacture of *paints*, pigments, *fertilisers*, detergents, *plastics* and many uses in the chemical, metallurgical and petrochemical industries
Nitrogen from the air, natural gas and oil fractions	Haber process for *ammonia manufacture* and catalytic oxidation of ammonia for *nitric acid manufacture*	Manufacture of fertilisers, dyes, pigments, detergents, *explosives*, plastics and *fibres*
Calcium phosphate rock	Treatment of phosphate rock with concentrated sulfuric acid to make *phosphoric(v) acid* and phosphates	Fertiliser industry, manufacture of washing powders, toothpaste, food industry, enamels and glazes
Fluorite (calcium fluoride)	Concentrated sulfuric acid on fluorite to make hydrogen fluoride; electrolysis of fluorides in hydrogen fluoride to make fluorine	Etching and polishing glass and integrated circuits, manufacture of fluorocarbons and hydrofluorocarbons to replace *CFCs*), pharmaceuticals and the polymer PTFE

chemotherapy is the use of chemicals to treat disease. Chemotherapy began with the work of Paul Ehrlich (1854–1915) who had the idea that it might be possible to find chemicals which kill the micro-organisms which cause disease without harming healthy living cells. Ehrlich was particularly interested in selective dyes. Some dyes, for example, take well on cotton but not on wool. Ehrlich found that methylene blue would dye nerve cells well but not other parts of the body. This inspired him to search for chemical 'magic bullets' to target micro-organisms and diseased cells.

The high hopes raised by Ehrlich's work led to many disappointments, until 1932 when Gerhard Domagk was involved in testing the medical effects of new dyes produced by a German chemical company. His work led to the development of sulfonamide drugs which were used to treat bacterial diseases until the discovery of *antibiotics*.

Today chemotherapy is widely used to treat cancer but few drugs are 'magic bullets'. If they are effective in destroying cancerous cells they are generally toxic and damage other parts of the body too (see *cisplatin*).

chiral compounds are composed of *asymmetric molecules* so that they have mirror-image forms which are not identical and cannot be superimposed on each other. The commonest chiral compounds have a carbon atoms attached to four different atoms or groups. The two mirror-image forms are known as enantiomers.

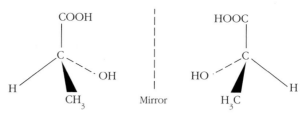

Mirror-image forms of 2-hydroxypropanoic acid (lactic acid)

Enantiomers behave identically in ordinary chemical reactions and their main physical properties are the same. They differ in their effect on *plane-polarised light* – they are optically active. One mirror-image form rotates the plane of polarised light in one direction. The other form has the opposite effect. They are *optical isomers*.

There are three common ways of naming compounds to distinguish between pairs of enantiomers. The symbols (+) and (−) indicate the direction in which the molecules rotate the plane of polarised light. The other two systems (*D/L* and *R/S*) distinguish the enantiomers according to the arrangement of atoms in three dimensions. The *D/L* system is commonly used for *amino acids*. The *R–S* system is based on a set of rules related to the *atomic numbers* of the atoms bonded to the chiral carbon atom.

Some complex ions are chiral.

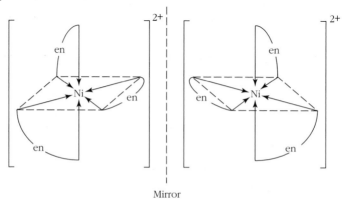

*Mirror-image forms of a complex between nickel(II) ions and 1,2-diaminoethane, en, (see **bidentate ligands**). Three-dimensional models make it easier to see that they are non-superimposable.*

chiral synthesis is based on a synthetic route that produces only one of the optical isomers of a *chiral compound*. This is of particular importance in the production of *pharmaceuticals*.

Chirality is very important in living organisms. Living cells are full of messenger and carrier molecules which interact selectively with the active sites and receptors in other molecules such as *enzymes*. The messenger and carrier molecules are all chiral and the body works with only one of the mirror-image forms. In most living things all the *amino acids,* for example, are *L*-amino acids.

The importance of chirality in living things was brought home to people by the impact of the thalidomide tragedy. Thalidomide was a sedative used in the early 1960s. Doctors believed that it was very safe and prescribed it widely. Sadly it was soon discovered that thalidomide could harm babies if taken by mothers during the early months of pregnancy.

Thalidomide is chiral. One isomer is the effective sedative. The other isomer causes malformations in babies (it is a *teratogen*). The testing of drugs and other pharmaceuticals is now much more thorough than it was in the 1950s and 1960s. When chiral molecules are involved, licensing authorities require pharmaceutical companies to carry out research with both isomers to identify their individual properties. Often, this leads to a requirement to produce a single *optical isomer*.

Scientists have developed techniques for making a pure single optical isomer of a pharmaceutical product including the use of:
* *enzymes* or bacteria to produce only one of the optical isomers and introduce the required stereoselectivity
* natural chiral molecules as starting molecules, such as *L*-amino acids or *D*-glucose
* chiral catalysts, possibly attached to a polymer support with reactants flowing over them.

chlor-alkali industry: see *chemical industry* and the *electrolysis of brine*.

chlorates(ɪ) and (v): see *chlorine oxoanions*.

chlorination is a reaction which replaces a hydrogen atom in an organic molecule with a chlorine atom. One example is the chlorination of *alkanes* – a *free-radical chain reaction* initiated by ultraviolet radiation.

Another example is the chlorination of *benzene* by chlorine in the presence of an iron(ɪɪɪ) chloride catalyst. This is an *electrophilic substitution* reaction (see also *methylbenzene*).

chlorine (Cl) occurs as a greenish, toxic gas consisting of Cl_2 molecules. It is the second element in the family of non-metals called the halogens (*group 7*). Its electron configuration is $[Ne]3s^23p^5$.

Chlorine consists of diatomic molecules with pairs of atoms held together by single covalent bonds. The molecules are non-polar so the *intermolecular forces* are relatively weak. Chlorine mixes freely with hydrocarbon solvents and the pale yellow-green colour of the solution is the same as chlorine vapour.

Chlorine is manufactured on a large scale by the *electrolysis of brine*.

Chlorine is a powerful *oxidising agent* which reacts directly with most elements. In its compounds chlorine is usually present in the −1 oxidation states but chlorine can be oxidised to positive oxidation states by oxygen and fluorine.

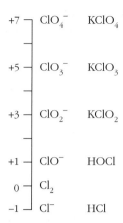

Oxidation states of chlorine

Chlorine forms ionic chlorides with metals:

Formation of ions in magnesium chloride when chlorine reacts with magnesium

Chlorine forms covalent, molecular chlorides with most non-metals but it does not react directly with carbon, oxygen or nitrogen. Hydrogen burns in chlorine to produce the colourless, acidic gas hydrogen chloride, HCl.

Reaction of hydrogen with chlorine to produce hydrogen chloride molecules

Hydrogen chloride is very soluble in water, forming **hydrochloric acid**.
Chlorine dissolves in water. It reacts reversibly with water forming a mixture of weak chloric(I) acid and strong hydrochloric acid. This is an example of a *disproportionation reaction* (see *chlorine oxoanions*):

$$Cl_2(aq) + H_2O(l) \rightleftharpoons HOCl(aq) + Cl^-(aq) + H^+(aq)$$

Chlorine oxidises a range of ions and molecules in solution:
- iron(II) ions to iron(III) ions
- bromide ions to bromine
- iodide ions to iodine
- sulfite ions to sulfate ions
- thiosulfate ions to sulfate ions
- hydrogen sulfide to sulfur

chlorine oxoanions form when chlorine reacts with water and alkalis. In these compounds chlorine is oxidised to positive oxidation states through a series of *disproportionation reactions*.

When chlorine dissolves in potassium (or sodium) hydroxide solution at room temperature it produces chlorate(I) and chloride ions. A solution of chlorine in alkali is used for *bleaching* and *chlorine water treatment*.

$$Cl_2(aq) + 2OH^-(aq) \rightarrow ClO^-(aq) + Cl^-(aq) + H_2O(l)$$

On heating the chlorate(I) ions disproportionates to chlorate(v) and chloride ions:

$$3ClO^-(aq) \rightarrow ClO_3^-(aq) + 2Cl^-(aq)$$

Potassium chlorate(v) can be crystallised from the solution. Careful heating just above the melting point converts potassium chlorate(v) to potassium chlorate(VII) and potassium chloride.

$$4KClO_3(s) \rightarrow 3KClO_4(s) + KCl(s)$$

chlorine water treatment: a life-saving application of *chlorine* chemistry to kill micro-organisms in drinking water and swimming pools. Since the nineteenth century, the treatment of drinking water with chlorine has helped to control the spread of diseases such as typhoid and cholera. In Europe today it is chlorine which makes almost all drinking water safe.

Chlorine disinfects tap water by forming chloric(I) acid, HOCl.

$$Cl_2(aq) + H_2O(l) \rightleftharpoons HOCl(aq) + H^+(aq) + Cl^-(aq)$$

Choric(I) acid is a powerful oxidising agent and a weak acid. It is an effective disinfectant because the molecule can pass through the cell walls of bacteria, unlike the negatively charged ClO$^-$ ions. Once inside the cell, the HOCl molecules break it open and kill the organism by oxidising and chlorinating within the cell.

There have been reports that the side effects of using chlorine can be harmful. Chlorine reacts with naturally occurring substances such as decomposing plant and animal materials in water to form trichloromethane and related compounds that are collectively known as trihalomethanes (THCs). Some research studies have shown a slightly increased risk of cancer in people who regularly drink chlorinated water.

cholesterol is a *steroid* which plays an important part in metabolism. It is a part of cell membranes and the precursor of *steroid* hormones such as testosterone and progesterone.

Structure of cholesterol

Cholesterol (a *lipid*) is insoluble in water but it is carried in the blood surrounded by protein molecules which keep it dispersed. There are two types of lipoprotein: low-density lipoproteins (LDL) and high-density lipoproteins (HDL). LDL cholesterol is often called 'bad' cholesterol because high levels of this form of the lipid are linked to fatty deposits in arteries and heart disease. HDL cholesterol is called 'good' cholesterol because it carries the lipid to the liver where it is broken down for excretion.

chromatography is a method for separating and identifying the chemicals in a mixture. Chromatography can be used to:
- separate and identify the chemicals in a mixture
- check the purity of a chemical product
- identify impurities in a product
- purify a chemical product (on a laboratory or industrial scale).

All types of chromatography have a stationary phase (a solid or a liquid held by a solid) and a mobile phase (a liquid or gas). The components of a mixture separate as the mobile phase moves through the stationary phase. Components which tend to mix with the mobile phase move faster. Components which tend to be held by the stationary phase move slower.

The basic principle underlying the separation is:
- *adsorption* when the stationary phase is a solid
- *partition* when the stationary phase is a liquid held as a thin layer on the surface of a solid.

There are several types of chromatography including *column chromatography*, *high-performance liquid chromatography* (HPLC), *paper chromatography*, *thin-layer chromatography* (TLC) and *gas chromatography* (GC).

chromium (Cr) is a hard, silvery metal, a *d-block element* with the electron configuration [Ar]$3d^5 4s^1$. This electron configuration is an exception to the normal [Ar]$3d^5 4s^2$ pattern for the first
series of d-block elements. Energetically it is more favourable to have one electron in each *d orbital* and thus to half-fill the d-subshell.

In solution chromium forms ions in the +2, +3 and +6 *oxidation states*.

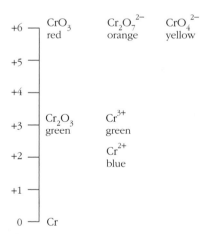

Oxidation states of chromium

There is an equilibrium between yellow chromate(VI) ions and orange dichromate(VI) ions in aqueous solution. The position of equilibrium depends on the pH.

$$2CrO_4^{2-}(aq) + 2H^+(aq) \rightleftharpoons Cr_2O_7^{2-}(aq) + H_2O(l)$$

In acid the hydrogen ion concentration is high so the solution is orange because dichromate ions predominate. Adding alkali removes hydrogen ions and turns the solution yellow as chromate ions form. These shifts in the position of equilibrium are as predicted by *Le Chatelier's principle*. Note that this is not a redox reaction. Chromium is in the +6 state on both sides of the equation.

Dichromate ions in acid solution are used to oxidise *alcohols* and *aldehydes*. Paper moistened with dichromate(VI) solution is used in the *gas test* for sulfur dioxide. The paper turns from orange to green as dichromate(VI) ions are reduced to chromium(III) ions. Potassium dichromate(VI) is used as a *primary standard* in *redox titrations*.

$$Cr_2O_7^{2-}(aq) + 14H^+(aq) + 6e^- \rightarrow 2Cr^{3+}(aq) + 7H_2O(l)$$

Chromium(III) ions in solution exist as aquo complexes, $[Cr(H_2O)_6]^{3+}$. The hydrated ions are acidic and behave in a similar way to aluminium(III) ions.

The removal of successive protons from the hydrated chromium(III) ion, $[Cr(H_2O)_6]^{3+}$, produces a neutral complex, $[Cr(H_2O)_3(OH)_3]$, which is sparingly soluble and precipitates. The removal of further protons produces a negative ion, $[Cr(OH)_6]^{3-}$, and the ionic compound redissolves. This is possible because of the high polarising power of the small, highly charged chromium(III) ions (see *hydrolysis of metal aqua ions*).

Under alkaline conditions *hydrogen peroxide* oxidises chromium(III) to chromium(VI).

Zinc reduces a green solution of chromium(III) to a blue solution of chromium(II) ions. Chromium(II) is a powerful *reducing agent* and is quickly converted to chromium(III) by oxygen in the air.

chromophore: the part of a molecule which gives rise to colour. Chromophores in *coloured compounds* of carbon have an extended system of *delocalised electrons*. Families of dyes with the same chromophore are made by attaching different *functional groups* to modify the properties of the dyes.

Functional groups linked to a chromophore may act as:
- auxochromes which change the colour of the dye
- solubilising groups which make the dye more soluble in water and help it stick to the fibres of a textile.

chrysoidine   alizarin yellow

Two dyes based on the same chromophore

cisplatin is an anticancer drug consisting of a complex ion of platinum, $Pt(NH_3)_2Cl_2$. The ion is planar and can shows *cis–trans isomerism*. The *cis* isomer inhibits cell division but does not prevent cell growth. This makes it a useful treatment for cancer. Unfortunately cisplatin is toxic and has unpleasant side effects.

cis isomer   *trans* isomer

Structure of the cis and trans isomers of Pt(NH$_3$)$_2$Cl$_2$

***cis–trans* isomers** are molecules with the same molecular and structural formulae but different shapes because of the lack of free rotation about double bonds. *Alkenes* and other compounds with C=C *double bonds*, such as *fatty acids*, may have *cis–trans* isomers because there is no rotation about the double bond. In the *cis* isomer, similar functional groups are on the same side of the double bond. In the *trans* isomer similar functional groups are on opposite sides of the double bond. In more complex molecules it is not possible to identify the *cis* and *trans* forms unambiguously and so chemists have an alternative naming system (see *E–Z isomers*)

cis-but-2-ene   trans-but-2-ene

Cis and trans isomers of but-2-ene. They are distinct compounds with different melting points, boiling points and densities.

Some inorganic *complexes* also have geometrical isomers.

cis isomer   *trans* isomer

Cis and trans isomers of the planar complex [Ni(NH$_3$)$_2$Cl$_2$]. The cis form is blue-violet. The trans form is green.

The anti-cancer drug, ***cisplatin*** is the *cis* isomer of a flat (planar) complex.

citric acid is a white crystalline solid which occurs widely in nature. Citric acid is present in citrus fruits, other fruits, vegetables and plants. The acid has a vital role in the biochemical citric-acid cycle which releases energy from sugars by *respiration*.

$$\begin{array}{c} H \\ | \\ H-C-COOH \\ | \\ HO-C-COOH \\ | \\ H-C-COOH \\ | \\ H \end{array}$$

Structure of citric acid

Citric acid has many uses:

- as an *acid* it is used to remove calcium carbonate limescale
- as a *ligand* it is added to washing powders to form stable chelates with calcium or magnesium ions in *hard water*
- mixed with its salts, it is used to make *buffer* solutions to control the pH in *pharmaceuticals*
- as a *food additive* (E330) with its salts (E322 and E333) it is used as a flavouring and preservative.

clathrate compounds: materials consisting of molecules of one element or compound trapped within a cage-like structure of another compound. *Ice*, for example, has a very open structure. So much so that it can trap molecules within the spaces between the water molecules. One example of this is methane hydrate which is a solid form of water with large amounts of methane imprisoned in cages made by hydrogen-bonded water molecules.

climate change describes long-term trends in the average climate, such as changes in average temperatures or rainfall. The term is often used to describe changes that are the direct or indirect results of human activities and that alter the composition of the Earth's atmosphere. In particular the term is linked to the emission of *greenhouse gases* and the resulting *global warming*.

clinical trials – see *drug development*.

clock reaction: a variant of the *initial-rate method* which is so-called because the reaction is set up to produce a sudden colour change after a certain time once it has produced a fixed amount of one substance.

The reaction of peroxodisulfate(VI) ions with iodide ions can be set up as a clock reaction:

$$S_2O_8^{2-}(aq) + 2I^-(aq) \rightarrow 2SO_4^{2-}(aq) + I_2(aq)$$

A small, known amount of sodium thiosulfate ions is added to the reaction mixture which also contains starch indicator. At first the thiosulfate ions, $S_2O_3^{2-}$, reacts with any iodine, I_2, as soon as it is formed turning it back to iodide ions, so there is no colour change. At the instant when all the thiosulfate has been used up, free iodine is produced and this immediately gives a deep blue-black colour with the starch. If t is the time for the blue colour to appear after mixing the chemicals, then $1/t$ is a measure of the initial rate of reaction.

close-packed structures are found in metal crystals. In a layer of close-packed spheres each atom has six other spheres touching it. In three dimensions, layers of close-packed atoms stack up in two possible ways. In hexagonal close-packing the third layer is directly over the first layer (aba). This is a common crystal structure for metals including magnesium, titanium and beryllium. In cubic close-packing it is the fourth layer which corresponds with the first layer (abca). This is also a common structure for metals and is found in aluminium, nickel, copper, silver and platinum.

In the two structures, each atom touches 12 nearest neighbours so for both the *coordination number* is 12. The *unit cell* of the cubic close-packed structure is a face-centred cube.

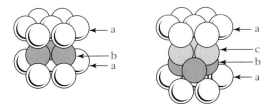

Hexagonal and cubic close-packing of metal atoms

coal is a fossil fuel formed by the action of heat and pressure on the remains of plants buried under sediments. Heating coal in the absence of air at 1000°C drives off coal tar, ammonia and coal gas leaving a residue of coke. Coke is required for a range of industrial processes including *iron extraction* in blast furnaces. Coal tar was a major source of organic chemicals until crude oil took over in the 1940s. Coal tar is a rich source of arenes and was the main source of *benzene* when William Perkin was pioneering the production of synthetic dyes.

cobalt (Co) is a hard, silvery metal, a *d-block element* which is less reactive than iron. It has the electron configuration $[Ar]3d^7 4s^2$.

Cobalt is an ingredient of alloys *steels* such as the ferromagnetic alloy alnico which makes excellent permanent magnets.

In solution cobalt forms ions in the +2 and +3 oxidation states. Cobalt(II) is the more stable state. Anhydrous cobalt(II) chloride is blue but it turns pink on adding water as the cobalt ions are hydrated to the $[Co(H_2O)_6]^{2+}$ ion. This is used as a test to detect water.

A dilute solution of cobalt chloride is pink because the cobalt(II) ions are hydrated. A concentrated solution is blue. A dilute solution also turns blue on adding concentrated hydrochloric acid. The colour change is due to a *ligand* substitution reaction as chloride ions replace the water molecules. The reaction is reversible:

$$[Co(H_2O)_6]^{2+}(aq) + 4Cl^-(aq) \rightleftharpoons [CoCl_4]^{2-}(aq) + 6H_2O(l)$$

It is normally very difficult to oxidise aqueous cobalt(II) to cobalt(III) but the reaction goes readily if the cobalt(II) ions are complexed with ammonia molecules. The Co(III) complex with ammonia is more stable than the Co(II) complex. The value for the *standard electrode potential* shows that aqueous Co(III) is a stronger oxidising agent than potassium manganate(VII) in acid solution:

$$[Co(H_2O)_6]^{3+}(aq) + e^- \rightleftharpoons [Co(H_2O)_6]^{2+}(aq) \qquad E^\ominus = +1.82 \text{ V}$$

When the two states are complexed with ammonia the standard electrode potential shifts to a value that shows that the Co(III) state is much more stable. Cobalt(II) is now a reducing state and can be oxidised to Co(III) by oxygen or hydrogen peroxide.

$$[Co(NH_3)_6]^{3+}(aq) + e^- \rightleftharpoons [Co(NH_3)_6]^{2+}(aq) \qquad E^\ominus = +0.10 \text{ V}$$

collision theory accounts for the effects of concentration, temperature and *catalysts* on *rates of reaction*. The idea is that a chemical reaction happens when the molecules or ions of reactant collide, making some bonds break and allowing new bonds to form.

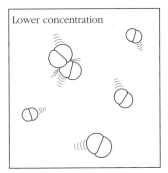
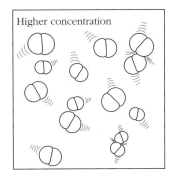

Raising the concentration means that the reacting particles are closer together. There are more collisions and reactions are faster.

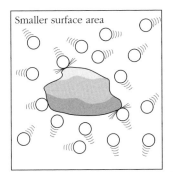
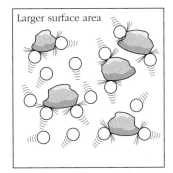

In a heterogeneous reaction of a solid with either a liquid or gas, the reaction is faster if the solid is broken up into smaller pieces. Crushing the solid increases its surface area, collisions can be more frequent and the rate of reaction is bigger.

It is not enough for the molecules just to collide. In soft collisions the molecules simply bounce off each other. Molecules are in rapid random motion and if every collision led to reaction all reactions would be explosively fast. Only pairs of molecules which collide with enough energy to stretch and break chemical bonds can lead to new products. Reactant molecules have to overcome the *activation energy*.

Even if the colliding molecules have enough energy, they may still not react if their orientation is not correct. In quantitative accounts of collision theory this is taken into account by including a steric factor in the equations.

Collision theory refers to the *Maxwell–Boltzmann distribution* of energies of molecules to explain the effects of temperature and catalysis on reaction rates.

colloids consist of fine particles (the disperse *phase*) of one substance dispersed in another (the continuous phase). The dispersed particles in liquids and gases are large enough to scatter light but they do not settle out and they show *Brownian motion*. The diameter of colloid particles is typically around 10 to 1000 nm (much larger than atomic diameters which are around 0.2 nm), so colloid chemistry is related to *nanotechnology*. Examples of colloids are *aerosols*, *emulsions*, *foams* and *gels*.

95

colorimetry is a method for measuring the *concentration* of compounds in solution. It can be used with chemicals which are themselves coloured or which give a colour when mixed with a suitable reagent.

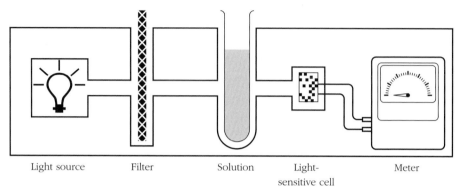

| Light source | Filter | Solution | Light-sensitive cell | Meter |

Diagram of a colorimeter

The filter lets through light which is strongly absorbed by the solution. The extent of absorption depends on the path length of the light through the solution (which is a constant for the sample chamber used) and on the concentration of the solution. The instrument is *calibrated* with a series of *standard solutions*. It can then be used to measure the concentration of unknown samples.

coloured compounds in many instances get their colour by absorbing some of the radiation in the visible region of the *electromagnetic radiation* spectrum with wavelengths between 400 nm and 700 nm.

Colour of compound	Wavelength absorbed/nm	Colour of light absorbed
greenish yellow	400–430	violet
yellow to orange	430–490	blue
red	490–510	blue-green
purple	510–530	green
violet	530–560	yellow-green
blue	560–590	yellow
greenish blue	590–610	orange
blue-green to green	610–700	red

It is the electrons in coloured compounds that absorb radiation as they jump from their normal state to a higher excited state. According to *quantum theory* there is a fixed relationship between the size of the energy jump and the wavelength of the radiation absorbed. In many compounds the jumps are so big that they absorb in the ultraviolet part of the spectrum. Such compounds are colourless.

colour of transition metal complex ions – this depends on:
- the metal
- the oxidation state of the metal
- the ligand
- the coordination number.

Colour in transition metal ions arises from electronic transitions between *d orbitals*. In a free atom all of the five d orbitals have the same energy. When a transition metal ion forms complex ions the d orbitals split into two groups with different energies. The size of the split depends on the number and nature of the ligands in the complex. This helps to account for colour changes.

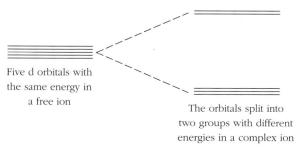

Five d orbitals with the same energy in a free ion

The orbitals split into two groups with different energies in a complex ion

Energy difference between d orbitals in a transition metal complex ion

If all the d orbitals are full there is no possibility of electronics transitions between them. This explains why Zn^{2+} and Cu^+ ions are colourless. Sc^{3+} ions are also colourless because the ion has no d electrons.

colour in organic compounds is often associated with *delocalised electrons*. The energy jumps between molecular orbitals absorb visible radiation in molecules with extended systems of alternating double and single bonds (see *conjugated systems*).

Skeletal formula of β-carotene, the main orange colour in carrots, used as a colourant in foods, drugs and cosmetics

other sources of colour – some colours are caused by physical effects rather than electronic transitions. Examples of these effects are:
- light scattering – moonstones, blue skies
- interference – soap bubbles and oil films on water
- diffraction – opal, liquid crystals
- refraction of some wavelengths more than others – 'fire' in gemstones.

column chromatography was the first type of *chromatography* developed. The Russian botanist Michel Tswett developed this type of chromatography in columns to separate plant pigments. In 1903 he separated leaf colours (chlorophylls, carotenes and xanthophylls) by dissolving a leaf extract in a hydrocarbon solvent. He added the extract to the top of a column of powdered chalk and then passed more pure solvent through the column to complete the separation. Running solvent through the column to separate the mixture is called 'elution' and the liquid which flows out of the bottom is the eluate.

This is an example of *adsorption* chromatography. Each compound in the mixture has its own equilibrium between adsorption on the solid and solution in the solvent. Compounds which are strongly adsorbed by the stationary phase move slower; compounds which are more soluble in the solvent move faster.

The liquid leaving the bottom of the column is collected in a series of tubes. Tubes containing parts of the mixture can be detected by their colour or some other method such as their appearance in ultraviolet light. Components of the mixture can be recovered by evaporating the solvent from these tubes. *High–performance liquid chromatography* is a development of column chromatography.

combinatorial chemistry is a technique of chemical synthesis which speeds up *drug development*. The method takes advantage of automation to prepare very large numbers of compounds very rapidly. In this way chemists build up 'libraries' of closely related compounds.

The starting materials for synthesis are attached to polymer beads. After each step of a synthesis, the beads with the product attached can be separated from the other reagents by filtering and washing. The final step removes the polymer beads to leave the products.

Initially the compounds are tested for their biological activity as mixtures. This is much quicker than the traditional method of making, purifying and testing individual compounds. It then only becomes necessary to synthesise the compounds individually if the tests show that a particular mixture shows promising biological activity.

combustion is the reaction of a *fuel* with oxygen to release energy. If there is plenty of air or oxygen, the carbon in the fuel turns to carbon dioxide and the hydrogen turns to steam (water):

$$CH_4(g) + 2O_2(g) \rightarrow CO_2(g) + 2H_2O(l)$$

methane
in natural gas

Incomplete combustion in a limited supply of oxygen is hazardous because the products are then soot and the toxic gas *carbon monoxide*. Combustion of fuels can give rise to other *atmospheric pollutants*.

combustion analysis is a method for determining the *empirical formula* of an organic compound. A weighed sample of the compound is burned in excess oxygen mixed with helium. This converts the carbon to carbon dioxide and the hydrogen to water. A catalyst ensures that combustion is complete. The inert helium carries the products of combustion and the excess oxygen through a tube which contains chemicals to remove any volatile halogen, sulfur or phosphorus compounds. Oxides of nitrogen are converted to nitrogen gas and the excess oxygen combines with copper.

The water vapour is absorbed in magnesium chlorate(vii). Carbon dioxide is absorbed in soda lime. In a modern instrument, measurements of the thermal conductivity of the helium before and after absorption make it possible to determine the masses of water, carbon dioxide and nitrogen formed by burning the sample.

From the results it is possible to calculate the *percentage composition* of the compound. Any mass of the sample not accounted for is assumed to be due to oxygen.

Worked example

Complete combustion of 0.15 g of a liquid compound produced 0.22 g of carbon dioxide and 0.09 g water. What is the empirical formula of the compound?

Notes on the method

The *molar mass* of carbon dioxide, CO_2 = 4 g mol^{-1} of which carbon is 12 g mol^{-1}

The molar mass of water, H_2O = 18 g mol^{-1} of which hydrogen is 2 g mol^{-1}

Answer

The mass of carbon in the sample = $\dfrac{12}{44} \times 0.22$ g = 0.06 g

The mass of hydrogen in the sample = $\dfrac{2}{18} \times 0.09$ g = 0.01 g

The total mass of carbon and hydrogen = 0.07 g in a sample with mass 0.15 g

So the difference gives the mass of oxygen in the sample which is 0.08 g

These are the amounts of the elements in the sample:

carbon: 0.06 g ÷ 12 g mol^{-1} = 0.005 mol

hydrogen: 0.01 g ÷ 1 g mol^{-1} = 0.01 mol

oxygen: 0.08 g ÷ 16 g mol^{-1} = 0.005 mol

The ratio C:H:O is 1:2:1

The empirical formula of the compound is CH_2O.

common ion effect: this describes the effect on an ionic equilibrium of adding another compound containing one of the same ions as the equilibrium system. The effect is a consequence of *Le Chatelier's principle*. It accounts for the fact that a salt such as barium sulfate is less soluble in 0.1 mol dm^{-3} sulfuric acid (1×10^{-9} mol dm^{-3} at 298 K) than it is in pure water (1×10^{-5} mol dm^{-3} at 298 K):

$$BaSO_4(s) \rightleftharpoons Ba^{2+}(aq) + SO_4^{2-}(aq)$$

In this example the common ion is the sulfate ion. The added sulfate ions from the strong acid, sulfuric acid, move the equilibrium to the left and reduce the solubility of the salt.

complex ions consist of a central metal ion linked to a number of molecules or ions with lone pairs of electrons. The surrounding molecules or ions are *ligands* which use lone pairs of electrons to form a *coordinate bond* with the metal ion. The number of ligands in a complex ion is typically two, four or six.

Transition metal ions form aquo complexes when dissolved in water. Examples are $[Fe(H_2O)_6]^{2+}(aq)$ and $[Cr(H_2O)_6]^{3+}(aq)$.

The overall charge of a complex ion is the sum of the charges on the metal ion and its ligands. (See also *names of complex ions, shapes of complex ions, chelates, coloured compounds, coordination compounds, coordination number, complexometric titrations* and *stability constants*.)

complexes with ammonia molecules are useful in *qualitative analysis* in both anion tests and *cation tests*. Example are the diamminesilver(I) ion and the tetraammine copper(II) ion. The *lone pair of electrons* on the nitrogen atom of each ammonia molecule forms a *coordinate bond* with the central metal ion.

$$[Ag(NH_3)_2]^+ \qquad\qquad [Co(NH_3)_6]^{2+}$$

diamminesilver(I) ion   hexaamminecobalt(II) ion

Complexes between metal ions and ammonia

Adding ammonia solution to a precipitate of a *silver halide* can distinguish the chloride from the bromide and the iodide. Only silver chloride, the least insoluble, will freely dissolve in dilute ammonia solution to form the diammine complex.

Formation of a complex may produce a colour change or redissolve a precipitate. Adding ammonia solution to a solution with copper(II) ions initially produces a pale blue precipitate of the hydroxide, but with excess ammonia the precipitate dissolves to give a deep blue solution as the tetraammine complex forms.

complexometric titration: a practical technique used to determine the *concentration* of metal ions. A *titration* measures the volume of a *standard solution* of a complex-forming reagent needed to react exactly with a measured volume of the unknown solution of metal ions, such as zinc ions.

Complex-forming titrations often use a standard solution of *EDTA* which forms very stable *complex ions* with metal ions. The procedure is the same as for any other titration.

To find the end-point the analyst adds an indicator and a *buffer solution* which forms a coloured but unstable complex with the metal ion in the mixture. A suitable indicator is Eriochrome black T which is blue in solution at pH 10. At the start of the titration it produces a wine-red complex. EDTA from a burette forms a more stable complex with metal ions and so takes the metal ions from the indicator. At the end-point all the metal ions have been complexed by titration with EDTA. The last drop of EDTA leaves no metal ions to form the red complex with the indicator so the indicator turns blue again.

composites are made by combining two or more materials to create a new material. A composite combines the desirable properties of its constituents and compensates for their disadvantages. Steel-reinforced concrete is a composite material, as is galvanised steel. Kitchen worktops are composites consisting of chipboard covered with paper impregnated with a *thermosetting polymer* such as a melamine–methanal resin.

Many important composites consist of fibres of one material, such as *glass*, aramids or graphite, embedded in a *polymer, metal* or *ceramic* matrix. The combinations are designed to give new materials with better properties, especially high stiffness per unit weight which is important in road and rail transport, aircraft, sporting goods and many other applications.

Glass fibres in a *polyester* matrix (so-called fibreglass) are used to make boat hulls and car bodies. Parts of aircraft and some sports gear are made from carbon fibres in an epoxy matrix.

computational chemistry takes advantage of powerful computers and ingenious software to investigate chemical behaviour without using test tubes. Chemists use computer models to explore reaction mechanisms, to predict the behaviour of novel catalysts and polymers, and to visualise the interaction between drugs and their target molecules in living organisms. Central to all the software is a theory (such as quantum mechanics) for calculating the energy of each possible state of a set of interacting atoms or molecules. The software uses its calculations to predict the arrangement of atoms or molecules that gives rise to the lowest energy for the system.

concentration of a solution is usually measured in moles per litre of solution (mol dm^{-3}). There are small volume changes when chemicals dissolve in water so it is important to note that concentrations normally refer to litres of solution and not to litres of the solvent:

$$\text{concentration/mol dm}^{-3} = \frac{\text{amount of solute/mol}}{\text{volume of solution/dm}^3}$$

Writing the formula of a chemical in square brackets is the usual shorthand for concentration in mol dm^{-3}. This is common in equilibrium and rate chemistry.

When ionic crystals dissolve, the ions separate and become independent:

$$CaCl_2(s) + aq \rightarrow Ca^{2+}(aq) + 2Cl^-(aq)$$

So if $[CaCl_2] = 0.1$ mol dm^{-3}, then $[Ca^{2+}] = 0.1$ mol dm^{-3} but $[Cl^-] = 0.2$ mol dm^{-3}.

Other ways of measuring concentrations are to use *mole fractions* or *parts per million* (ppm). (See also *dilution*.)

Worked example

What is the concentration of a solution of potassium manganate(VII) made by dissolving 3.95 g of the solid in water and making the solution up to 500 cm³?

Answer

The molar mass of potassium manganate(VII), $(KMnO_4) = 158.0$ g mol^{-1}

Amount of $KMnO_4$ in solution $= \dfrac{3.95 \text{ g}}{158.0 \text{ g mol}^{-1}} = 0.025$ mol

Volume of the solution $= \dfrac{500}{1000}$ dm³ $= 0.5$ dm³

Concentration of the solution $= \dfrac{0.025 \text{ mol}}{0.5 \text{ dm}^3} = 0.05$ mol dm^{-3}

---

concentration–time graph: see *zero order reaction*, *first order reaction* and *second order reaction*.

condensation polymers are produced by a series of condensation reactions splitting off water between the functional groups of the *monomers*. Examples of condensation polymers are *polyamides* and *polyesters*. Natural condensation polymers include *proteins* and the *carbohydrates* starch and cellulose.

Where each monomer has two functional groups this type of polymerisation produces chains. Condensation polymers make good *fibres* because they form long, straight-chain molecules with few side chains and relatively strong intermolecular forces between polar bonds in neighbouring chains. *Cross-linking* is possible if one of the monomers has three functional groups.

condensation reaction: a reaction in which molecules join together by splitting off a small molecule such as water. The *addition–elimination reactions* of carbonyl compounds are examples of condensation reactions. The formation of an ester from an acid and an alcohol is also a condensation reaction (see also *polyamides* and *polyesters*).

| acid | alcohol | ester | water |

Condensation reaction to form the ester ethyl ethanoate

conditions for organic reactions specify concisely but accurately the reagents, catalysts, temperatures and pressures needed for good *yields*. Factors to consider include:
● whether a solvent is needed – if so, whether the solvent is water or a non-aqueous solvent such as ethanol
● the concentration of the reagents – whether acids, for example, are dilute or concentrated
● the temperature – whether the mixture has to be cooled, heated or kept at room temperature
● the pressure – many industrial process take place at several times the pressure of the atmosphere
● whether a catalyst is required.

conducting polymers consist of long-chain molecules with delocalised electrons that have been treated so that they conduct electricity. Poly(ethyne) contains alternating double and single bonds. In the stable form of the polymer all the double bonds have the *cis* configuration. Forming poly(ethyne) in the presence of a *Ziegler–Natta catalyst* and then doping the product with iodine produces a material which conducts electricity.

Some polymers conduct only when exposed to light. Films of polymers with this property are used to make the light-sensitive surface of photocopiers. An example of a photoconducting polymer is poly(vinyl carbazole).

conjugate acid–base pairs: see *acid–base equilibria*.

conjugated system: a system of alternate double and single bonds in a molecule. Organic molecules with extended conjugated systems absorb radiation in the visible part of the spectrum and are therefore coloured. Examples are the orange β-carotene molecule in carrots and the indicator phenolphthalein in alkali (see *coloured compounds*).

colourless   red

Structures of phenolphthalein in acid and alkali

conservation of mass (law of): the law which chemists take for granted every time they write a balanced chemical equation. The law states that matter is neither created nor destroyed during a chemical reaction.

constant-boiling mixture: see *azeotropic mixture*.

Contact process: see *sulfuric acid manufacture*.

continuous processes manufacture chemicals on a large scale in industrial plants which operate for 24 hours a day. Raw materials are constantly fed into the plant and products are continuously removed. Examples of continuous processes are the *fractional distillation of oil*, the Haber process for *ammonia manufacture*, *iron extraction* in a blast furnace and the Contact process for *sulfuric acid manufacture*.

conversion of units is often needed to make sure that the units are consistent before carrying out calculations. Typically the units of measurement (such as cm^3) have to be converted to different units for calculation (in this case dm^3).

Worked example

In a titration, the volume of acid added from the burette was 23.5 cm^3. What is the volume in litres (dm^3)?

Notes on the method

Multiply by an appropriate factor to convert from the unit used for measurement to the unit needed for the calculation. Find the factor by:
- writing down the relationship between the two units, then by
- writing the relationship as a ratio, so that the units cancel when it multiplies the measurement, such that the original unit is replaced by the unit required.

Answer

$1 \ dm^3 = 1000 \ cm^3$

Hence $23.5 \ cm^3 \times \dfrac{1 \ dm^3}{1000 \ cm^3} = 0.0235 \ dm^3$

---

coordinate bond is an alternative name for a *dative covalent bond*. The term coordinate bond is often used when describing the dative bonding between ligands and a metal ion in *complex compounds*.

coordination compounds contain complexes which may be cations, anions or neutral molecules. In a coordination compound, *ligand* molecules or negative ions form coordinate bonds (*dative covalent bonds*) with a metal ion.

Examples of coordination compounds:
- $K_3[Fe(CN)_6]$ containing the negatively charged complex ion $[Fe(CN)_6]^{3-}$
- $[Ni(NH_3)_6]Cl_2$ containing the positively charged complex ion $[Ni(NH_3)_6]^{2+}$
- $Ni(CO)_4$ a neutral complex between nickel atoms and carbon monoxide molecules.

coordination number: the number of nearest neighbours to an atom or ion in a crystal structure or the number of *coordinate bonds* formed between the ligands and the metal ion in a complex ion.

copolymerisation is used to modify the properties of *polymeric materials* by producing polymer chains from a mixture of monomers. ABS, for example, is a rigid, tough plastic widely used for the casing of domestic equipment and for parts of cars. It is a copolymer of **A**crylonitrile (CH_2=CH—CN), **B**utadiene (CH_2=CH—CH=CH_2) and **S**tyrene (C_6H_5—CH=CH_2).

copper (Cu) is a ductile metal with a reddish colour. It has the electron configuration [Ar]$3d^{10}4s^1$. This electron configuration is an exception to the normal [Ar]$3d^x4s^2$ pattern for the first series of *d-block elements*. Energetically it is more favourable to fill the d subshell and leave only one electron in the 4s.

Copper is relatively unreactive. It corrodes very slowly in moist air and is not attacked by dilute non-oxidising acids. Copper is a good conductor of electricity; it is widely used in electricity cables and for domestic water pipes.

Copper's mechanical properties are enhanced by making alloys such as *brass* and *bronze*.

Copper forms compounds in the +1 and +2 oxidation states. Under normal conditions copper(II) is the stable state in aqueous solution. Copper(I) *disproportionates* in aqueous solution. So when Cu_2O dissolves in dilute sulfuric acid the products are copper(II) sulfate and copper.

$$2Cu^+(aq) \rightleftharpoons Cu^{2+}(aq) + Cu(s)$$

The equilibrium lies well to the right. Copper(I) in the presence of water can exist as insoluble compounds such as Cu_2O, CuI or CuCl. Iodide ions, for example, reduce copper(II) to copper(I) ions which immediately precipitate with more iodide ions as white copper(I) iodide.

$$2Cu^{2+}(aq) + 4I^-(aq) \rightarrow 2CuI(s) + I_2(s)$$

Copper(I) can exist in aqueous solution as stable complexes such as $[Cu(NH_3)_2]^+$ and $[Cu(CN)_4]^{3-}$.

Fehling's solution and *Benedict's solution* contain deep blue copper(II) complexes in alkali. The reagents are used to detect *reducing sugars* and to distinguish *aldehydes* from *ketones*. A reducing sugar or aldehyde reduces the reagent to copper(I) oxide on heating. The blue colour goes and a reddish-brown precipitate forms.

(See *names of complex ions* and *ligand substitution reactions* for more examples of copper complexes. See *carbon monoxide* for an example of copper acting as a catalyst. See *cation tests* and *flame tests* for methods used to detect the presence of copper(II) ions.)

copper refining uses *electrolysis* to turn copper which is 99.5% pure into 99.99% pure metal. High purity is important especially when copper is to be used as an electrical conductor.

Impure copper is cast into anodes and the cathodes are thin sheets of pure copper. The electrolyte is a mixture of copper(II) sulfate and sulfuric acid.

At the anodes: $Cu(s) \rightarrow Cu^{2+}(aq) + 2e^-$

At the cathodes: $Cu^{2+}(aq) + 2e^- \rightarrow Cu(s)$

Valuable impurities such as gold and silver are recovered from the bottom of the electrolysis cell because they do not dissolve in the electrolyte.

corrosion of a metal is a redox process in which oxygen, water and acids attack metals. Most metals are extracted from oxides. Corrosion turns them back into oxides.

The most familiar and economically serious example of corrosion is the rusting of iron. Rusting is an electrochemical reaction.

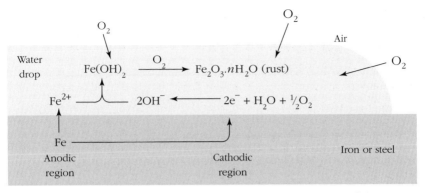

An electrochemical cell on the surface of iron in contact with water and the air

When iron is in contact with air and water, the regions rich in oxygen are cathodic. In these regions oxygen is reduced to hydroxide ions. Other parts of the metal surface are anodic. In these regions iron is oxidised to iron(II) ions. Iron(II) ions and hydroxide ions diffuse together and form a precipitate of iron(II) hydroxide which is then oxidised to rust, hydrated iron(III) oxide. Cathodic protection can stop iron rusting, so can painting or coating with plastic to provide a barrier to exclude water and air.

coulomb (symbol C) is the *SI unit* of electric charge.

electric charge (C) = current (C s^{-1}) × time (s)

A coulomb is the amount of electric charge flowing each second past a point in a circuit when the current is one ampere (1 A = 1 C s^{-1}).

covalent bonds form when atoms share electrons. The atoms are held together by the attraction between the positive charges on their nuclei and the negative charge on the shared electrons.

Covalent bonds are the strong bonds which hold together the atoms of non-metals in molecules and *giant structures*. Molecules have a definite shape because covalent bonds have a definite length and direction (see *shapes of molecules and ions*).

Dot-and-cross diagrams to show single covalent bonding in molecules. A simpler way of showing the bonding in molecules is also shown. A line between two symbols represents a covalent bond.

One shared pair of electrons gives rise to a single bond. *Double bonds* and *triple bonds* are also possible with two or three shared pairs.

The non-metals common in organic chemistry generally form a fixed number of covalent bonds. Knowing this helps to work out the structures of molecules.

Element	Number of covalent bonds	Examples		
carbon, C	4	$H-\underset{\underset{H}{\vert}}{\overset{\overset{H}{\vert}}{C}}-H$	$\underset{H}{\overset{H}{\diagdown}}C{=}O$	$O{=}C{=}O$
hydrogen, H	1	$\underset{H}{\overset{H}{\diagdown}}O$	$H-\underset{\overset{\vert}{H}}{\overset{\overset{H}{\vert}}{N}}-H$	$H-Cl$
oxygen, O	2	$O{=}O$	$\underset{H}{\overset{H}{\diagdown}}O$	$H-C\underset{O-H}{\overset{O}{\diagup\diagdown}}$
nitrogen, N	3	$N{\equiv}N$	$Cl-\underset{\overset{\vert}{Cl}}{\overset{\overset{Cl}{\vert}}{N}}-Cl$	$CH_3-N\underset{H}{\overset{H}{\diagdown\diagup}}$
halogens, F, Cl, Br, I	1	$H-Br$	$Cl-\underset{\underset{Cl}{\vert}}{\overset{\overset{Cl}{\vert}}{C}}-Cl$	$H-\underset{\underset{H}{\vert}}{\overset{\overset{H}{\vert}}{C}}-I$

covalent radius: the covalent radius of an element is half the **bond length** when two atoms of the element are linked by a single covalent bond.

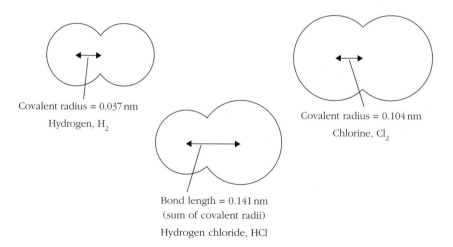

Covalent radius = 0.037 nm
Hydrogen, H_2

Covalent radius = 0.104 nm
Chlorine, Cl_2

Bond length = 0.141 nm
(sum of covalent radii)
Hydrogen chloride, HCl

The length of a single covalent bond between atoms of different elements can be estimated fairly accurately by adding the covalent radii for the two atoms

cracking: see *catalytic cracking* and *thermal cracking*.

critical temperature is the temperature above which it is impossible to liquefy a gas however high the pressure (see *vapours*).

cross-linking is the formation of chemical bonds between polymer chains to modify the properties of polymers such as *thermosetting polymers* as well as natural *rubber* during vulcanising.

Cross-linking determines the three-dimensional shape of *proteins*. Cross-links in protein molecules take various forms including:
- *hydrogen bonding* between amino acid side chains
- disulfide bridges formed by *covalent bonds* between cystine side chains.

crude oil is a complex mixture of *hydrocarbon* molecules formed over millions of years when the remains of microscopic sea creatures trapped in sediments were converted by heat and pressure to petroleum. Crude oil is now the main source of *fuels* and organic compounds. The composition of crude oil varies from one oilfield to another. Some crudes contain significant quantities of sulfur and nitrogen compounds as well as traces of various metals. *Fractional distillation of oil* is the first step in refining to produce fuels and *lubricants* as well as feedstocks for the *petrochemical industry*.

crystallinity of polymers arises when the long-chain molecules are regular enough to lie more or less parallel to each other in some regions of the solid. The rest of the solid where the chains are tangled together is an *amorphous solid*.

The crystallinity of *polymeric materials* with regular, unbranched chains is generally higher than in polymers with branched or irregular molecules. Relatively strong *intermolecular forces* between chains also favour crystallinity. Highly crystalline polymers are stronger and less flexible than more amorphous polymers.

crystal structures of ionic compounds include the *sodium chloride structure*, the *caesium chloride structure* and the *fluorite structure*. In an ionic crystal the ions behave like charged spheres in contact. The structures are only stable if each ion is in contact with its nearest neighbours. A caesium ion is large enough to have eight chloride ions around it, as in the caesium chloride structure. Sodium ions are smaller and only big enough to touch six neighbouring chloride ions, as in the sodium chloride structure.

Compounds with ionic *giant structures* are hard, melt at high temperatures and only conduct electricity when molten or dissolved in water.

The *Born–Haber cycle* makes it possible to decide whether or not the bonding in a crystal is purely ionic through the *electrostatic forces* between ions.

crystal structures of metals: the important metal structures are the two *close-packed structures* and the *body-centred cubic structure*.

crystal structures of non-metals: most solid *non-metals*, such as *iodine*, *sulfur* and white *phosphorus*, are molecular so the forces between the particles in the crystals are weak *intermolecular forces*. The crystals easily melt or turn to vapour on gentle heating.

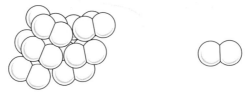

The crystal structure of iodine. Molecules of I₂ held in a regular lattice of intermolecular forces, and a separate iodine molecule.

Some non-metals consist of giant structures of atoms held together by *covalent bonds*. Examples are *carbon* (as graphite or diamond) and *silicon*. Covalent bonds are strong and point in a definite direction so these structures have very high melting points.

Diamond is very hard. Graphite, the other *allotrope* of carbon with a giant structure, has a layer lattice. It too has a high melting point but it can be used as a *lubricant* because the forces between the layers are weak and they can slide over each other.

curly arrows describe the movement of electrons as bonds break and form in the steps of the mechanism of a reaction. A curly arrow with both halves of the arrowhead shows the movement of a pair of electrons. The tail of the arrow starts where the electron pair comes from. The head of the arrow points to where the electron pair ends up:

$$
\begin{array}{ccc}
\underset{\underset{CH_3}{|}}{\overset{\overset{CH_3}{|}}{CH_3-C-Br}} & \longrightarrow & \underset{\underset{CH_3}{|}}{\overset{\overset{CH_3}{|}}{CH_3-\overset{+}{C}}} \quad + \quad Br^-
\end{array}
$$

*Movement of an electron pair during **heterolytic bond breaking***

A curly arrow with only half an arrowhead indicates the movement of a single electron:

$$ Cl:Cl \longrightarrow Cl\bullet \ + \ Cl\bullet $$

*Movement of single electrons during **homolytic bond breaking***

current is a flow of electric charge and is measured in *amperes*. In a metal the flow of charge is the movement of electrons. In an electrolyte, the charge carriers are negative ions moving towards the anode and positive ions moving towards the cathode.

cyclic hydrocarbons are *hydrocarbons* with rings of carbon atoms (but no *benzene* rings). Examples are cycloalkanes (*saturated compounds*) and cycloalkenes (*unsaturated compounds*).

cyclohexane C_6H_{12} cyclohexene C_6H_{10}

Structures of cyclohexane and cyclohexene

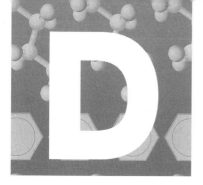

Dalton's law of partial pressures states that the total pressure of a mixture of gases is the sum of the *partial pressures* of the individual gases. This means that the gases in a mixture do not interfere with each other. Each gas makes its own independent contribution to the total pressure.

dative covalent bond: a covalent bond formed when one atom contributes both of the shared pair of electrons. Chemists call this a 'dative covalent bond' because the word 'dative' means giving, and one atom gives both the electrons to make the bond. Once formed there is no difference between a dative bond and any other *covalent bond*.

Ammonia forms a dative covalent bond with a hydrogen ion when it is acting as a base.

Formation of an ammonium ion

An ammonia molecule can form a dative bond with an electron deficient compound such as $AlCl_3$ or BF_3, thus forming a donor-acceptor compound.

Two representations of the donor-acceptor compound formed by ammonia with AlCl$_3$

Dative covalent bonding also accounts for the structures of *carbon monoxide*, the Al_2Cl_6 molecules in *aluminium chloride* vapour and the solid chloride of *beryllium*.

oxonium ion   nitric acid   carbon monoxide

Examples of dative covalent bonds

The alternative name for dative bonding is 'coordinate bonding'. Chemists often use this alternative name when describing the bonding in **complex ions** and other **coordination compounds**.

d-block elements are the elements in the three horizontal rows of elements in periods 4, 5 and 6 for which the last electron added to the atomic structure goes into a **d orbital**. In period 4, the d-block elements run from scandium ($1s^2 2s^2 2p^6 3s^2 3p^6 3d^1 4s^2$) to zinc ($1s^2 2s^2 2p^6 3s^2 3p^6 3d^{10} 4s^2$).

The changes in the properties across a series of d-block elements are much less marked than the big changes across a **p-block element** series. This is because from one element to the next, as the **atomic number** of the nucleus increases by one, the extra electron goes into the inner d subshell. In period 4, the outer shell is always the 4s **orbital** which is filled before the 3d starts to fill.

	3d	4s		3d	4s
Sc [Ar]	↑	↑↓	Fe [Ar]	↑↓ ↑ ↑ ↑ ↑	↑↓
Ti [Ar]	↑ ↑	↑↓	Co [Ar]	↑↓ ↑↓ ↑ ↑ ↑	↑↓
V [Ar]	↑ ↑ ↑	↑↓	Ni [Ar]	↑↓ ↑↓ ↑↓ ↑ ↑	↑↓
Cr [Ar]	↑ ↑ ↑ ↑ ↑	↑	Cu [Ar]	↑↓ ↑↓ ↑↓ ↑↓ ↑↓	↑
Mn [Ar]	↑ ↑ ↑ ↑ ↑	↑↓	Zn [Ar]	↑↓ ↑↓ ↑↓ ↑↓ ↑↓	↑↓

*Electron configurations of d-block elements as free atoms. Note that orbitals fill singly before the electrons start to pair up. Note also that the configurations of **chromium** and **copper** do not fit the general pattern.*

The chemistry of an atom is to a large extent determined by its outer electrons because they are the first to get involved in reactions. So the elements Sc to Zn in period 4 are similar in many ways.

All the d-block elements are **metals** with useful properties for engineering and construction. Most have high melting points. A plot of **physical properties** against proton number often has two peaks corresponding to the half-filling and then filling of the d shell.

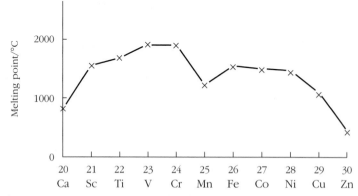

Plot of melting point against atomic number for the elements Ca to Zn

Some, but not all, of the d block elements are classified as **transition elements**.

decantation involves gently pouring off most of a liquid or solution from a solid after centrifuging, or simply allowing it to settle to the bottom of a container.

Filtering is often quicker if the solid is first allowed to settle and then most of the liquid decanted through the filter paper before pouring in the bulk of the solid. As a result most of the liquid passes through the filter paper before its pores are clogged by solid particles.

decarboxylation is a chemical reaction in which a carboxylic acid group (−COOH) is removed and replaced by a hydrogen atom. One method of decarboxylation is to heat the sodium salt of a carboxylic acid with soda lime (a solid reagent that is in effect a mixture of sodium hydroxide and calcium hydroxide). In equations, soda lime is normally represented as NaOH.

$$CH_3CH_2COONa(s) + NaOH(s) \rightarrow CH_3CH_3(g) + Na_2CO_3(s)$$

decomposition is a reaction in which compounds break down into simpler substances, which may be compounds or elements. Heating is often necessary for decomposition (see *thermal decomposition*).

dehydration removes the elements of water from a compound to form a new compound. Concentrated *sulfuric acid* is a powerful dehydrating agent. It dehydrates blue copper sulfate crystals, sucrose and ethanol.

$$CuSO_4 \cdot 5H_2O(s) \xrightarrow{-5H_2O} CuSO_4(s)$$
$$\text{blue} \qquad\qquad\qquad\qquad \text{white}$$

$$C_{12}H_{22}O_{11}(s) \xrightarrow{-11H_2O} 12C(s)$$
$$\text{sucrose, white} \qquad\qquad\qquad \text{black}$$

$$C_2H_5OH(l) \xrightarrow{-H_2O} CH_2{=\!=}CH_2(g)$$
$$\text{ethanol} \qquad\qquad\qquad\qquad \text{ethene}$$

deliquescent substances take up water vapour from the air and dissolve in it. Examples are calcium chloride, potassium hydroxide and sodium hydroxide. Deliquescent substances make good drying agents in *desiccators*.

Deliquescent substances cannot be used as *primary standards* in volumetric analysis because they cannot be weighed accurately.

delocalised electrons: bonding electrons which are not fixed between two atoms in a bond but are shared between three or more atoms. Electron delocalisation accounts for the shape, stability and properties of *benzene* rings, nitrate ions and other ions of *oxoacids*, the acidity of *carboxylic acids* and *phenols*, and the existence of organic *coloured compounds*.

Electron delocalisation takes place in molecules where the conventional structure shows alternating double and single bonds. Delocalisation also affects ions where an atom with a lone pair of electrons and a negative charge is separated by a single bond from a double bond.

The extreme example of delocalisation is *metallic bonding* where electrons are shared between all the atoms in a crystal. Extended delocalisation over the planes of carbon atoms explains the electrical conductivity of graphite.

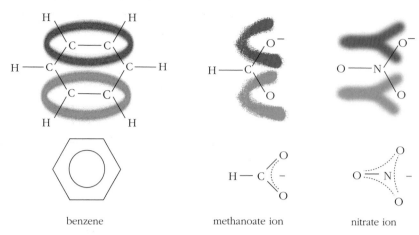

benzene   methanoate ion   nitrate ion

Examples of delocalisation in molecules and ions. The bond lengths of the bonds affected by delocalisation in each of these examples are the same – shorter than single bonds but longer than double bonds.

denaturation of proteins happens when the three-dimensional shape of a *protein* is disrupted by heating, by extremes of pH, or by the presence of *heavy metal* ions. The chemical activity of protein molecules is linked to their shape. Changing the structure of proteins means that they lose their normal activity. *Enzymes*, for example, cease to act as *catalysts* when denatured.

density is the mass per unit volume of a material.

$$\text{density} = \frac{\text{mass}}{\text{volume}}$$

The symbol for density is ρ and the SI unit is kg m^{-3}. In chemistry densities are generally given in g cm^{-3}. The densities for most solids and liquids are in the range 0.5 to 10 g cm^{-3}.

The density of water is 1.0 g cm^{-3} and in approximate laboratory work it is common to assume that dilute aqueous solutions have the same density as water. Solids and immiscible liquids float if they are less dense than water. The density of *ice* is 0.92 g cm^{-3} so ice floats on water.

At room temperature and pressure the densities of gases are about a thousand times smaller than those of solids and liquids. Measurement of gas densities can be used to calculate the *molar masses* of gases.

deprotonation is the removal of a *proton* from a molecule or ion.

derivatives are used by chemists to identify unknown organic compounds. Converting a compound to a crystalline derivative produces a substance which can be purified by recrystallisation and then identified by measuring its melting point. Chemists identify carbonyl compounds by converting them to crystalline derivatives with *2,4-dinitrophenylhydrazine*.

desalination is a process for obtaining pure water from seawater or other sources of water containing dissolved salts. Methods of desalination include *distillation* at low pressure, freezing, reverse *osmosis*, electrodialysis and *ion exchange*.

desiccator: a container with a *drying agent* used to remove the moisture (or other liquids) from chemical products which decompose if warmed in an oven. Chemicals which can be oven dried may also be stored in a desiccator as they cool to stop them picking up moisture from the air. A desiccator is a useful place to store chemicals which must be kept dry.

Vacuum desiccators are fitted with a tap so that air can be pumped out. The partial vacuum speeds up evaporation and diffusion of the vapour to the drying agent.

desulfurisation is a process used to remove sulfur. Desulfurisation processes are used to reduce emissions of sulfur dioxide which is an *atmospheric pollutant*. This reduces *acid deposition*. Examples of desulfurisation are the removal of sulfur from *diesel fuel* and the removal of sulfur dioxide from the flue gases of power stations that burn coal or oil. Sulfur dioxide is an acidic oxide that can be absorbed in a slurry of either *calcium carbonate* or calcium hydroxide.

detergents clean things by removing dirt from surfaces. They are included in *washing powders or liquids*. The chemicals which act as detergents are surface active agents or *surfactants*. Water is a polar solvent with a high *surface tension*. This means that water alone it is not good at removing dirt and grease. Detergents help to clean by:

- lowering the surface tension of water so that is spreads out and wets the surface
- separating grease and particles from the surface
- suspending the dirt in water so that it can be rinsed away.

There are two main types of detergent:

- soap detergents (usually just called *soaps*) made from animal *fats* or *vegetable oils*
- soapless detergents (usually just called detergents) which are made using chemicals from oil.

deuterium is the *isotope* of hydrogen with atomic number 1 but mass number 2. The symbols used for deuterium are ^2_1H or D. About 0.015% of natural hydrogen is the deuterium isotope. Deuterium oxide (heavy water) is D_2O.

The relative mass of D is twice that of H and this difference means that deuterium compounds react more slowly than their normal hydrogen equivalents. During *electrolysis* of acidified water H_2 is formed at the cathode more readily than D_2 so the concentration of deuterium increases as electrolysis continues. Eventually it is possible to make almost pure heavy water; this is used as a moderator in some nuclear power stations.

Deuterated solvents, such as $CDCl_3$, are used in *proton NMR spectroscopy*. While deuterium oxide can be used in proton NMR to detect functional groups such as —OH or —NH_2 which contain *labile protons*.

dialysis is used to separate dissolved ions and small molecules from *colloid* particles. Dialysis uses a *selectively permeable membrane* which lets through the small ions and molecules but traps larger colloid particles. Dialysis is used in medicine to treat the blood of patients with kidney failure. Dialysis removes the waste products of metabolism from blood and helps to adjust the concentration of ions. Blood cells and colloidal-sized protein molecules cannot pass through the membrane.

diamond is one of the *allotropes* of carbon; it consists of a giant structure of atoms. Larger diamonds are valued as gemstones because of the mineral's high refractive index and dispersive power. Diamond is also an important industrial material because it is the hardest natural substance and the best solid conductor of heat energy. Its thermal conductivity is five times as great as that of copper. This makes diamonds very useful for cutting, grinding and polishing other hard materials. Diamond tipped tools are hard wearing and they do not overheat. Industrial diamonds are manufactured by heating graphite to 1800°C at very high pressure in the presence of a metal such as nickel or iron.

diastereoisomers (diastereomers) are *stereoisomers* that are not *enantiomers*. *Chiral compounds* with two chiral centres, such as 2,3-dihydroxybutanedioic acid (tartaric acid), can exist as three stereoisomers. Two of the isomers are asymmetric; they are mirror images of each other and so they are enantiomers. The third stereoisomer has a plane of symmetry at right angles to the bond joining the two central carbon atoms. This is the meso form of the acid which is not optically active and is not an enantiomer.

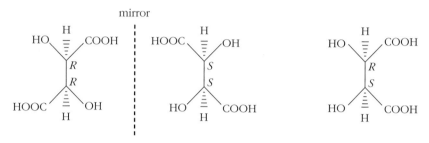

*Stereoisomers of 2,3-dihydroxybutanedioic acid. The chiral centres are labelled with **R–S** names. The meso form has a plane of symmetry so it does not have distinct mirror image forms and is not optically active.*

diatomic molecule: strictly a molecule with two atoms such as N_2 or HCl but the term is generally used for the molecules of elements with two identical atoms. Examples of diatomic elements are oxygen, O_2, nitrogen, N_2 and the halogens, F_2, Cl_2, Br_2 and I_2.

diazonium salts: salts formed when aryl amines, such as *phenylamine*, react with *nitrous acid* (HNO_2) below about 10°C.

Diazonium salts are unstable so they are made as needed and kept cold. Above 10°C, benzene diazonium chloride decomposes to phenol and nitrogen.

The commercial importance of diazonium salts is based on their coupling reactions to form *azo dyes*. Diazonium salts are also useful intermediates which make it possible to form derivatives of arenes.

Diazotisation – the formation of benzene diazonium chloride

dienes are hydrocarbons with two double bonds. Buta-1,3-diene is a product of *thermal cracking* which is a useful intermediate for making other organic compounds. Synthetic *rubbers* are copolymers of butadiene, or one of its derivatives, with other unsaturated compounds such as styrene.

diesel fuel is heavier and oilier than *petrol*. The hydrocarbon molecules in diesel typically contain 14 carbon atoms ($C_{14}H_{30}$), while the molecules in petrol are generally smaller. Less *refining* is needed to make diesel. Diesel *fuel* also has a higher energy density than petrol.

Diesel engines emit *atmospheric pollutants* that include oxides of nitrogen (NO_x), particulate matter and unburnt hydrocarbons, as well as carbon monoxide from their exhausts. The emission of *particulates* is much more of a problem with diesel than with petrol engines. There are technologies based on catalysts to trap particulates and adsorb oxides of nitrogen which can cut emissions by up to 90% but they only work with very low-sulfur fuel.

diffusion is a spreading out and mixing process in gases or solutions as molecules or ions mingle with each other. Molecules diffuse from a region where they are more concentrated to a region where their concentration is lower. Eventually diffusion evens out differences in concentration. The smaller the molecules or ions, the faster they diffuse. Diffusion is a consequence of the rapid, random motion of molecules in gases and liquids. Diffusion through membranes is important in *dialysis* and *osmosis*.

dilution is the process of adding more solvent to a solution to lower the concentration. Quantitative dilution is an important procedure in analysis. The purpose is to make a solution with known concentration by accurately diluting a *standard solution*. Successive dilutions provide a series of solutions which can be used to *calibrate* instruments such as *colorimeters*.

The procedure is to take a measured volume of the more concentrated solution with a pipette and run it into a graduated flask. The flask is then carefully filled to the mark with purified water.

The key to calculating the volumes to use when diluting a solution is to remember that the amount in moles of the reagent in the final solution must equal the amount in moles of the sample taken from the concentrated solution. If c is the concentration in mol dm^{-3} and V is the volume in dm^3:

- the amount in moles of the reagent in the concentrated solution $= c_A V_A$
- the amount in moles of the reagent in the diluted solution $= c_B V_B$

So: $c_A V_A = c_B V_B$

Worked example

What volume of a 1.00 mol dm^{-3} solution of copper(II) sulfate is needed to prepare 100 cm^3 of a 0.1 mol dm^{-3} solution?

Notes on the method

Start by a conversion to give the volume of the diluted solution in dm^3.

V_A can be calculated because all the other terms in the relationship $c_A V_A = c_B V_B$ are known.

Answer

Final volume of the diluted solution is to be $100 \text{ cm}^3 \times \dfrac{1 \text{ dm}^3}{1000 \text{ cm}^3} = 0.1 \text{ dm}^3$

$1.0 \text{ mol dm}^{-3} \times V_A = 0.1 \text{ mol dm}^{-3} \times 0.1 \text{ dm}^3$

$V_A = \dfrac{(0.1 \text{ mol dm}^{-3} \times 0.1 \text{ dm}^3)}{1.0 \text{ mol dm}^{-3}} = 0.01 \text{ dm}^3 = 10 \text{ cm}^3$

Pipetting 10 cm^3 of the concentrated solution into a 100 cm^3 graduated flask and making it up to the mark gives the required dilution.

dimerisation: two molecules linking together. Dimers may be held together by *covalent bonds* (see *aluminium chloride*) or by *hydrogen bonding*.

$$CH_3 - C \overset{O \cdots H-O}{\underset{O-H \cdots O}{}} C - CH_3$$

An ethanoic acid dimer in a non-polar solvent. In aqueous solution, ethanoic acid does not dimerise because there are so many water molecules with which they can form hydrogen bonds.

2,4-dinitrophenylhydrazine: in acid solution this is used to detect and identify *carbonyl compounds*. Mixing an aldehyde or ketone with the reagent produces an yellow–orange precipitate. The *addition–elimination reaction* produces a 2,4-dinitrophenylhydrazone. The solid *derivative* can be filtered off, *recrystallised* and identified by measuring its melting point. This makes it possible to identify the original carbonyl compound.

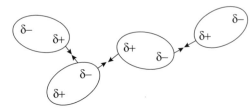

Formation of the 2,4-dinitrophenylhydrazone derivative of propanone

diols are alcohols with two —OH groups. An example is ethane-1,2-diol used as *antifreeze*.

dioxins are a family of stable compounds which are classified as *persistent organic pollutants* (POPs). Chemically they contain chlorine atoms bonded to benzene rings. Some dioxins are toxic and can cause skin disease, cancer and birth defects. Traces of dioxins may form during the manufacture of chlorinated phenols, the bleaching of wood pulp for paper and during the incineration of wastes containing chlorine compounds such as PCBs and the addition polymer PVC.

dipole moments: see *polar molecules*.

dipole–dipole interactions are *intermolecular forces* involving *polar molecules* with permanent dipoles. These cohesive forces account for the fact that compounds such as propanone are liquids at room temperature while comparable non-polar hydrocarbons are gases.

Attractions between molecules with permanent dipoles

diprotic acid: an *acid* such as sulfuric acid (H_2SO_4) and ethanedioic acid ($C_2O_4H_2$), which can give away two *protons* (H^+). A diprotic acid has two values of K_a (and pK_a): one for the first ionisation and another for the second. There are two *end-points* when a diprotic acid is titrated with a strong base.

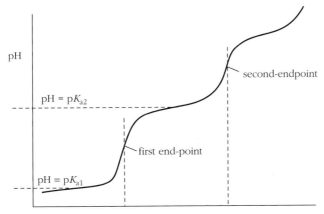

pH changes during acid–base titration *of a weak diprotic acid with a strong base*

disaccharides: see *carbohydrates*.

disinfectants are chemicals which destroy micro-organisms. Chlorine and oxygen *bleaches* are used as disinfectants. Unlike *antiseptics*, disinfectants cannot be used on skin and other living tissues.

displacement reactions are redox reactions which can be used to compare the relative strengths of metals as *reducing agents* and non-metals as *oxidising agents*.

A more reactive metal displaces a less reactive metal from one of its salts. Iron, for example, displaces copper from a copper(II) sulfate solution. The iron atoms reduce the copper ions.

$$Fe(s) + Cu^{2+}(aq) \rightarrow Fe^{2+}(aq) + Cu(s)$$

This reaction is used to extract copper from solutions obtained from low-grade ores by *leaching* with acid.

In *group 7*, a more reactive halogen displaces a less reactive halogen. The order of reactivity for the halogens is Cl > Br > I. The more reactive halogen oxidises the ions of a less reactive halogen.

$$Br_2(aq) + 2I^-(aq) \rightarrow 2Br^-(aq) + I_2(s)$$

Standard electrode potentials make it possible to predict the direction of change in a displacement reaction (see E_{cell} and *electrochemical series*).

displayed formulae show all the atoms and bonds in a molecule (see also *structural formulae*).

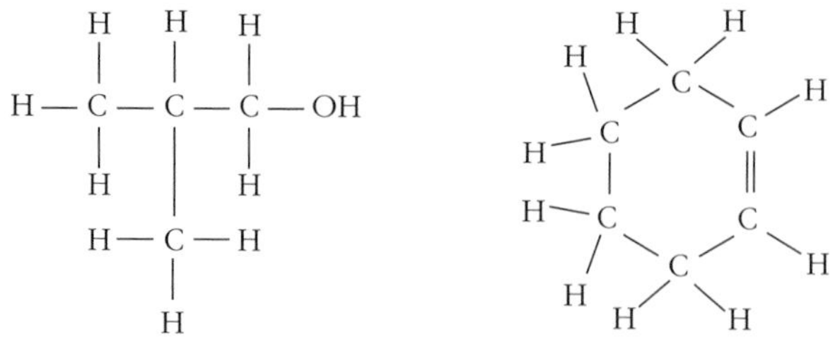

2-methylpropan-1-ol   cyclohexene

Examples of displayed formulae

disproportionation reaction: a reaction in which the same element both increases and decreases its *oxidation number*. *Copper*(I) ions disproportionate in aqueous solution to a mixture of Cu^{2+}(aq) ions and Cu atoms.

A series of disproportionation reactions is used to make *chlorine oxoanions*. The oxygen in *hydrogen peroxide* disproportionates when it decomposes. Note that disproportionation refers to a particular element in a compound, not to the compound as a whole.

(See the *electrochemical series* for the use of standard electrode potential values to predict whether or not disproportionation will take place.)

dissociation: a reaction in which a compound splits into two or more smaller products which may recombine to form the original compound if the conditions change.

Some compounds dissociate on heating. Examples are ammonium chloride and dinitrogen tetroxide. These *thermal dissociation* reactions are *reversible*. On cooling the original compounds reform.

$$NH_4Cl(s) \underset{cool}{\overset{heat}{\rightleftharpoons}} NH_3(g) + HCl(g)$$

$$N_2O_4(g) \underset{cool}{\overset{heat}{\rightleftharpoons}} 2NO_2(g)$$

Acids dissociate into ions in solution in water. The extent to which they ionise distinguishes strong from weak acids and this is measured by the acid dissociation constant, K_a.

distillation: a technique for separating and purifying a liquid by heating to vaporise the liquid and then cooling to condense it. Simple distillation is used to purify water, to recover pure solvents and to separate a liquid product from reaction mixtures. Substances which do not evaporate are left behind in the distillation flask.

One of the easiest ways to determine the boiling point of a liquid is to distil a pure sample of the liquid in an apparatus with a thermometer.

(See also *fractional distillation*, *steam distillation* and *vacuum distillation*.)

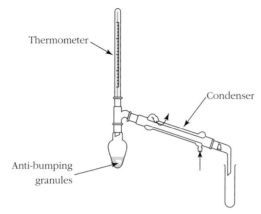

Apparatus for simple distillation

distribution constant: see *partition*.

disulfide bridges: see *cross-linking*.

DNA (deoxyribonucleic acid) is a double helix made up of two poly*nucleotide* chains. Both chains have a sugar phosphate backbone. Every sugar unit has one of four bases linked to it. The four bases are adenine (A), cytosine (C), guanine (G) and thymine (T). The links between the chains are formed by **hydrogen bonding** between pairs of bases: A pairs with T and C pairs with G.

Link to one
strand of DNA
double helix

Link to the other
strand of the DNA
double helix

adenine (A)   thymine (T)

Hydrogen bonding between adenine and thymine in a DNA double helix

Genes are sections of DNA molecules found in the chromosomes of cells.

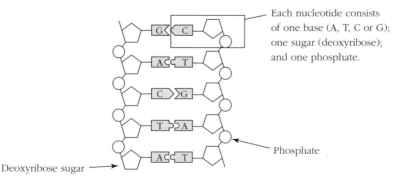

Each nucleotide consists
of one base (A, T, C or G);
one sugar (deoxyribose);
and one phosphate.

Phosphate

Deoxyribose sugar

Representation of part of a DNA double helix

During protein synthesis in cells, transcription produces messenger *RNA* from the genes of DNA. The ribosomes in cells then translate the genetic code into a sequence of *amino acids* in a *protein*.

dolomite is a carbonate mineral with the composition $MgCO_3.CaCO_3$. It is a raw material for the production of magnesium and its salts.

d orbitals are the five atomic orbitals in a d subshell. In a free atom or ion the five d orbitals are at the same energy level. The five orbitals are not the same shape and they split into two groups with different energies when a *d-block element* ion is surrounded by molecules or ions in a complex ion. This helps to account for *coloured compounds* containing complex ions.

dot-and-cross diagrams show the way in which the outer electrons of atoms are shared or transferred when *covalent bonds* and *ionic bonds* form. Dots, crosses and small circles help to keep count of the number of electrons contributed by each atom to bonding. These diagrams derive from the Lewis theory of chemical bonding and are also called Lewis electron-dot diagrams. In the Lewis theory the diagrams are in line with the *octet rule*.

Dot-and-cross diagrams for molecules show both bonding and lone pairs of electrons; they help to predict the *shapes of molecules*.

Despite the use of dots and crosses for the electrons coming from different atoms, all the electrons are the same. Once bonds form it is impossible to say which electron came from which atom.

double bond: two covalent bonds between two atoms as in *oxygen, alkenes* and *ketones*. With two electron pairs involved in the bonding, there is a region of high electron density between the two atoms joined by a double bond.

oxygen	carbon dioxide	ethene
$O{=}O$	$O{=}C{=}O$	

Examples of molecules with double bonds. The double bond holds the atoms of ethene in a plane and prevents rotation.

The molecular orbital model for a double bond shows that one of the bonds is a normal *sigma bond* while the second bond is a *pi bond*.

double salts are solid ionic salts with two different metal cations in a lattice of negative ions. Crystalline double salts are not mixtures but distinct compounds with a definite structure. An example is ammonium iron(II) sulfate, $(NH_4)_2SO_4.FeSO_4.6H_2O$.

When dissolved in water, double salts behave just like a mixture of two simple *salts*. The usual way of making a double salt is to mix equal amounts (in moles) of the two single salts in solution and then to crystallise the double salt. Other examples of double salts are the *alums*.

drugs are substances which affect natural chemical processes in the body.

In medicine, drugs are substances used for the treatment, relief, diagnosis or prevention of disease. A drug is what a doctor prescribes. A *medicine* is the whole formulation.

There are many classes of medical drug which include:

- drugs which act on the central nervous system - such as hypnotics, sedatives, tranquillisers, antidepressants, *narcotics, anaesthetics* and *analgesics*
- drugs which kill bacteria, such as *antibiotics* and sulfonamides (see *chemotherapy*)
- anti-viral drugs, such as those used to combat flu or suppress HIV
- cardiovascular drugs, which help to lower blood pressure, stimulate the heart or control the heartbeat
- drugs such as *cisplatin* used in cancer treatment
- drugs used for the digestive system, such as *antacids*, laxatives and drugs to treat ulcers
- *hormones* such as *insulin*, growth hormone and *steroids*.

The human body has a chemical communication system which is based on *hormones* and neurotransmitters. The chemicals act on the receptor sites on body cells. Faults in this communication system can cause diseases either because too many messenger molecules are released or too few of them. Some drugs act by blocking the receptors for messenger molecules when there are too many of them. Drugs of this type are antagonists. Others drugs make up for a lack of natural messenger molecules and these drugs are called agonists.

People also take drugs for pleasure, stimulation and relaxation. Some recreational drugs such as alcohol and nicotine are legal. Others such as ecstasy, cannabis, cocaine and opiates are illegal. Drugs used to treat disease, such as steroids, can also be exploited to enhance performance in sport. Sensitive methods of chemical analysis are widely used to detect traces of drugs in blood, breath and urine (see *breathalyser* and *gas chromatography*.)

drug development is the research and development programme which leads to the launch of a new drug. Scientists in the industry that produces *pharmaceuticals* use the term 'new chemical entity' (NCE) to describe a newly-discovered chemical for treating disease. Discovering an NCE can be hugely profitable, so pharmaceutical companies spend large sums on research and development.

Before a new compound can become a marketable drug the manufacturer has to show that:

- the drug meets a definite need of patients and their doctors, and is better than existing treatments
- it is technically possible to make it on a large scale
- there is a big enough market to make it commercial
- it is safe.

Claims for the medical benefits of a new medicine and for its use in treatments have to be supported by scientific evidence. The pharmaceutical industry has to carry out a rigorous programme of testing. After screening thousands of chemicals produced by *combinatorial chemistry*, a few chemicals of interest are tested on preparations of cells and tissues. This is *in vitro* testing.

Cells *in vitro* do not have the complex interactions between cells and tissues found in living bodies. *In vivo* testing is needed and this begins in animals. If the drug proves to be safe and effective in animal tests, it can then be approved for clinical trials with humans. In the first phase of a clinical trial, the scientists will typically run a pilot study with a small group of healthy volunteers. As confidence in the drug grows, the trials move on to larger-scale studies in patients that compare the new product with the currently prescribed treatments.

drying agents remove water from moist liquids and gases. Liquid products of organic preparations are often dried with small amounts of anhydrous sodium sulfate or anhydrous magnesium sulfate before the final distillation step. Gases made with laboratory reagents are often moist and can be dried by bubbling them through concentrated sulfuric acid, or more safely by passing them though a U-tube containing *silica gel*. Anhydrous calcium chloride is frequently used in *desiccators* to absorb water, alcohols or amines.

drying oils are oils such as linseed oil which polymerise and set when exposed to air. As a drying oil, linseed oil is used in the manufacture of *paints* and varnishes.

ductility is a property of materials, especially metals, which can be drawn out under tension without breaking. *Copper* is a ductile metal which can be pulled through a die (narrow hole) to make thin wire.

dyes are coloured materials used to dye fabrics and other materials such as paper, hair, leather and food. A good dye is a fast dye which means that it is not only coloured but can also attach itself strongly to a material so that it does not wash out – nor should it fade in the light.

Plants were the main source of natural dyes until William Perkin chanced on the first synthetic dye, mauveine in 1856. Alizarin and indigo were plant dyes which were the basis of large-scale industries. The discovery of synthetic routes to these dyes in the late nineteenth century were early triumphs for *organic chemistry*.

Examples of synthetic dyes are *azo dyes*, *vat dyes* and *fibre-reactive dyes*. Each type of *coloured compound* used in dyes is based on a *chromophore* in a range of chemical structures which include functional groups that can modify the colour, affect the solubility of the dye and help the dye to bond to fibres in fabrics.

dynamic equilibrium: the state of balance in a *reversible* process when neither the forward change nor the backward change is complete; both changes are still going on at equal rates so that they cancel each other out and there is no overall change.

Examples of systems which can be in dynamic equilibrium include:
- water and ice at 0°C, or more generally the liquid and solid states of a substance at its melting point
- a liquid with its saturated *vapour* in a closed container
- a solid with a *saturated solution* of the solid
- a solute distributed between two immiscible solvents
- a *weak acid* in aqueous solution, or any other reversible reaction at constant temperature and pressure.

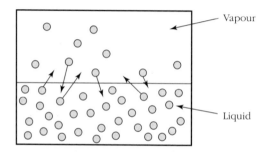

Dynamic equilibrium between a liquid and its vapour

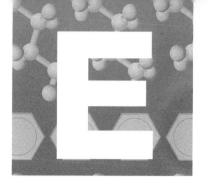

m

$E^{\ominus}_{cell}$ is the standard cell potential. It is the voltage between the terminals of an *electrochemical cell* under *standard conditions* when no current is being taken from the cell.

There is a convenient shorthand for describing cells. The standard emf of the cell, $E^{\ominus}_{cell}$ is written alongside the cell diagram. The agreed convention is that the sign of $E^{\ominus}_{cell}$ gives the polarity of the right-hand electrode.

$$Zn(s) \mid Zn^{2+}(aq) \mathbin{\vdots} Cu^{2+}(aq) \mid Cu(s) \qquad\qquad E^{\ominus}_{cell} = +1.14 \text{ V}$$

If $E^{\ominus}_{cell}$ is positive the reaction tends to go according to the cell diagram read from left to right. As a current flows in a circuit connecting the two electrodes in the cell diagram, zinc atoms turn into zinc ions and dissolve, as copper ions turn into copper atoms and deposit on the copper electrode:

$$Zn(s) + Cu^{2+}(aq) \rightarrow Zn^{2+}(aq) + Cu(s)$$

$E^{\ominus}_{cell}$ values can be worked out for any redox reaction with the help of a table of *standard electrode potentials*.

$$E^{\ominus}_{cell} = E^{\ominus}_{\text{right-hand electrode}} - E^{\ominus}_{\text{left-hand electrode}}$$

$E^{\ominus}_{cell}$ is measured in volts (V). The voltage measures the energy transferred in *joules* per *coulomb* of charge flowing through a circuit connected to a cell. The definition of the volt as a 'joule per coulomb' makes it possible to calculate the standard *free energy change*, $\Delta G^{\ominus}$, for reactions from the values of $E^{\ominus}_{cell}$.

$$\Delta G^{\ominus} = -zFE^{\ominus}_{cell}$$

where z is the number of electrons transferred according to the equation for the reaction and F is the *Faraday constant*.

EDTA is the common abbreviation for a particular ion which binds so firmly with metal ions that it holds them in solution and effectively makes them chemically inactive. A small amount of EDTA can be used as a *food additive* to preserve foods such as salad dressing to trap traces of metal ions which otherwise would catalyse the oxidation of oils. Another practical application is the inclusion of EDTA in bathroom cleaners to help remove limescale by dissolving deposits of calcium carbonate left by hard water. Thanks to this ability to form *chelates*, EDTA can be used to treat lead poisoning.

In laboratories EDTA is the disodium salt of **e**thylene **d**iamine**t**etr**a**acetic **a**cid (or as it is now called 1,2-bis[bis(carboxymethyl)amino]ethane). The ion can fold itself round metal ions so that four oxygen atoms and two nitrogen atoms can present lone pairs to form coordinate bonds with the metal ion. It is a hexadentate *ligand*. This is an example of a *polydentate ligand*.

Complex ion formed by EDTA with a metal ion. For simplicity the disodium salt of EDTA is often represented in equations as Na$_2$Y.

efficiency in an energy system is the percentage of the energy available from the burning fuel (or other energy resource) that ends up where it is wanted, for example as electrical energy.

efflorescence is the loss of *water of crystallisation* from a hydrated salt kept in an open container. Hydrated crystals of sodium carbonate (washing soda, Na$_2$CO$_3$.10H$_2$O) turn powdery in an unsealed container as they gradually lose most of their water.

elastomers are *polymeric materials* which can be stretched to several times their normal length and recover their original size and shape when released. They are elastic materials. Natural rubber is an elastomer. So are synthetic rubbers such as the *copolymer* of butadiene and styrene.

electric arc furnace: a furnace for making *steel* by *recycling* scrap iron and steel. The heat of an arc from carbon electrodes melts the metal. Added lime combines with impurities to form a *slag*. Other ingredients are added as necessary. The process can be precisely controlled to give small batches of steel matched to a specified composition.

electrical conductors are solids, liquids or gases which conduct electricity because they contain charged particles which are free to move. They include:
- metals and graphite in which the charge carriers are *delocalised electrons*
- electrolytes in *electrolysis* cells in which the charge carriers are positive and negative ions.

electrical insulators are solids, liquids or gases which do not conduct electricity because they contain no free-moving charged particles. Increasing the applied voltage may turn an insulator into a conductor if the voltage becomes high enough to ionise the atoms or molecules.

electrochemical cells produce an electric potential difference (voltage) from a redox reaction. Some electrochemical cells are designed for practical use. Examples include *alkaline cells* and rechargeable *lithium polymer cells, nickel–cadmium cells* and *lead–acid cells.*

The reaction of zinc metal with aqueous copper(II) ions is a redox reaction which can be described by two half-equations. Zinc is oxidised to zinc ions as copper(II) ions are reduced to copper metal.

$$Zn(s) \rightarrow Zn^{2+}(aq) + 2e^-$$

$$Cu^{2+}(aq) + 2e^- \rightarrow Cu(s)$$

In an electrochemical cell the two half-reactions happen in separate *half-cells*. The electrons flow from one cell to the other through a wire connecting the electrodes. The electric circuit is completed by a *salt bridge* connecting the two solutions.

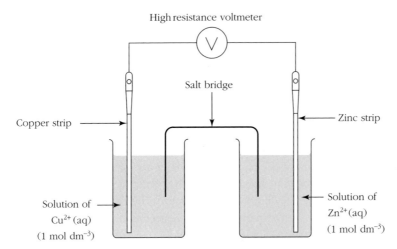

Diagram of an electrochemical cell. The copper electrode is positive and the zinc electrode is negative.

This is chemical convention for representing this cell:

$$Cu(s)\,|\,Cu^{2+}(aq)\,\|\,Zn^{2+}(aq)\,|\,Zn(s) \qquad E^{\ominus}_{cell} = -1.14\ V$$

The tendency for the current to flow is measured with a high-resistance voltmeter which measures the value of $E^{\ominus}_{cell}$ when no current is flowing. In the example shown, electrons tend to flow out of the zinc electrode (negative) round the circuit and into the copper electrode (positive).

electrochemical series: a series showing half-reactions for oxidation and reduction.

The order of the half-reactions is determined by their *standard electrode potentials*. In the list below, the *half-cell* with the most negative electrode potential is at the top of the list. The most negative electrode has the greatest tendency to give up electrons, so it is powerfully reducing.

The half-cell with the most positive electrode potential is shown at the bottom of the list. The most positive electrode has the greatest tendency to gain electrons, so it is the most powerfully oxidising.

(Note that standard electrode potentials may also be listed in the opposite order, in which case the most powerfully reducing electrode is at the bottom and the most powerfully oxidising electrode is at the top.)

Half cell	Half-reaction	$E^{\ominus}/V$
$Na^+(aq)\,|\,Na(s)$	$Na^+(aq) + e^- \rightarrow Na(s)$	-2.71
$Mg^{2+}(aq)\,|\,Mg(s)$	$Mg^{2+}(aq) + 2e^- \rightarrow Mg(s)$	-2.37
$Zn^{2+}(aq)\,|\,Zn(s)$	$Zn^{2+}(aq) + 2e^- \rightarrow Zn(s)$	-0.76
$2H^+(aq)\,|\,H_2(g)$	$2H^+(aq) + 2e^- \rightarrow H_2(g)$	0.00 (by definition)
$Cu^{2+}(aq)\,|\,Cu(s)$	$Cu^{2+}(aq) + 2e^- \rightarrow Cu(s)$	$+0.34$
$[I_2(aq),2I^-(aq)]\,|\,Pt$	$I_2(aq) + 2e^- \rightarrow 2I^-(aq)$	$+0.54$
$[Br_2(aq),2Br^-(aq)]\,|\,Pt$	$Br_2(aq) + 2e^- \rightarrow 2Br^-(aq)$	$+1.09$
$[Cl_2(aq),2Cl^-(aq)]\,|\,Pt$	$Cl_2(aq) + 2e^- \rightarrow 2Cl^-(aq)$	$+1.51$

The top part of the above electrochemical series shows half-reactions for metal ions and metals. Sodium is the most reactive of the metals when it reacts as a reducing agent forming metal ions; copper is the least reactive. This broadly corresponds to the reactivity series for metals and the order of reaction shown by metal/metal ion *displacement reactions*.

The bottom part of the above electrochemical series shows half-reactions for halogens and halide ions. Chlorine is the most reactive of these halogens when it acts as an oxidising agent forming chloride ions. Iodine is the least reactive. This corresponds to the order of reactivity of the halogens as shown by their displacement reactions.

An electrochemical series based on electrode potentials can be used to predict the direction of change in redox reactions. First identify the two half-equations. Write them down one above the other. The more positive half-reaction tends to go from left to right taking in electrons, while the more negative half-reaction goes from right to left.

Using half-reactions and $E^{\ominus}$ values to predict the direction of change

In this way we can predict from electrode potentials whether or not a *disproportionation reaction* is likely to occur. Copper(I) ions disproportionate in aqueous solution, but iron(II) ions tend not to disproportionate.

Standard electrode potentials predict the direction of change but say nothing about the rate of change. A reaction which is *feasible* may not in fact happen because it is so slow. Also, the predictions apply only under standard conditions. Changing the concentrations can alter the direction of change.

More
negative
electrode

$Cu^{2+}(aq) + e^- \rightleftharpoons Cu^+(aq)$ $E^{\ominus} = +0.15$ V

Copper(I)
disproportionates

More
positive
electrode

$Cu^+(aq) + e^- \rightleftharpoons Cu(s)$ $E^{\ominus} = +0.52$ V

More
negative
electrode

$Fe^{2+}(aq) + 2e^- \rightleftharpoons Fe(s)$ $E^{\ominus} = -0.44$ V

Iron(III) reacts
with iron metal
to form iron(II)

More
positive
electrode

$Fe^{3+} + e^- \rightleftharpoons Fe^{2+}$ $E^{\ominus} = +0.77$ V

Using standard electrode potential values to decide whether or not ions are likely to disproportionate

electrode: an electrical conductor that carries an electric current into or out of an *electrolysis* cell, an *electrochemical cell* or a *fuel cell* (see *anode* and *cathode*).

electrode potential: see *standard electrode potential*.

electrolysis is a process which uses an electric current to decompose a molten ionic compound, or a solution of ions, into elements. Ions can conduct electricity only when they are free to move, which is why electrolytes are molten compounds or solutions.

Large-scale manufacturing processes which use electrolysis include *aluminium extraction* and the *electrolysis of brine*. *Electroplating*, *copper refining* and *anodising* are also electrolytic processes.

During the electrolysis of molten salts such as lead bromide:

- metals appear at the *cathode*, e.g. $Pb^{2+}(l) + 2e^- \rightarrow Pb(l)$ (reduction)
- non-metals appear at the *anode*, e.g. $2Br^-(l) \rightarrow Br_2(g) + 2e^-$ (oxidation)

During the electrolysis of a solution of a salt in water, the change at the cathode depends on the type of metal ion in the salt. If the metal is:

- low in the *electrochemical series* (unreactive) it forms at the cathode, e.g.
  $Cu^{2+}(aq) + 2e^- \rightarrow Cu(s)$
- high in the electrochemical series (reactive) the product at the cathode is hydrogen, from hydrogen ions produced by the ionisation of water, e.g. $2H^+(aq) + 2e^- \rightarrow H_2(g)$

During the electrolysis of a solution of a salt in water, the change at the anode may depend on the type of electrode. If the anode is made of carbon or platinum the products are:

- halogen molecules if the ions in the solution are chloride, bromide or iodide ions, e.g.
  $2Cl^-(aq) \rightarrow Cl_2(g) + 2e^-$
- otherwise, oxygen from the water, $4OH^-(aq) \rightarrow O_2(g) + 2H_2O(l) + 4e^-$

The concentration of the solution can affect the product at the anode. Chlorine forms at the anode during electrolysis of a concentrated solution of sodium chloride, but as the solution gets more dilute increasing amounts of oxygen form.

If the anode is a metal such as copper or silver, the atoms in the electrode turn into ions, e.g.

$$Cu(s) \rightarrow Cu^{2+}(aq) + 2e^-$$

The amount of change at an electrode depends on:

- the charge on the ions
- the amount of electric charge that flows.

electrolysis of brine is the basis of the chlor-alkali industry. Electrolysis of brine is used to manufacture chlorine, hydrogen and sodium hydroxide. The process is also the source of chlorine bleach, made by allowing the chlorine and sodium hydroxide to mix and react forming sodium chlorate(I).

Brine is a solution of sodium chloride in water. During electrolysis, chlorine forms at the positive electrode (anode). Hydrogen bubbles off the negative electrode (cathode) while the solution turns into sodium hydroxide.

The cell used for the electrolysis of brine has to be carefully designed because chlorine reacts with sodium hydroxide. The cell has to keep the chlorine and sodium hydroxide apart.

The two main types of cell used for the process in the UK are the flowing mercury cell and the membrane cell. The mercury cell is being phased out because mercury is expensive and hazardous. Traces of mercury escaping into the environment cause pollution.

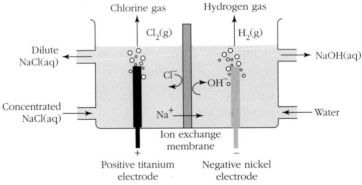

Main inputs and outputs of a membrane cell. The membrane allows the solution to pass through but stops the chlorine mixing with the alkali. The membrane has ion exchange properties; it lets positive ions through but not negative ions.

electromagnetic radiation is radiation such a X-rays, ultraviolet radiation, visible light and infrared rays which can travel through space as oscillating electric and magnetic waves.

There is a spectrum of types of electromagnetic radiation which differ only by their *wavelength*, λ, and *frequency*, v. In space all radiation travels at the same speed, that of light, $c = 2.99 \times 10^8$ m s^{-1}.

$$\lambda = \frac{c}{v} = \frac{2.99 \times 10^8 \text{ m s}^{-1}}{v}$$

	radio frequency		microwave		infrared			ultraviolet	X-rays		γ-rays

	NMR spectroscopy	heating	detecting organic functional groups	initiating chemical reactions	ionising radiations

Frequency v/Hz	10^5	10^6	10^7	10^8	10^9	10^{10}	10^{11}	10^{12}	10^{13}	10^{14}	10^{15}	10^{16}	10^{17}	10^{18}	10^{19}	10^{20}
Wavelength λ/m		10^3		1			10^{-3}			10^{-6}			10^{-9}			

The electromagnetic spectrum

When electromagnetic radiation interacts with matter it behaves as if made up of packets of energy. These energy quanta are also called photons. The higher the frequency, v, of the radiation the higher the energy of the quanta where $E = hv$. The higher energy photons of ultraviolet radiation and X-rays can have a much greater chemical effect than lower energy photons of infrared and radio waves.

electron: one of the *fundamental particles* that make up atoms.

electron affinity: the enthalpy change when gaseous atoms of an element gain electrons to become negative ions. Electron affinities are precisely defined and only used in thermo-chemical cycles such as the *Born–Haber cycle*. Chemical reactions in laboratories do not normally involve free gaseous atoms and ions.

The first electron affinity of an element is the enthalpy change when one mole of gaseous atoms gains electrons to form one mole of gaseous ions. These two equations define the first and second electron affinities for oxygen:

$$O(g) + e^-(g) \rightarrow O^-(g) \qquad \Delta H^\ominus = -141 \text{ kJ mol}^{-1}$$

$$O^-(g) + e^-(g) \rightarrow O^{2-}(g) \qquad \Delta H^\ominus = +798 \text{ kJ mol}^{-1}$$

The gain of the first electron is exothermic but adding the second electron to a negatively charged particle is an endothermic process. Overall, adding two electrons to a gaseous oxygen atom is endothermic.

(Note that chemists use *electronegativity* values and **not** electron affinities to account for polar covalent bonds.)

electron configuration: the number and arrangement of electrons in an atom of an element. Electrons fill energy levels according to the *aufbau principle*.

The electron configuration helps to make sense of the chemistry of an element. It is the electrons in the outer shell which largely determine the chemical properties of an element. Elements in the same *group* of the periodic table have similar properties because they have the same outer electron configuration. There are trends in properties down a group because of the *shielding* effect of the increasing number of inner full shells.

There are several common conventions for writing electron configurations.

A shortened form of electron configuration uses the symbol of the previous *noble gas* to stand for the inner full shells. According to this convention the electron configuration of sodium is [Ne]3s^1.

Representations of the electron configuration of sodium, $1s^2 2s^2 2p^6 3s^1$ including the electron-in-boxes notation (right)

electron density maps show the whereabouts of electrons in crystals. The lines in the maps connect points with equal electron densities. Each map shows the pattern of electron density for a two-dimensional slice through the crystal. The maps are an interpretation of the information from X-ray diffraction studies. This is possible because it is the electrons in a crystal which scatter X-rays (see **X-ray crystallography**).

Electron density maps for an ionic crystal provide evidence for electron transfer from the metal to the non-metal. The ions show up as distinct particles. Electron density maps for molecules show electrons shared between atoms in *covalent bonds*.

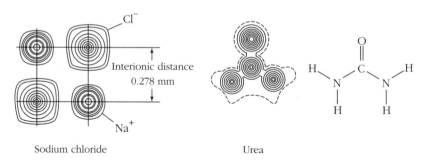

Sodium chloride   Urea

Electron density maps for an ionic crystal and for a molecule

electron-in-boxes notation: see *electron configuration*.

electron-pair repulsion theory: see *shapes of molecules*.

electron transfer takes place during *redox reactions*. An atom, molecule or ion loses electrons when it is oxidised (**o**xidation **is l**oss). The electrons are transferred to the atom, molecule or ion being reduced (**r**eduction **is g**ain). Hence the memory aide: OIL RIG. *Oxidation numbers* and *half-equations* for redox reactions help to keep track of electron transfer reactions.

electronegativity measures the pull of an atom of an element on the electrons in a chemical bond. The stronger the pulling power of an atom, the higher its electronegativity. There are two quantitative scales of electronegativity; one devised by Linus Pauling and the other by Robert Mulliken. However, electronegativity is generally used to compare one element with another qualitatively so it is enough to know the trends in values across and down the periodic table.

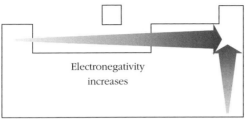

Trends in electronegativity for s- and p-block elements. Values are low to the left of each period and rise towards the right, giving rise to a repeating (periodic) pattern.

The highly electronegative elements, such as fluorine and oxygen, are at the top right of the periodic table. The least electronegative elements, such as caesium, are at the bottom left.

The bigger the difference in the electronegativities of the elements forming a bond, the more polar the bond. Oxygen is more electronegative than hydrogen so an O—H bond is *polar* with a slight negative charge on the oxygen atoms and a slight positive charge on the hydrogen atom.

$$\overset{\delta+}{H} - \overset{\delta-}{Cl} \qquad \overset{H}{\underset{H}{\diagdown}} \overset{\delta+}{C} = \overset{\delta-}{O} \qquad \overset{\delta+}{H} - \overset{2\delta-}{O} \underset{\underset{\delta+}{H}}{|}$$

Examples of polar covalent bonds between atoms with different electronegativities

The bonding in a compound becomes ionic if the difference in electronegativity is large enough for the more electronegative element to remove electrons completely from the other element. This happens in compounds such as sodium chloride, magnesium oxide and calcium fluoride.

electronic structure: the arrangement of electrons in an atom in its main energy levels and sublevels. (See *electron configuration*.)

electrophiles are reactive ions and molecules that attack parts of molecules which are rich in electrons. They are 'electron-loving' reagents. Electrophiles form a new bond by accepting a pair of electrons from the molecule attacked during a reaction.

Examples of electrophiles are the H^+ and the NO_2^+ ions. Other electrophiles are the atoms at the $\delta+$ end of a *polar covalent bond*. See *electrophilic addition* and *electrophilic substitution*.

electrophilic addition takes place when *alkenes* undergo *addition reactions*. The *electrophile* attacks the electron-rich region of the double bond between two carbon atoms. Electrophiles which add to alkenes include hydrogen bromide, bromine, sulfuric acid and water (in the presence of an acid catalyst).

Hydrogen bromide molecules are *polar*. The hydrogen atom, with its $\delta+$ charge is the electrophilic end of the molecule.

*Electrophilic addition of hydrogen bromide to ethene. The reaction takes place in two steps. The intermediate has a positive charge on a carbon atom. It is a carbocation. Curly arrows show the movement of a pair of electrons. Bond breaking is **heterolytic** and the intermediates are ions.*

Bromine molecules are not polar but they become polarised as they approach the electron-rich double bond. Electrons in the double bond repel electrons in the bromine molecule. The δ^+ end of the molecule is electrophilic.

Electrophilic addition of bromine to ethene

(See also *Markovnikov's rule*.)

electrophilic substitution is the characteristic reaction of *arenes* such as benzene. The electrophile attacks the electron-rich benzene ring with its six *delocalised electrons*. When describing the mechanism of this type of reaction it is easier to keep track of what is happening to the electrons using the *Kekulé structure* for *benzene*.

Electrophilic substitution of bromine in benzene is rapid in the presence of iron. Iron reacts with some bromine to form iron(III) bromide, which is the catalyst. Iron(III) bromide is an electron-pair acceptor (Lewis acid) which produces the electrophile, Br^+. An alternative catalyst is *aluminium chloride* (another Lewis acid). The catalysts for this reaction are sometimes described as *halogen carriers*.

Formation of the electrophile

The first step of electrophilic substitution in benzene is very similar to the first step of *electrophilic addition* to alkenes. The intermediate could then complete the reaction by adding a bromide ion but in fact what happens is the loss of H^+ leading to substitution. Loss of H^+ is preferred because it keeps the stable benzene ring stabilised by six delocalised electrons.

The mechanism of electrophilic substitution. Note that an alternative second step (addition of a bromide ion) is not favoured because it gives a product which does not have a stable benzene ring.

Other examples of electrophilic substitution are the **nitration of benzene**, the **sulfonation of benzene** and **Friedel–Crafts reactions**.

electrophoresis is a technique for separating and identifying organic compounds. It can be used to separate compounds such as **amino acids** which become electrically charged when dissolved in water at a particular pH. Electrophoresis is used in hospitals to diagnose disease. It is important in genetic profiling to separate out the bands which form the 'DNA fingerprint'. It is also used in research to study the structures of **proteins** and **nucleic acids**.

Electrophoresis takes place on strips of a gel soaked in a buffer solution to keep the pH constant. Small spots of several samples can be put on each gel. The charged molecules move when an electric voltage is set up between the ends of the gel and they separate into bands according to their size and charge. Most samples are colourless so, once the separation is complete, the gel is dipped into a stain to show up the bands.

electroplating uses **electrolysis** to coat one metal with another. The process is often used to coat metals with thin layers of copper, nickel, chromium, silver or zinc. The object to be plated forms the cathode in an electrolysis cell containing a solution of ions of the metal to be plated. Good results depend on careful control of the conditions, including the composition and concentration of the electrolyte, the temperature and the size of the current. Often the metal ion is present as a **complex ion**. During silver plating, for example, the silver ions are complexed with cyanide ions.

electrostatic forces are the forces between charged particles. Unlike charges attract each other. Positive ions, for example, attract negative ions in ionic compounds. Like charges repel each other. Positive ions in an electrolyte are repelled by the positive anode while being attracted to the negative cathode.

The size of the electrostatic force between two charges varies according to Coulomb's law. The bigger the charges, the stronger the force – the force is proportional to the size of the charges, Q. The greater the distance, d, between the two charges, the smaller the force – the force is inversely proportional to the distance squared:

electrostatic force $\alpha \dfrac{Q_1 \times Q_2}{d^2}$

electrovalent bonding: see **ionic bonding**.

elements are chemically the simplest substances. All the atoms of an element have the same number of protons in the nucleus. The atomic number identifies an element and fixes its position in the **periodic table**. When an atom turns into an ion by gaining or losing electrons it becomes a charged atom of the same element. The **isotopes** of an element have different numbers of neutrons in the nucleus but they have the same chemical properties because they have the same number of protons, and therefore the same **electron configuration**.

133

elimination reaction: a reaction which splits off a simple compound from a molecule while forming a double bond.

Elimination of a *hydrogen halide* from a *halogenoalkane* produces an *alkene*. The conditions are to heat the halogenoalkane with a solution of potassium hydroxide in ethanol. Using ethanol as the solvent instead of water makes the alternative *nucleophilic substitution* reaction less likely.

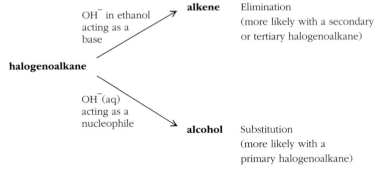

halogenoalkane

OH⁻ in ethanol acting as a base → **alkene** Elimination (more likely with a secondary or tertiary halogenoalkane)

OH⁻ (aq) acting as a nucleophile → **alcohol** Substitution (more likely with a primary halogenoalkane)

Competition between elimination and substitution

The elimination reaction is favoured if the halogen atom is in the middle of the carbon chain or at a branch in the chain.

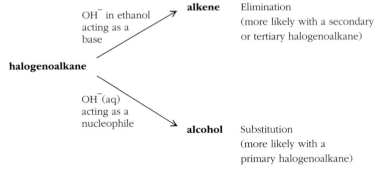

Elimination of hydrogen bromide from 2-bromopropane

Another example of an elimination reaction is the removal of water from an *alcohol* to form an alkene. This is a useful reaction in synthesis for introducing double bonds into molecules. The conditions for reaction are either to pass the vapour of the alcohol over a hot, solid catalyst such as aluminium oxide, or to dehydrate the alcohol by heating it with phosphoric acid or sulfuric acid.

Formation of an alkene from an alcohol

The acid catalyst protonates the —OH group so that the **leaving group** is a water molecule – a better leaving group than a hydroxide ion. The proton is reformed in the second step of the reaction, so overall the acid catalyst is unchanged.

Ellingham diagrams explain the conditions for metal extraction from oxide ores. The diagrams are graphs showing how the free energy changes of formation (ΔG) of oxides vary with temperature.

emf (electromotive force) measures the 'voltage' of an electrochemical cell (see $E^{\ominus}_{cell}$).

emission spectrum: see *atomic spectrum* and *atomic emission spectroscopy*.

empirical formula: the formula of a compound found by calculation from the combining masses of elements. The empirical formula shows the simplest ratio of the amounts of elements in the compound; it therefore gives the ratio of the numbers of atoms.

For molecular compounds, the *relative molecular mass* shows whether or not the empirical formula is the same as the *molecular formula*.

m

Worked example

Analysis of a sample of a salt shows that it consists of 1.30 g zinc combined with 0.24 g carbon and 0.96 g oxygen. What is the formula of the salt?

Notes on the method

Look up the molar masses of the elements in a book of data.

Recall that: amount of substance/mol $= \dfrac{\text{mass of substance/g}}{\text{molar mass/g mol}^{-1}}$

Answer	**zinc**	**carbon**	**oxygen**
Combining masses	1.30 g	0.24 g	0.96 g
Molar masses of elements	65 g mol^{-1}	12 g mol^{-1}	16 g mol^{-1}
Amounts combined	$\dfrac{1.30\,\text{g}}{65\,\text{g mol}^{-1}}$	$\dfrac{0.24\,\text{g}}{12\,\text{g mol}^{-1}}$	$\dfrac{0.96\,\text{g}}{16\,\text{g mol}^{-1}}$
	= 0.02 mol	= 0.02 mol	= 0.06 mol
Simplest ratio of amounts	1	1	3

The formula is $ZnCO_3$.

---

emulsions consist of droplets of one liquid finely dispersed in another liquid. Emulsions are *colloids*. Milk is an emulsion of fat droplets in water. Salad cream is an emulsion of vegetable oil in vinegar. An emulsion is often 'thicker' (more *viscous*) than either liquid on its own. Many cosmetic and medical creams are emulsions.

Just shaking two immiscible liquids together generally produces droplets which quickly separate back into two layers on standing. It is usually necessary to add an emulsifying agent to stabilise an emulsion. Many creamy foods are emulsions and the lists of their ingredients include *food additives* that are emulsifying agents such as lecithin and alginates. Emulsifying agents are *surfactants*.

enantiomers: see *chiral compounds*.

endothermic changes take in energy from their surroundings. Melting and evaporation are endothermic *changes of state*.

For an endothermic reaction, the enthalpy change, ΔH, is positive. During an endothermic reaction more energy is taken up in breaking bonds in the reactants than is given out as new bonds form in the products.

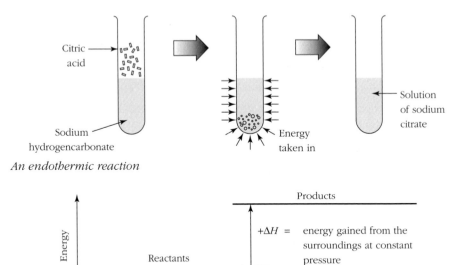

An endothermic reaction

Energy level diagram for an endothermic reaction

The fact that there are *spontaneous* endothermic reactions shows that it is not safe to predict that a reaction will not happen just because the *enthalpy change*, ΔH, is positive. Even if ΔH is positive, the *free energy change* (ΔG) can be negative if the *entropy* change (ΔS_{system}) is large enough and positive.

end-point: the point during a titration when a colour change shows that enough of the solution in the burette has been added to react exactly with the amount of chemical in the flask. Ideally the end-point corresponds with the *equivalence point*. In *acid–base titrations* it is common to use a coloured *acid–base indicator* to detect the end-point.

In some *redox titrations* the end-point is shown by a permanent change in the colour of the solution in the flask once a drop of excess reagent has been added. In *potassium manganate(vii)* titrations the solution in the flask becomes permanently pink with only a small excess of the purple-coloured solution of manganate(vii) ions.

energy conservation is a consequence of the first law of *thermodynamics*. In the sciences, energy conservation helps to keep account of *energy transfers* because the total quantity of energy before and after any change is the same.

Public campaigns to 'save energy' have a different meaning. They refer to the importance of conserving fuels and other useful sources of energy. Energy is conserved in the scientific sense when fuels burn but the energy is spread around at a relatively low temperature in the environment so it can no longer be used for heating and to drive engines.

energy levels in atoms are the energies of the electrons in the available *atomic orbitals*. According to *quantum theory* each electron in an atom has a definite energy. When atoms gain or lose energy, the electrons jump from one energy level to another.

energy (enthalpy) level diagrams are a convenient way of displaying and comparing energy changes. The diagrams show the difference in energy between one state and another. Chemists use these diagrams to show the *energy levels* in atoms, and to summarise *exothermic* and *endothermic* processes.

energy recovery is one of the ways of dealing with plastic waste, and other combustible waste. The energy from *incineration* of the waste is recovered to generate electricity.

energy transfers from one *system* to another system, or between a system and its surroundings, happen in chemistry by heating, by the emission and absorption of radiation (such as light) or by the flow of electric charges.

enthalpy change: the exchange of energy between a reaction mixture (*system*) and its surroundings when the reaction takes place at constant *pressure*. The enthalpy change is calculated when the starting temperature of the reactants and final temperature of the products are the same. The symbol for an enthalpy change is ΔH and the units are kJ mol^{-1}.

The sign of ΔH is decided by noting what happens to the reaction mixture.
- If the reaction is *exothermic* the system loses energy to its surroundings and ΔH is negative.
- If the reaction is *endothermic* the system gains energy from its surroundings and ΔH is positive.

Standard enthalpy changes for reactions are calculated for carefully specified conditions and for precise amounts of chemicals (in moles). The standard conditions for enthalpy changes are 298 K and 1 bar (= 100 kPa). The symbol for a standard enthalpy change is $\Delta H^{\ominus}_{298}$

Books of data list standard enthalpy changes but it is often impossible to measure them directly under the stated conditions. Many values are determined indirectly using *Hess's law*. Values measured under non-standard conditions are corrected to give the value under standard conditions.

Enthalpy changes for changes of state (melting, vaporising or subliming) are usually measured at the melting or boiling point.

enthalpy change of atomisation is the *enthalpy change* to produce one mole of gaseous atoms from an element. The element at the start has to be in its *standard state* (normal, stable state) at 298 K and 1 bar *pressure*.

$$\text{element in its standard state} \xrightarrow{\Delta H^{\ominus}_{at}} \underset{1 \text{ mol}}{\text{gaseous atoms}}$$

Values for the standard enthalpy of atomisation of elements are used in the *Born–Haber cycle* to determine *lattice energies*. They are also used to calculate *bond enthalpies*.

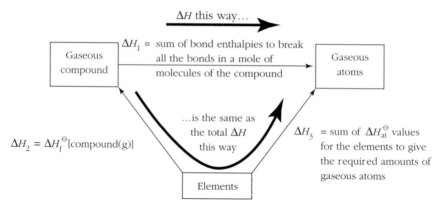

ΔH this way...

ΔH_1 = sum of bond enthalpies to break all the bonds in a mole of molecules of the compound

| Gaseous compound | | Gaseous atoms |

...is the same as the total ΔH this way

$\Delta H_2 = \Delta H_f^{\ominus}[\text{compound(g)}]$

ΔH_3 = sum of $\Delta H_{at}^{\ominus}$ values for the elements to give the required amounts of gaseous atoms

Elements

Thermochemical cycle to determine average bond enthalpies, $\Delta H_1 = -\Delta H_2 + \Delta H_3$

Like all thermochemical quantities the precise definition is important. The standard enthalpy change of atomisation of an element, $\Delta H^{\ominus}_{at,\,298}$, is the enthalpy change when one mole of gaseous atoms forms from the element in its **standard state** under **standard conditions**.

enthalpy change of combustion is the *enthalpy change* when one mole of a compound burns completely in oxygen. The compound and the products of burning must be in their normal stable states. For a carbon compound, complete combustion means that all the carbon burns to carbon dioxide and that there is no soot or carbon monoxide. When burning a compound containing hydrogen, the water formed must end up as a liquid, not gas.

Values of enthalpies of combustion are much easier to measure than many other enthalpy changes. The importance of these values is that, with the help of *Hess's law*, they can be used to calculate *enthalpy changes of formation*.

Like all thermochemical quantities the precise definition of standard enthalpy of combustion is important. The standard enthalpy change of combustion of a substance, $\Delta H^{\ominus}_{c,\,298}$, is the enthalpy change when one mole of the substance burns completely in oxygen under **standard conditions** with the reactants and products in their **standard states**.

enthalpy change of formation is the *enthalpy change* when one mole of a compound forms from the elements in their normal stable states. The more stable state of an element is chosen where there are **allotropes**. For carbon, perhaps surprisingly, graphite turns out to be more stable thermochemically than diamond.

Tables of standard enthalpies of formation are very useful because, with a *Hess's law* cycle, they can be used to calculate the enthalpy changes for reactions.

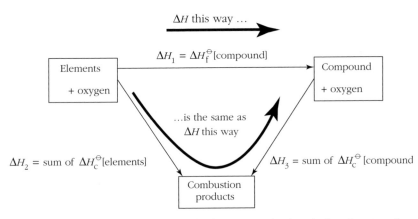

ΔH this way ...

$\Delta H_1 = \Delta H_f^{\ominus}$[compound]

| Elements |
| + oxygen |

→

| Compound |
| + oxygen |

...is the same as ΔH this way

ΔH_2 = sum of $\Delta H_c^{\ominus}$[elements]

ΔH_3 = sum of $\Delta H_c^{\ominus}$ [compound

| Combustion |
| products |

*Outline of a thermochemical cycle for calculating standard enthalpy changes of formation from standard **enthalpy changes of combustion**, $\Delta H_1 = \Delta H_2 - \Delta H_3$*

Like all thermochemical quantities, the precise definition of standard enthalpy of formation is important. The standard enthalpy change of formation of a compound, $\Delta H_{f,98}^{\ominus}$, is the enthalpy change when one mole of the compound forms from its elements under ***standard conditions*** with the elements and the compound in their ***standard states***.

Worked example

Calculate the enthalpy of formation of propane, C_3H_8, at 298 K given these standard enthalpies of combustion:

- propane, $\Delta H_c^{\ominus}$ [C_3H_8] = – 2220 kJ mol^{-1}
- carbon, $\Delta H_c^{\ominus}$ [C] = – 393 kJ mol^{-1} v
- hydrogen, $\Delta H_c^{\ominus}$ [H_2] = – 286 kJ mol^{-1}

Notes on the method

Draw up a thermochemical cycle. Use Hess's law to produce an equation for the enthalpy changes. All the enthalpy changes are given except $\Delta H_f^{\ominus}$[C_3H_8].

Pay careful attention to the signs. Put the value and sign for a quantity in brackets when multiplying, adding or subtracting enthalpy values.

Answer

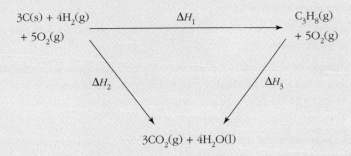

$3C(s) + 4H_2(g)$
$+ 5O_2(g)$

ΔH_1

$C_3H_8(g)$
$+ 5O_2(g)$

ΔH_2

ΔH_3

$3CO_2(g) + 4H_2O(l)$

According to Hess's law $\Delta H_1 = \Delta H_2 - \Delta H_3$

$\Delta H_1 = \Delta H_f^{\ominus} [C_3H_8]$

$\Delta H_2 = 3 \times \Delta H_c^{\ominus} [C] + 4 \times \Delta H_c^{\ominus} [H_2] = 3 \times (-393 \text{ kJ mol}^{-1}) + 4 \times (-286 \text{ kJ mol}^{-1})$
$$= -2323 \text{ kJ mol}^{-1}$$

$\Delta H_3 = \Delta H_c^{\ominus} [C_3H_8] = -2220 \text{ kJ mol}^{-1}$

Hence $\Delta H_f^{\ominus} [C_3H_8] = (-2323 \text{ kJ mol}^{-1}) - (-2220 \text{ kJ mol}^{-1})$
$$= -2323 \text{ kJ mol}^{-1} + 2220 \text{ kJ mol}^{-1}$$
$$= -103 \text{ kJ mol}^{-1}$$

enthalpy change of hydration is the *enthalpy change* for the hydration of one mole of gaseous under standard conditions to produce a solution in which the concentration of ions is 1 mol dm^3.

$$Na^+(g) + aq \rightarrow Na^+(aq) \qquad \Delta H^{\ominus}_{\text{hydration}} = -390 \text{ kJ mol}^{-1}$$

In this equation '+ aq' is short for adding water. The enthalpy change of hydration is a specific example of *enthalpy change of solvation*.

enthalpy change of melting (fusion) is the *enthalpy change* when one mole of a solid turns to a liquid at its melting point.

Physicists generally use 'specific latent heat' values when studying the energy changes when solids melt. Specific latent heats of melting are measured in kilojoules per kilogram rather than kilojoules per mole.

enthalpy change of neutralisation is the *enthalpy change of reaction* when an acid and an alkali neutralise each other.

$$HCl(aq) + NaOH(aq) \rightarrow NaCl(aq) + H_2O(l) \qquad \Delta H^{\ominus}_{\text{neutralisation}} = -57.1 \text{ kJ mol}^{-1}$$

The size of enthalpy of neutralisation for dilute solutions of **strong acid** with **strong base** is always close to 57.1 kJ mol^{-1}. The reason is that these acids and alkalis are fully ionised, so in every instance the reaction is the same:

$$H^+(aq) + OH^-(aq) \rightarrow H_2O(l) \qquad \Delta H^{\ominus} = -57.1 \text{ kJ mol}^{-1}$$

The enthalpy changes of neutralisation for reactions involving weak acids and weak bases differ from this value because there are other energy changes, such as the energy needed to remove a proton from a weak acid.

Enthalpies of neutralisation can be measured approximately by mixing solutions of acids and alkalis in a *calorimeter*.

enthalpy change of reaction is the enthalpy change when the amounts shown in the chemical equation react. The standard enthalpy change for reaction is defined at 298 K, 1 bar pressure and with the reactants and products in their normal, stable states under these conditions. The concentration of any solutions is 1 mol dm^{-3}.

The enthalpy change for a reaction can easily be calculated from tabulated values for standard *enthalpy changes of formation* using *Hess's law*.

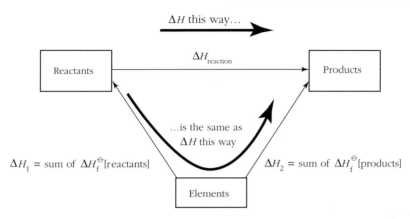

Outline of a thermochemical cycle for calculating standard enthalpies of reaction from standard enthalpies of formation

According to Hess's law:

$$\Delta H^{\ominus}_{\text{reaction}} = -\Delta H_1 + \Delta H_2 = -\Delta H_2 - \Delta H_1$$

$$\Delta H^{\ominus}_{\text{reaction}} = \text{sum of } \Delta H^{\ominus}_f \text{ [products]} - \text{sum of } \Delta H^{\ominus}_f \text{ [reactants]}$$

Worked example

Calculate the enthalpy change for the reduction of iron(III) oxide by carbon monoxide.

$\Delta H^{\ominus}_f [Fe_2O_3] = -824 \text{ kJ mol}^{-1}$

$\Delta H^{\ominus}_f [CO] = -110 \text{ kJ mol}^{-1}$

$\Delta H^{\ominus}_f [CO_2] = -393 \text{ kJ mol}^{-1}$

Notes on the method

Write the balanced equation for the reaction. All the enthalpy changes are given except ΔH for the reaction. Note that by definition $\Delta H^{\ominus}_f$ [element] $= 0$ kJ mol^{-1}.

Pay careful attention to the signs. Put the value and sign for a quantity in brackets when multiplying, adding or subtracting enthalpy values.

Answer

$$Fe_2O_3(s) + 3CO(g) \rightarrow 2Fe(s) + 3CO_2(g)$$

So $\Delta H^{\ominus}_{\text{reaction}} = \{\text{sum of } \Delta H^{\ominus}_f \text{ [products]}\} - \{\text{sum of } \Delta H^{\ominus}_f \text{[reactants]}\}$

$\Delta H^{\ominus}_{\text{reaction}}$ $\{2 \times \Delta H^{\ominus}_f [Fe] + 3 \times \Delta H^{\ominus}_f [CO_2]\} - \{\Delta H^{\ominus}_f [Fe_2O_3] + 3 \times \Delta H^{\ominus}_f [CO]\}$

$= \{0 + 3 \times (-393 \text{ kJ mol}^{-1})\} - \{(-824 \text{ kJ mol}^{-1}) + 3 \times (-110 \text{ kJ mol}^{-1})\}$

$= (-1179 \text{ kJ mol}^{-1}) - (-1154 \text{ kJ mol}^{-1}) = -25 \text{ kJ mol}^{-1}$

enthalpy change of solution is the *enthalpy change* when one mole of a substance dissolves in a stated amount of water under standard conditions,

$$NaCl(s) + aq \rightarrow Na^+(aq) + Cl^-(aq) \qquad \Delta H^{\ominus}_{\text{solution}} = +4 \text{ kJ mol}^{-1}$$

In this equation '+ aq' is short for adding water.

The enthalpy change of solution is the difference between the energy needed to separate the ions from the crystal lattice (– *lattice energy*) and the energy given out as the ions are hydrated (sum of the *enthalpy changes of hydration*).

$$\Delta H^{\ominus}_{\text{solution}} = \Delta H^{\ominus}_{\text{hydration cation}} + \Delta H^{\ominus}_{\text{hydration anion}} - \Delta H^{\ominus}_{\text{lattice}}$$

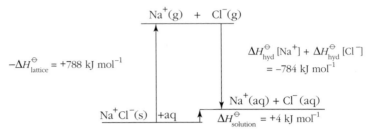

Energy (enthalpy) level diagram for sodium chloride dissolving in water. The enthalpy change for going from the crystal to the gaseous ions is (–lattice enthalpy). This is the second type of Born–Haber cycle for analysing enthalpy changes involving ionic compounds.

Note that the enthalpy change of solution is a small difference between two large enthalpy changes. Also note that the lattice energy and the hydration enthalpies tend to be affected in the same way by changes, to the sizes of the ions and their charges. The smaller the ions and the larger the charges, the greater the lattice energy but also the greater the hydration enthalpies. This means that it is not easy to make use of enthalpy cycles to account for trends in solubilities of ionic compounds down group 1 or group 2 in the periodic table.

Also note that an ionic salt may well dissolve in water even if $\Delta H^{\ominus}_{\text{solution}}$ is slightly positive. Dissolving makes a significant difference to the numbers of ways twhat particles and energy can be arranged. The *entropy* change of the system, $\Delta S^{\ominus}_{\text{system}}$ may be positive and large enough to outweigh a negative entropy change in the surroundings.

enthalpy change of solvation is the enthalpy change when one mole of gaseous ions reacts with solvent molecules under standard conditions to produce a solution in which the concentration of ions is 1 mol dm^{-3}. Ions are hydrated when the solvent is water (see *enthalpy change of hydration*).

enthalpy change of vaporisation is the *enthalpy change* when one mole of a liquid turns to a *vapour* at its boiling point. Evaporation separates the particles in a liquid, so values for the enthalpy change of vaporisation give a measure of the strength of the bonding between particles in liquids. Substances with strong ionic or metallic bonding have much higher boiling points and enthalpies of vaporisation than substances consisting of molecules with weak *intermolecular forces*.

Physicists generally use specific latent heat values in studying the energy changes when liquids evaporate. Specific latent heats of vaporisation are measured in joules per kilogram rather than joules per mole.

entity: a term which can refer to any distinct particle such as an atom, molecule, ion, electron or free radical. The term is used in precise definitions such as the definition of *amount of substance* in moles. It is important to be precise when specifying a particular entity. These, for example, are four distinct chemical entities: a hydrogen atom, H; a hydrogen molecule, H_2; a hydrogen ion, H^+; and a hydride ion, H^-.

entropy *(S)* is a thermochemical quantity which makes it possible to predict the direction of changes (see also *standard molar entropy*). Change happens in the direction which leads to a total increase in entropy. When considering chemical reactions, it is convenient to calculate the total entropy change in two parts: the entropy change of the system and the entropy change of the surroundings.

$$\Delta S_{total} = \Delta S_{system} + \Delta S_{surroundings}$$

The entropy of a system measures randomness, or degree of disorder, in a system based on the number of ways, W, of arranging the molecules and sharing out the energy between the molecules. Nothing seems to be happening in a closed flask of gas kept at a constant temperature. The *kinetic theory of gases* tells us, however, that the molecules are in rapid motion, colliding with each other and with the walls of the container. The gas stays the same while constantly rearranging its molecules and redistributing the energy. The 'number of ways', W, is large for a gas. Gases generally have higher entropies than liquids which in turn have higher entropies than solids.

The relationship between S and W was derived by Ludwig Boltzmann (1844–1906), the Austrian physicist who was the first person to explain the laws of thermodynamics in terms of the behaviour of atoms and molecules in motion:

$$S = k \ln W$$

where S is the entropy of the system, k is the Boltzmann constant and $\ln W$ is the natural logarithm of the number of ways of arranging the particles and energy in the system.

The entropy change of the surroundings during a chemical reaction is determined by the size of the *enthalpy* change, ΔH, and the temperature, T. The relationship is:

$$\Delta S_{surroundings} = -\frac{\Delta H}{T}$$

The minus sign is included because the entropy change is bigger the more energy is transferred to the surroundings. For an exothermic reaction, which transfers energy to the surroundings, ΔH is negative, so $-\Delta H$ is positive.

For many exothermic reactions at about room temperature, the value of $-\Delta H/T$ is much larger than ΔS_{system} which means that ΔS_{total} is positive. This explains why exothermic reactions generally tend to go (see *entropy changes and feasibility*), but there are exceptions.

Chemists often find it more convenient to work with *free energy changes* which can be regarded as 'total entropy changes in disguise'.

entropy changes and feasibility: a reaction is only *feasible* if the total *entropy* change, ΔS_{total}, is positive:

$$\Delta S_{total} = \Delta S_{system} + \Delta S_{surroundings}$$

It turns out that:
- most exothermic reactions tend to go because at about room temperature the value of $-\Delta H/T$ is much larger and more positive than ΔS_{system} which means that ΔS_{total} is positive.
- an endothermic reaction can be feasible so long as the increase in the entropy of the system is greater than the decrease in the entropy of the surroundings.
- a reaction which does not tend to go at room temperature may become feasible as the temperature rises because $\Delta S_{surroundings}$ decreases in magnitude as T increases.

Entropy change of the system	Enthalpy change	Entropy change of the surroundings	Is the reaction feasible?
increase (ΔS_{system} positive)	exothermic ($\Delta H_{reaction}$ negative)	increase ($\Delta S_{surroundings}$ positive)	Yes: ΔS_{total} is positive
decrease (ΔS_{system} negative)	exothermic ($\Delta H_{reaction}$ negative)	increase ($\Delta S_{surroundings}$ positive)	Maybe: ΔS_{total} is positive if $\Delta S_{surroundings}$ is more positive than ΔS_{system} is negative
increase (ΔS_{system} positive)	endothermic ($\Delta H_{reaction}$ positive)	decrease ($\Delta S_{surroundings}$ negative)	Maybe: ΔS_{total} is positive if ΔS_{system} is more positive than $\Delta S_{surroundings}$ is negative
decrease (ΔS_{system} negative)	endothermic ($\Delta H_{reaction}$ positive)	decrease ($\Delta S_{surroundings}$ negative)	No: ΔS_{total} is negative

environmental chemistry is the study of chemical changes in the environment. Chemists use a range of analytical methods to estimate the amounts of elements in the environment. They apply the theory of *reaction kinetics, equilibrium law* and *thermochemistry* to explain the changes they observe.

Environmental chemists use models to summarise their findings and to make predictions. The models show how the various elements cycle through the environment (see *carbon cycle* and *nitrogen cycle*). The models show the main reservoirs for an element and the size of the flows of an element from one reservoir to another.

Knowledge of the scale of natural changes shows whether or not human activity is likely to disturb natural changes locally, regionally or globally. So environmental chemistry helps scientists to assess the seriousness of various types of pollution (see *atmospheric pollutants, acid deposition, incineration, landfill, CFCs, ozone in the stratosphere, ozone in the troposphere, greenhouse effect, persistent organic pollutants, steady state systems* and *waste*).

enzymes are *protein* molecules which are the *catalysts* for biochemical reactions that would otherwise be very slow. Saliva contains the enzyme amylase which aids digestion by speeding up the *hydrolysis* of starch to *sugar*. Catalase is an enzyme which protects living cells from a build up of hydrogen peroxide by speeding up its decomposition to water and oxygen. Some washing powders contains proteases – enzymes which break down proteins such as blood on dirty clothes.

An enzyme molecule consists of a coiled protein chain. The coiling gives rise to an active site with just the right shape and size for the *substrate* molecules. The enzyme can act only on molecules with the right shape to fit into the active site. This accounts for the specificity of enzymes.

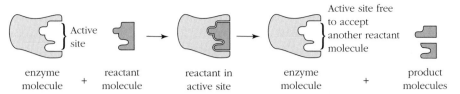

enzyme molecule + reactant molecule → reactant in active site → enzyme molecule + product molecules

A model illustrating an enzyme with an active site splitting a molecule into two smaller molecules. This is sometimes called the lock-and-key model because of the way that the active site fits the reactant molecule can be likened to a key fitting a lock.

The substrate is converted to products at the active site. These products then separate from the enzyme leaving the active site free to accept another molecule of the substrate. *Enzyme kinetics* can help scientists to explain the process.

Each enzyme works best at a particular temperature and pH and is less effective under other conditions. (See also *inhibitor.*) If the temperature is too high, or the pH too extreme, the tertiary structure of the protein molecules may be disrupted. This denatures the protein and destroys the active site of the enzyme making it inactive.

enzyme kinetics is the study of the factors affecting the rates of reactions catalysed by enzymes. The results can often be interpreted in terms of this three-stage *mechanism of a reaction*:

enzyme (E) + substrate (S) → enzyme-substrate complex (ES)

→ enzyme-product complex (EP) → enzyme (E) + products (P)

The hydrolysis of the substrate *urea* ($H_2N—CO—NH_2$), for example, forms carbon dioxide and ammonia when catalysed by the enzyme, urease (E). When the substrate (S) concentration is low, the reaction is a *first order reaction* with respect to urea. The rate equation is:

rate = k[urease][urea], or rate = k[E][S]

At low substrate concentrations both E and S appear in the rate equation. The first step is rate determining:

E + S → ES

When the substrate concentration is high, the reaction becomes a *zero order reaction* with respect to urea. The *rate equation* is then:

rate = k[urease], or rate = k[E]

At high substrate concentrations all the enzyme *active sites* are occupied and so [ES] is a constant. The first step is no longer rate determining. The rate of reaction now depends only on the concentration of the ES complex, which is determined by [E].

epoxy resins are used as adhesives. The *resin* is a sticky liquid *polymer* with relatively short chains.

Mixing the resin with a hardener starts a chemical reaction which *cross-links* the chains to make a hard, strong, glass-like polymer. The higher the temperature, the faster the reaction.

equilibrium law is a quantitative law for predicting the amounts of reactants and products present when a *reversible* reaction reaches a state of *dynamic equilibrium*.

In general, for a reversible reaction at equilibrium:

$$aA + bB \rightleftharpoons cC + dD$$

$$K_c = \frac{[C]^c [D]^d}{[A]^a [B]^b}$$

This is the form for the equilibrium constant, K_c when the *concentrations* of the reactants and products are measured in moles per litre. [A], [B] and so on are the equilibrium concentrations in mol dm^{-3}.

The concentrations of the chemicals on the right-hand side of the equation appear on the top line of the expression. The concentrations of reactants on the left appear on the bottom line. Each concentration term is raised to the power of the number in front of its formula in the equation.

For gas reactions it is often more convenient to use partial pressures as the measure of concentrations. The equilibrium constant is then K_p.

Equilibrium constants are constant for a particular temperature. The variation of K with temperature accounts for the *temperature effect on equilibria*.

Worked example

Calculate the value of K_c for the reaction: $PCl_5(g) \rightarrow PCl_3(g) + Cl_2(g)$ given that when 8.4 mol of $PCl_5(g)$ is mixed with 1.8 mol $PCl_3(g)$ and allowed to come to equilibrium in a 10 dm^3 container, the amount of $PCl_5(g)$ at equilibrium is 7.2 mol.

Notes on the method

Write down the equation. Underneath write first the initial amounts, then write the amounts at equilibrium. Use the equation to calculate the amounts not given.

Calculate the equilibrium concentrations given the volume of the container. Substitute in the expression for K_c.

Answer

Equation	$PCl_5(g)$	$\rightleftharpoons$	$PCl_3(g)$	+	$Cl_2(g)$
Initial amounts/mol	8.4		1.8		0
Equilibrium amounts/mol	7.2 (so 1.2 mol reacts)		1.8 + 1.2 = 3.0		1.2
Equilibrium concentrations/ mol dm^{-3}	7.2 ÷ 10 = 0.72		3.0 ÷ 10 = 0.3		1.2 ÷ 10 = 0.12

$$K_c = \frac{[PCl_3(g)][Cl_2(g)]}{[PCl_5(g)]} = \frac{0.3 \text{ mol dm}^{-3} \times 0.12 \text{ mol dm}^{-3}}{0.72 \text{ mol dm}^{-3}} = 0.05 \text{ mol dm}^{-3}$$

---

equivalence point is the point during any *titration* when the amount in moles of one reactant added from a burette is just enough to react exactly with all of the measured amount of chemical in the flask as shown by the *balanced equation*.

In a well-planned titration the colour change observed at the *end-point* corresponds exactly with the equivalence point.

errors of measurement are not mistakes but unavoidable differences between measured values and the true values. Often the true value is not known, so analysts have to assess the *measurement uncertainty*.

There are random errors which cause repeat measurements to vary and to scatter around a mean value. Averaging a number of readings helps to take care of random errors. The smaller the random errors, the higher the *precision of data* is ensured.

There are systematic errors which affect all measurements in the same way, making them all lower or all higher than the true value. Systematic errors do not average out. Identifying and eliminating systematic errors is important for increasing the *accuracy of data*. Systematic errors can be corrected by *calibration* with another instrument that is known to be more reliable, but may be too expensive for general use.

There are also errors arising from the choice of measuring instrument. The accuracy of an instrument such as a burette, pipette or measuring cylinder is limited by its quality and by how accurately the chemist can take a reading from the scale. Instrumental errors can be reduced by choosing a suitable measuring device.

esters are sweet-smelling compounds found in perfumes and in fruit *flavours*. The ester 3-methylbutylethanoate gives pear drops their taste and smell. Ethyl butanoate has the odour of pineapples. Ripe fruits contain mixtures of esters. Esters are also used as *plasticisers* and *solvents*.

The general formula for an ester is RCOOR′, where R and R′ are *alkyl* or *aryl groups*. Ethyl ethanoate is only slightly soluble in water.

methyl ethanoate   ethyl ethanoate   ethyl benzoate

Names and structures of esters

An ester forms when a carboxylic acid reacts with an alcohol in the presence of a drop of concentrated sulfuric acid to act as a catalyst. The reaction is reversible. It takes an excess of the alcohol to convert most of the acid to ester.

Formation of ethyl ethanoate from ethanoic acid and ethanol

Another way to make an ester is to mix an alcohol with either an *acyl chloride* or an *acid anhydride*.

Hydrolysis splits an ester into an acid and an alcohol. Acids or bases can catalyse the hydrolysis. Base catalysis is generally more efficient because it is not reversible. The acid is produced as its negative ion which does not react with the alcohol.

$$CH_3C \underset{OCH_2CH_3}{\overset{O}{\diagup}} + OH^- \xrightarrow{\text{heat}} CH_3C \underset{O^-}{\overset{O}{\diagup}} + CH_3CH_2OH$$

ester   salt

Base-catalysed hydrolysis of ethyl ethanoate

Lithium tetrahydridoaluminate(III) (LiAlH$_4$) in dry ether reduces esters to alcohols more easily than it reduces carboxylic acids.

$$CH_3C \underset{OCH_2CH_3}{\overset{O}{\diagup}} \xrightarrow[\text{dry ether}]{\text{LiAlH}_4 \text{ in}} CH_3CH_2OH$$

Reduction of an ester to an alcohol by lithium tetrahydridoaluminate(III)

Compounds with more than one ester link include **triglycerides** in fats and oils as well as **polyester** fibres. Alkaline hydrolysis of fats and oils produces **soaps**.

Several compounds used in medicine are esters including **aspirin, paracetamol** and the local **anaesthetics** novocaine and benzocaine. The insecticides malathion and pyrethrin are also esters.

ethanoic acid manufacture produces over six million tonnes of this bulk chemical each year. The **carboxylic acid** is a key intermediate on the way to making a very wide range of other chemicals including polymers, drugs, herbicides, dyes, adhesives and cosmetics. Changes to the manufacture of the acid illustrates the applications of the principles of **green chemistry**.

Until the 1970s, the main method of making ethanoic acid was to oxidise hydrocarbons from crude oil with oxygen in the presence of a cobalt(II) ethanoate **catalyst**. This process ran at 180–200°C and at 40–50 times atmospheric pressure. The reaction produced a mixture of products which had to be separated by fractional distillation. The economics of the process depended on finding markets for all the products.

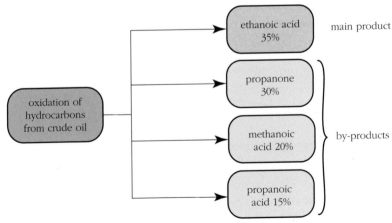

Percentages of the product and by-products from the manufacture of ethanoic acid by oxidising hydrocarbons. The atom economy of the process was about 35%

In 1970, Monsanto introduced a new process. In this process methanol and carbon monoxide combine to make ethanoic acid as the only product in the presence of a catalyst made from rhodium and iodide ions.

$$CH_3OH(l) + CO(g) \rightarrow CH_3COOH(l)$$

This process runs at 150–200°C and 30–60 times atmospheric pressure. This is a very efficient process with high *yields* and a 100% *atom economy*. Much less energy is needed than in the older process based on hydrocarbons. The reaction is fast and the *catalyst* has a long life.

Subsequently, the oil company BP bought and developed a variant of the Monsanto method called the Cativa process. The reaction of methanol with carbon monoxide is the same but it uses an iridium catalyst instead of a rhodium catalyst. The catalyst is made even more effective by adding ruthenium compounds as catalyst *promoters* which speed up the reaction by a factor of three.

The Cativa process has several advantages over the Monsanto process:
- iridium is much cheaper than rhodium
- the process is much faster and so more product can be made without increasing the size of the plant
- less carbon monoxide reacts with water so the yield from CO is over 94%
- the catalyst is more selective in producing ethanoic acid so less energy is needed to purify the product
- the catalyst is more stable so it lasts longer
- water concentrations in the plant can be lower, so less energy is needed to dry the product.

ethanol is the best-known member of the family of *alcohols*. It is the alcohol in beer, wine and spirits.

The chemical industry makes ethanol by the *hydration* of ethene in the presence of an acid catalyst. An alternative source of the alcohol is *fermentation* and this is the source of ethanol for use as *bioethanol*.

Methylated spirit consists of ethanol (alcohol) mixed with about 5% methanol to make it undrinkable. Industrial methylated spirit is used as a *solvent* commercially and in laboratories. Surgical spirit is industrial methylated spirit with other additives including castor oil. The methylated spirit ('meths') sold in hardware stores as a solvent and fuel is additionally coloured blue by a dye.

ethanoyl chloride: see *acyl chlorides*.

ethene ($CH_2\!=\!CH_2$) is the simplest *alkene*. It is a reactive gas. The petrochemical industry makes ethene on a very large scale by *thermal cracking* of natural gas or naphtha from the *fractional distillation* of oil.

Ethene is the starting material for the manufacture of a vast range of petrochemicals. The main products are polymers made by *addition polymerisation* such as polythene, PVC and polystyrene.

ethers are organic compounds with the general formula R—O—R', where R and R' are *alkyl* or *aryl groups*. The most familiar ether is ethoxyethane which has a sweet smell and is an *anaesthetic*.

$$CH_3-O-CH_3 \qquad CH_3CH_2-O-CH_2CH_3 \qquad CH_3-O-CH_2CH_3$$

methoxymethane   ethoxyethane   methoxyethane

Names and structures of ethers

Ethers are more *volatile substances* than alcohols because they have no hydrogen bonds. The vapour of ethoxyethane can be hazardous because it forms explosive mixtures in air.

Ethers are also much less soluble in water than alcohols. Ethers are good solvents for covalent compounds but not for salts so an ether such as ethoxyethane is one of the liquids used for *solvent extraction*.

One of the reactions that forms ethers is the nucleophilic substitution reaction of a halogenoalkane with a sodium alkoxide.

$$CH_3CH_2Br + CH_3CH_2O^-Na^+ \rightarrow CH_3CH_2OCH_2CH_3 + NaBr$$

Ethers are relatively unreactive because they do not have O—H bonds. They do not react with acids or alkalis, oxidising agents, sodium or phosphorus pentachloride. This makes ethers, such as ethoxyethane, useful as non-aqueous solvents for reactions such as *lithium tetrahydridoaluminate (iii)* ($LiAlH_4$) reductions and the formation and use of *Grignard reagents*.

ethyne is the simplest member of the series of hydrocarbons called *alkynes*.

$$H-C\equiv C-H$$

The structure of ethyne

Alkynes such as ethyne undergo *addition reactions*. Hydrogen adds to ethyne at room temperature in the presence of a platinum or palladium catalyst, or at 150°C in the presence of nickel. The reaction first forms ethene and then ethane.

Electrophilic addition of hydrogen halides takes place in two stages. Ethyne adds hydrogen chloride to give first chloroethene and then 1,1-dichloroethane. Note that Markovnikov's rule applies to the second addition.

$$H-C\equiv C-H \xrightarrow{HCl} CH_2{=}CHCl \xrightarrow{HCl} CH_3CHCl_2$$

A mixture of ethyne and chlorine explodes, but the two gases combine smoothly in the presence of a catalysts such as aluminium chloride or iron(III) chloride. The reaction produces first 1,2-dichloroethene and then 1,1,2,2-tetrachloroethane.

Ethyne polymerises to form poly(ethyne) which can be treated to produce a *conducting polymer*.

eutectic mixture: a mixture of two or more chemicals with the lowest melting (or freezing) point of all the possible mixtures. Metals can form eutectic mixtures as can solutions of salts such as sodium nitrate in water.

For pairs of metals, such as tin and lead, the eutectic mixture is the *alloy* with the lowest melting point of all the possible alloys. An alloy with the eutectic composition freezes (or melts) at a specific temperature like a pure metal. The melting point of the eutectic, however, is lower than both melting points of the pure metals.

eutrophication happens when the water in rivers or lakes is enriched by fertilisers from farmland or by nutrients from sewage and this makes it possible for algae to multiply rapidly. Thick layers of algae block out the light from plants growing below the surface so that they cannot produce oxygen as fast as usual. Then bacteria start to break down the mass of algae, using up the remaining oxygen in the water. Other organisms such as fish die because they are starved of oxygen.

evaporation happens at the surface of a liquid as it turns to a gas. This is an example of a *change of state*. As a liquid like water evaporates, the faster-moving molecules near the surface escape from the pull of intermolecular forces and break away to become a vapour. Liquids tend to cool as they evaporate because the faster-moving molecules break away so that the average kinetic energy of the remaining molecules falls.

Evaporation is an **endothermic** process. Energy must enter from the surroundings to keep a substance at a constant temperature as it evaporates. So volatile liquids feel cold as they evaporate on the skin.

excited state: the state of an atom or molecule when one or more of its electrons is raised to a higher energy above the stable *ground state*. Heating, electricity or electromagnetic radiation can provide the energy to excite atoms or molecules. Energy is released back to the surroundings when the electrons fall back to the stable energy levels of the ground state. This happens during *flame tests*. Heat from a Bunsen flame excites the electrons in metal atoms from the sample. As the electrons drop back to the lower energy levels they give off radiation with a wavelength determined by the size of the energy jump (see *quantum theory*).

exothermic reactions give out energy to their surroundings. Freezing and condensing are exothermic *changes of state*.

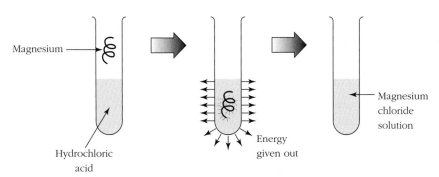

Magnesium

Hydrochloric acid

Energy given out

Magnesium chloride solution

Diagrams to illustrate an exothermic reaction

For an exothermic reaction the enthalpy change, ΔH, is negative. During an exothermic reaction, more energy is given out as new bonds form in the products than is needed to break bonds in the reactants.

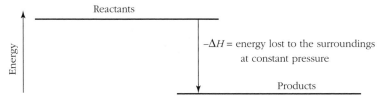

Reactants

Energy

$-\Delta H$ = energy lost to the surroundings at constant pressure

Products

Energy level diagram for an exothermic reaction

151

Most *spontaneous reactions* are exothermic but there are exceptions, so the *free energy change* (or total *entropy* change) must be examined to make reliable predictions about the direction and extent of change.

explosions are reactions that go with a bang. Gunpowder explodes if ignited in a confined space. The explosion is caused by the rapid build up of pressure as the reaction produces a large volume of hot gas from a small volume of solid.

The 'pop' during the test-tube test for hydrogen is a small explosion. This reaction explodes because it is a *free-radical chain reaction* with a step in which one radical reacts to produce two radicals which can then both react. The branching chain reaction produces more and more free radicals so the reaction goes faster and faster.

Explosives are compounds or mixtures of chemicals designed to produce explosions. Explosions are used in quarrying, mining, road building and to demolish old buildings. They are also used in warfare and as rocket propellants.

extraction of metals involves the reduction of compounds separated from ores (see also *metal extraction*). Sulfide ores are converted to oxides by *roasting* before extraction, as in *zinc extraction*. Carbon and carbon monoxide are cheap reducing agents used for *iron extraction* and to reduce the oxides of other relatively unreactive metals such as *manganese* from manganese(IV) oxide.

$$MnO_2(s) + C(s) \rightarrow Mn(s) + CO_2(g)$$

Tungsten cannot be extracted with carbon because tungsten carbide (WC) forms rather than tungsten metal. Tungsten(VI) oxide (WO_3) is heated with hydrogen at 900°C.

$$WO_3(s) + 3H_2(s) \rightarrow W(s) + 3H_2O(g)$$

Other reducing agents used to extract metals are sodium or magnesium for *titanium extraction*.

Scrap iron can be used in a *displacement reaction* to obtain copper from solutions containing $Cu^{2+}(aq)$ obtained by *leaching* oxide ores. Electrolysis is used for *aluminium extraction* and the extraction of other very reactive metals such as sodium.

extrusion is a process for shaping a metal, such as *aluminium*, or a *plastic*, such as PVC, by forcing the material through a shaped nozzle (die). Cooks use extrusion in a similar way when decorating a cake with icing sugar. A plastics extruder works by melting pellets of the polymer and then using a rotating screw to push the molten material through a die.

E–Z isomers are compounds with double bonds in which the same groups are arranged differently because of the lack of rotation about the double bond. This naming system applies unambiguously to all such isomers, unlike the traditional description of *cis–trans isomers*.

These three steps lead to the correct *E–Z* name for simpler examples.
1. Look at the two atoms bonded to the first carbon atom in the C—C bond. Apply the *Cahn–Ingold–Prelog priority rules* to decide which atom has the higher priority.
2. Similarly, identify the group with the higher priority of the two attached to the second carbon atom in the C—C bond.

3. If the two groups of higher priority are on the same side of the double bond, the isomer is designated *Z*– but if the two groups of highest priority are on opposite sides of the double bond, the isomer is designated *E*–.

Z comes from the German word *zusammen* meaning together. *E* comes from the word *entgegen* which is the German for opposite. In many cases a compound classed as *trans* in one system is *E* in the other and a compound classed as *cis* is *Z* in the other, but there are examples where this is not the case.

E-pent-2-ene (*trans*)   *Z*-pent-2-ene (*cis*)

E-2-chlorobut-2-ene (*cis*)   *Z*-2-chlorobut-2-ene (*trans*)

Structures of E and Z isomers compared to their cis–trans names. Chlorine atoms have a higher priority than carbon atoms. $\Delta H^{\ominus}_{reaction}$ = {sum of $\Delta H^{\ominus}_f$ [products]} – {sum of $\Delta H^{\ominus}_f$ [reactants]}

$\Delta H^{\ominus}_{reaction}$ {2 × $\Delta H^{\ominus}_f$ [Fe] + 3 × $\Delta H^{\ominus}_f$ [CO_2]} – {$\Delta H^{\ominus}_f$ [Fe_2O_3] + 3 × $\Delta H^{\ominus}_f$ [CO]}

face-centred cubic structure is one of the two *close-packed* crystal structures of metals. Metals with the face-centred cubic structure are calcium, aluminium, cobalt, nickel and copper.

Faraday constant *(F)* is the charge on one mole of electrons. The Faraday constant is used in calculations to determine the amount of change during *electrolysis*.

$$F = eL = 96\ 480\ \text{C mol}^{-1}$$

where *e* is the charge on one electron and *L* is the *Avogadro constant*. It is possible to measure *F* and *e* experimentally so this relationship can be used to arrive at a value for the Avogadro constant.

Worked example

In the electrolysis of molten lead(II) bromide, what mass of lead forms at the cathode when a current of 0.5 A flows for 1930 s.

Notes on the method

A current of one *ampere* means a flow of charge of one coulomb per second.

Work to 3 significant figures using the value of 96 500 C mol^{-1} for the Faraday constant. Write the half-equation for the process at the electrode where lead forms.

Answer

Amount of electric charge passed through the electrolyte = 0.5 C s^{-1} × 1930 s = 965 C

Amount of electrons passed = $\dfrac{965\ \text{C}}{96\ 500\ \text{C mol}^{-1}}$ = 0.01 mol

At the cathode: $Pb^{2+} + 2e^- \rightarrow Pb$

2 mol electrons are needed to form 1 mol of lead.

So a flow of 0.01 mol electrons produces 0.005 mol lead.

The mass of 0.005 mol lead = 0.005 mol × 207 g mol^{-1} = 1.04 g

---

fats belong to the family of compounds call *lipids*. Fats are esters of propane-1,2,3-triol *(triglycerides)* which are solid at around room temperature (below 20°C) because they contain a high proportion of saturated *fatty acids*. Solid triglycerides are generally found in animals. The triglycerides from plants are mostly liquids at around room temperature; these are *vegetable oils*.

fatty acids are carboxylic acids with long hydrocarbon chains which are combined with propane-1,2,3-triol to make *triglycerides* in *fats* and *vegetable oils*. *Saturated* fatty acids do not have double bonds in the hydrocarbon chain. There is one or more double bonds in an **unsaturated** fatty acid.

Fatty acid	Chemical name	Formula	Type
palmitic	hexadecanoic	$CH_3(CH_2)_{14}COOH$	saturated
stearic	octadecanoic	$CH_3(CH_2)_{16}COOH$	saturated
oleic	octadec-9-enoic	$CH_3(CH_2)_7CH{=}CH(CH_2)_7COOH$	unsaturated
linoleic	octadec-9,12-dienoic	$CH_3(CH_2)_4CH{=}CHCH_2CH{=}CH(CH_2)_7COOH$	unsaturated

Animal fats have a higher proportion of saturated fats in their triglycerides. In lard, for example, the main fatty acids are palmitic acid (28%), stearic acid (8%) and only 56% oleic acid. In olive oil the main fatty acids are oleic acid (80%) and linoleic acid (10%).

Polyunsaturated fats contain fatty acids with two or more double bonds in the chain.

Some fats and oils provide fatty acids which are essential to the human diet, including linoleic and linolenic acids.

f-block elements are the *lanthanide elements* and *actinide elements* in periods 6 and 7 of the periodic table. For these elements, the last electron added to the atomic structure goes into one of the seven f orbitals in the fourth or fifth shells. For convenience, the two rows of f-block elements are usually shown underneath the *periodic table* to cut down its overall width.

feasibility: a reaction is feasible if it tends to go – even if the reaction is very, very slow. In this sense the reaction between methane (in natural gas) and oxygen is feasible at room temperature but, because of a high *activation energy*, the reaction is so slow that it does not happen. Put a match to the mixture of air and gas and the reaction is rapid.

Chemists can decide whether or not a reaction is feasible by considering either the total *entropy* change for the reaction or its *free energy change*. If ΔS_{total} is positive or ΔG is negative, the reaction tends to go.

For many chemical reactions, ΔG approximately equals ΔH, so chemists often use the sign of ΔH as a guide to feasibility. This can be misleading if the *entropy* change for the reaction system is large.

For redox reactions, *standard electrode potentials* offer an alternative way of deciding whether or not reactions tend to go. $\Delta G \propto -E_{cell}$, so if the cell emf is positive the redox reaction in the cell will tend to go.

For other reactions, such as acid–base reactions, the easiest guide to the direction and extent of change is the equilibrium constant, K_c, where $\Delta G \propto -\ln K_c$.

What this shows is that ΔS_{total}, ΔG, E_{cell} and K_c values can all answer the same questions for a reaction:

- Will the reaction go?
- How far will it go?

In practice chemists use the quantity which is most convenient to measure. Knowing the value of one of the three quantities it is possible to calculate the other two because they are all related.

It is always important to bear in mind that even if a reaction is feasible it may be very slow (see also *spontaneous reaction*).

feedstock is the raw material for a process in the *chemical industry*.

Fehling's solution is used to distinguish *aldehydes* from *ketones* and to identify *reducing sugars*. Aldehydes and reducing sugars give a positive result. Ketones and other sugars do not. The deep blue solution turns greenish and then loses its blue colour as an orange-red precipitate of copper(I) oxide appears.

Fehling's reagent does not keep so it is made when required by mixing two solutions. One solution is copper(II) sulfate in water. The other is a solution of 2,3-dihydroxybutanedioate (tartrate) ions in strong alkali. The 2,3-dihydroxybutanedioate salt is included to form a complex with copper(II) ions so that they do not precipitate as copper(II) hydroxide with the alkali.

fermentation converts sugars to alcohol (ethanol) and carbon dioxide. Fermentation is an example of *anaerobic respiration*. Fermentation is catalysed by a group of enzymes from yeast, referred to as zymase. In brewing the aim is to produce alcohol in beer. In baking fermentation produces the gas bubbles which makes the dough rise.

$$C_6H_{12}O_6(aq) \rightarrow 2CO_2(g) + 2C_2H_5OH(aq)$$

glucose   carbon   ethanol
                    dioxide

ferromagnetism: the property of materials which can be used to make permanent magnets. The three ferromagnetic metals are iron, cobalt and nickel. Alnico is a magnetic alloy of these three metals.

fertilisers supply plants with the mineral salts they need for growth. Plants need three major nutrients – nitrogen, phosphorus and potassium – which must be available in the soil in a soluble form so that they can be taken up by roots.

Manure and compost are traditional fertilisers used in 'organic farming'. They release nutrients slowly as they rot down through the action of bacteria and fungi in the soil.

Intensive agriculture relies on fertilisers made from minerals and the air. 'Straight fertilisers' contain one of the three elements nitrogen (N), phosphorus (P) or potassium (K). 'Compound fertilisers' contain two or more of these elements.

Organic fertilisers, such as *urea*, have advantages over inorganic nitrates and ammonium salts. Organic fertilisers:
● release nitrogen more slowly by hydrolysis in the soil
● do not affect the pH of the soil to the same extent when spread on fields.

The use of manufactured fertilisers greatly increases crop yields. However the use of large amounts of fertiliser increases the risk of nutrients leaching from the soil into rivers, lakes and groundwater. This can lead to *eutrophication* (see also *ammonia manufacture* and the *nitrogen cycle*).

fibres are long thin threads which make up materials such as asbestos, wool and polyester fabrics. Most of the fibres used for textiles consist of natural or synthetic polymers. Wool, silk and hair are natural fibres made of *protein*. Cotton, linen and kapok are plant fibres made of the *carbohydrate* cellulose.

Examples of synthetic polymers used to make fibres are *polyester, polyamides* and polypropylene. Synthetic polymers are spun into fibres by forcing the hot molten polymer through a spinneret. Fibres form as the liquid cools and solidifies.

Fibres of inorganic materials are increasingly important for making *composite* materials. Examples are the fibres of glass and graphite.

fibre-reactive dyes are strongly bound to the fabrics they dye because they react with the molecules in the fibres, forming covalent bonds. They are fast *dyes* which do not fade during washing.

Structure of a fibre-reactive dye which forms covalent bonds with amino groups in the protein molecules of wool

fillers are materials mixed with *plastics* or *rubbers* to modify their properties. Fillers can make plastics stronger, tougher, harder or easier to mould. Fillers are also used to cut costs by bulking out plastics with something cheaper.

filtration separates an insoluble solid from a liquid. The liquid which passes through the filter is the filtrate. The solid retained on the paper is the residue.

The use of a *Buchner flask and funnel* speeds up filtration.

The use of a small plug of mineral wool or cotton wool in a small funnel cuts losses when filtering off a drying agent from an organic liquid before final distillation.

fire extinguishers put out fires by excluding air, by cooling to slow down burning or by starting a chemical reaction that stops the fire. The choice of fire extinguisher depends on the type of fire. Water is the cheapest way to put out burning wood, paper or rubbish. Water cools the fire as it evaporates and produces steam which helps to exclude air. Water is dangerous on electrical fires and useless on burning hydrocarbons or metals. Burning oil floats on water and continues burning so water just spreads the fire. Burning metals react vigorously with water. Alternative fire extinguishers use carbon dioxide gas, or non flammable liquids, or dry powders.

Halons such as CF_2ClBr are being phased out as fire-fighting chemicals because, like CFCs, they destroy ozone in the stratosphere. An alternative approach to fighting hydrocarbon fires is based on fire extinguishing foams (FEF) such as those developed by Pyrocool Technologies.

fire retardants are chemicals which delay or prevent burning. They can be used to treat materials to make them less combustible. A coating of aluminium hydroxide acts as a fire retardant. On heating, the hydroxide decomposes to aluminum oxide. This is an endothermic reaction which cools its surroundings while releasing steam that can help to exclude air. The residue of aluminium oxide forms a protective layer over the material.

first law of thermodynamics: see *thermodynamics (laws of)*.

first order reaction: a reaction is first order with respect to a reactant if the rate of reaction is proportional to the concentration of that reactant. The concentration term for this reactant is raised to the power one in the *rate equation*.

$$\text{rate} = k[X]^1 = k[X]$$

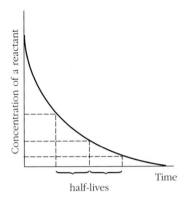

Variation of concentration of a reactant plotted against time for a first order reaction

The gradient of this graph at any point measures the rate of reaction. The **half-life** for a first order reaction is a constant so it is the same wherever it is read off the curve. It is independent of the initial concentration.

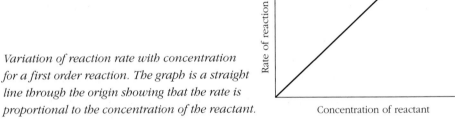

Variation of reaction rate with concentration for a first order reaction. The graph is a straight line through the origin showing that the rate is proportional to the concentration of the reactant.

The rate of reaction of 2-bromo-2-methylpropane with hydroxide ions is first order with respect to the tertiary halogenoalkane but *zero order* with respect to hydroxide ions.

$$\text{rate} = k[(CH_3)_3CBr]$$

fission means splitting. The word is used in two contexts in chemistry – bond breaking (**homolytic bond breaking** and **heterolytic bond breaking**) and the splitting of atomic nuclei (nuclear fission).

flame tests help to detect some metal ions in salts. An element gives a coloured flame if its *atomic spectrum* includes lines in the visible region of the spectrum.

Flame tests are particularly useful in qualitative analysis to distinguish metal ions which otherwise behave in a similar way. Both magnesium and calcium ions, for example, are precipitated by sodium hydroxide to give white precipitates which are insoluble in excess of the alkali. Calcium compounds give an orange-red flame but magnesium compounds do not colour a flame.

Metal ion	Colour
lithium	bright red
sodium	bright yellow
potassium	pale mauve
calcium	orange-red
strontium	scarlet
barium	yellowish-green
copper(II)	blue-green

flammable substances are *hazards* because they easily catch fire in air. Chemicals are labelled as highly flammable if they:

- may spontaneously catch fire in air
- are liquids with a flash point below 21°C but above 0°C
- are solids which may catch fire and keep burning after brief contact with a flame
- are gases which burn in air if ignited at normal pressure
- react to form flammable gases (such as hydrogen) in contact with water or water vapour.

flavourings are *food additives* which give food, confectionary, toothpaste and other products their flavours. Many natural fruit flavours are subtle blends of *esters* and other chemicals. Chemists can imitate fruit flavours by mixing esters. The *stereochemistry* of a molecule can affect its flavour.

fluids are materials which flow. Liquids or gases are fluids because the atoms or molecules are not held in fixed positions but are free to move around.

fluoridation involves adding traces of fluoride compounds to water supplies to prevent tooth decay. The benefits were identified in areas where water supplies naturally contain fluoride ions. Adding fluoride ions to drinking water is controversial and opposed by people who think that any kind of enforced treatment is wrong. Others point to possible harmful effects because fluoride ions become *toxic chemicals* at higher concentrations. There is a lack of firm scientific evidence to justify the fluoridation of public water supplies.

fluorine (F) is a pale yellow gas made up of F_2 molecules. It is the most reactive of the halogens in *group 7* of the periodic table. Its electron configuration is $[He]3s^23p^5$.

The *bond enthalpy* of fluorine is much lower than would be expected from the trend in values for group 7. The theory is that the bond is weakened by the repulsion between lone pairs on adjacent fluorine atoms. Lone pairs on larger atoms, such as chlorine atoms, are further apart and experience less repulsion. The relatively low bond energy for fluorine helps to account for the fact that most of its reactions are highly exothermic.

Fluorine is the most *electronegative* of all elements so it forms ionic compounds with metals. Fluorine is the most powerful *oxidising agent* and its oxidation state is –1 in all its compounds. It oxidises other elements to their highest positive oxidation state. Sulfur, for example, forms SF_6.

Uses of fluorine include the manufacture of a wide range of compounds consisting of only carbon and fluorine (see *fluorocarbons*). The most familiar of these is the very slippery, non-stick *addition polymer* poly(tetrafluoroethene), better known as PTFE.

fluorite structure: the cubic crystal structure of the ionic compound calcium fluoride, CaF_2. Each positive ion is surrounded by eight nearest neighbours at the corners of a cube and each negative ion is surrounded by four positive ions. So the *coordination numbers* are 8 for the positive ion and 4 for the negative ion. Other compounds with this structure are the fluorides of Sr, Ba, Cd, Pb and Hg.

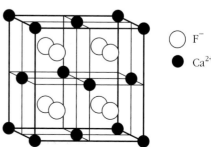

O F^-

● Ca^{2+}

Structure of calcium fluoride showing 8:4 coordination. The structure consists of a face-centred cubic array of positive ions with negative ions in the tetrahedral holes.

159

fluorescent substances can absorb energy from ultraviolet (or other) radiation and immediately re-emit the energy as visible light. The inside surfaces of fluorescent lamp bulbs or tubes are coated with compounds such as magnesium tungstate and zinc silicate. These substances fluoresce when irradiated by the ultraviolet light from the mercury vapour in the lamp.

The optical brighteners in *detergents* are fluorescent. They absorb ultraviolet light from the Sun and re-emit the energy as blue light. This compensates for any yellowing of the fabric with age.

Fluorescence is an example of *luminescence*.

fluorocarbons are compounds of fluorine and carbon only. They are very *inert*, thermally stable and non-flammable. They are also non-toxic. They are electrical and thermal insulators. They are used during the manufacture and testing of electronic components and also as refrigerants, coolants and lubricants. Some fluorocarbons are good solvents for oxygen and, in emergencies, can act as 'artificial blood'.

foams consist of gases finely dispersed in liquids or solids. Foams are *colloids*. The mass of bubbles in a liquid foam holds the liquid and gas in place. This is useful in foam fire extinguishers which can blanket a fire with carbon dioxide held by the foam. It is also useful in shaving foam which retains the liquid soap on the skin.

Solid foams can be rigid but have a low density. Expanded polystyrene and foamed polyurethanes are examples. These materials are excellent thermal insulators because they trap a large volume of gas, and gases are very poor thermal conductors.

food additives are natural or synthetic chemicals added to food, which can act as:
- colours to make food look more attractive
- preservatives to prevent the growth of micro-organisms
- *antioxidants* to stop oxygen from the air making food unfit to eat – oxidation, for example, turns fats and oils rancid
- emulsifiers and stabilisers to control food texture (see *emulsion*)
- sweeteners to improve the taste (see *sweetness*).

E numbers indicate that additives have been approved for use by the food industry in the European Union.

forensic science is the application of scientific techniques to gather evidence for a court of law. Forensic scientists make increasing use of sophisticated analytical techniques such as *mass spectrometry*, *chromatography* and various forms of *spectroscopy*.

formula (plural formulae) is used by chemists to represent the composition and structure of elements and compounds. (See *empirical formula*, *molecular formula*, *structural formulae*, *skeletal formulae* and *displayed formulae*.)

formula mass: see *formula unit*.

formulation: the list of ingredients for a mixture of chemicals, such as for a *medicine*. Medical drugs are not usually pure chemical compounds. The other additives in the formulation may have a range of purposes such as making the drug soluble, suspending it in water, thickening the mixture, preserving the medicine or improving its flavour.

formula unit: the symbols used in equations and calculations to represent elements and compounds with *giant structures*. Examples are:
- sodium chloride, a giant structure of ions, formula unit NaCl
- silicon dioxide, a giant covalent structure, formula unit SiO_2.

The molar mass of a substance with a giant structure is the relative mass of its formula unit. For silicon dioxide (SiO_2) the molar mass = $[28 + (2 \times 16)] = 60$ g mol^{-1}. This is sometimes called the formula mass.

fossil fuels are non-renewable energy resources which we now burn to utilise the energy from the Sun which was stored up millions of years ago by *photosynthesis*. The fossil fuels are coal, oil and natural gas.

Fossil fuels produce *atmospheric pollutants* as they burn and contribute to the formation of *acid deposition* when they burn. The fuels contain varying amounts of sulfur compounds. The proportion of sulfur compounds in crude oil, for example, varies from less than 1% up to 7%, depending on its origin. When crude oil is distilled, sulfur compounds tend to be concentrated in the heavy fractions such as fuel oils. Oil refining removes much of this *sulfur* and there is very little in petrol.

fractional distillation is a method for separating mixtures of liquids with different boiling points. On a laboratory scale, the process takes place with distillation apparatus fitted with a glass fractionating column between the flask and the still head. Separation is improved if the column is packed with inert glass beads or rings to increase the surface area where rising vapour can mix with condensed liquid running back to the flask. The column is hotter at the bottom and cooler at the top. The thermometer reads the boiling temperature of the compound passing over into the condenser.

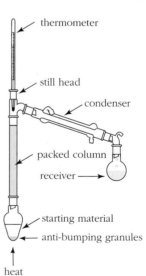

thermometer

still head

condenser

packed column

receiver

starting material

anti-bumping granules

heat

Apparatus for fractional distillation.

fractional distillation of oil is a large-scale, *continuous process* for separating crude oil into fractions. A furnace heats the oil which then flows into a fractionating tower containing about 40 'trays' pierced with small holes. Condensed vapour flows over the trays and runs down into the tray below. Rising vapour mixes with liquid on a tray as it bubbles up through the holes.

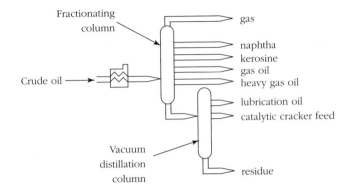

*Fractional distillation
of crude oil*

The column is hotter at the bottom and cooler at the top. Rising vapour condenses when it reaches the tray with liquid at a temperature below its boiling point. Condensing vapour releases energy which heats the liquid in the tray, which in turn evaporates the more volatile compounds in the mixture.

With a series of trays, the outcome is that the **hydrocarbons** with small molecules rise to the top of the column while larger molecules stay at the bottom. Fractions are drawn off from the column at various levels.

Some components of crude oil have boiling points too high for them to vaporise at atmospheric pressure. Lowering the pressure in a **vacuum distillation** column reduces the boiling points of the hydrocarbons and makes it possible to separate them.

Refining of oil fractions by **catalytic cracking, isomerisation** and **reforming** produces fuels such as petrol. Thermal cracking converts oil fractions into feedstocks for the **chemical industry.**

fragmentation patterns: see **mass spectrometry.**

fragrancies: see **perfumes.**

free energy change, ΔG: the thermochemical quantity used by chemists to decide whether a reaction tends to go and how far it will go. ΔG is the test for the *feasibility* of a reaction. If ΔG is negative the reaction is feasible.

The idea of *entropy* underlies the quantity 'free energy'. Any change tends to happen if the total entropy change is positive. Working with entropy values is generally less convenient because of the need to consider the entropy changes in the surroundings as well as in the reaction mixture. This was the reason why Willard Gibbs suggested the concept 'free energy', which he defined by $\Delta G = -T\Delta S_{total}$, where $\Delta S_{total} = \Delta S_{system} + \Delta S_{surroundings}$.
For a change at constant temperature and pressure:

$$\Delta S_{surroundings} = \frac{-\Delta H}{T}$$

And so $\Delta S_{total} = \Delta S_{system} + \dfrac{-\Delta H}{T}$

Hence $-T\Delta S_{total} = -T\Delta S_{system} + \Delta H$

From Gibb's definition this becomes $\Delta G = \Delta H - T\Delta S_{system}$.

This table summarises the implications of this important relationship.

Enthalpy change	Entropy change of the system	Is the reaction feasible?
exothermic (ΔH negative)	increase (ΔS positive)	yes, ΔG is negative
exothermic (ΔH negative)	decrease (ΔS negative)	yes, if the number value of ΔH is greater than the magnitude of $T\Delta S$
endothermic (ΔH positive)	increase (ΔS positive)	yes, if the magnitude of $T\Delta S$ is greater than the number value of ΔH
endothermic (ΔH positive)	decrease (ΔS negative)	no, ΔG is positive

The table shows that a change that is not feasible at a low temperature may become feasible if the temperature rises. Generally the values of ΔH and ΔS do not change markedly with temperature and so it is possible to estimate the temperature at which a reaction that is not feasible at room temperature becomes feasible at a higher temperature.

Full data is not available for all reactions so that it is not always possible to calculate the free energy change. At relatively low temperatures the $T\Delta S$ term is often relatively small compared to the enthalpy change so that $\Delta G \approx \Delta H$.

This means that chemists can often use the sign of ΔH as a guide to feasibility. This can be misleading if the magnitude of the entropy change for the reaction system is large. Also the approximation becomes less justified at higher temperatures when T is bigger and so the magnitude of $T\Delta S_{system}$ is bigger.

free radicals are reactive particles with unpaired electrons. Free radicals form when covalent bonds break in the way which leaves one electron on each of the atoms joined by the bond. This is *homolytic bond breaking*. The symbol for a free radical generally shows the unpaired electron as a dot. Other paired electrons in the outer shells are generally not shown.

$$Br \!:\! Br \longrightarrow Br^\bullet + {}^\bullet Br$$

Formation of free radicals

Free radicals are *intermediates* in reactions taking place:
- in the gas phase at high temperature or in ultraviolet light
- in a non-polar solvent, either when irradiated by ultraviolet light or with an initiator.

Examples of free-radical processes include the *cracking* of hydrocarbons, burning of *petrol* in an engine cylinder and the formation and destruction of the *ozone in the stratosphere* by *CFCs*. (See also *free-radical chain reactions*.)

free-radical chain reactions involve three stages:
- initiation – the step which produces free radicals
- propagation – steps giving products and more free radicals
- termination – steps which remove free radicals by turning them into molecules.

Free-radical substitution – the reaction of an alkane, such as methane, with bromine in sunlight in a free-radical chain reaction. The main products are bromomethane and hydrogen bromide. The presence of some ethane in the mixture of products is evidence for the termination step:

Initiation: $Br — Br \rightarrow Br\bullet + Br\bullet$

Propagation: $CH_4 + Br\bullet \rightarrow CH_3\bullet + HBr$

$CH_3\bullet + Br_2 \rightarrow CH_3Br + Br\bullet$

Termination: $CH_3\bullet + CH_3\bullet \rightarrow CH_3CH_3$

$CH_3\bullet + Br\bullet \rightarrow CH_3Br$

Free-radical addition polymerisation – free-radical chain reactions can also be used to make ***addition polymers*** from alkenes and from epoxy compounds.

The initiator can be an organic ***peroxide*** such as benzoyl peroxide, which acts as a source of free radicals, $R—O\bullet$.

$$R — O \text{—} O — R \longrightarrow 2R — O\bullet$$

This leads to the propagation steps which are repeated many times, producing a very long chain molecule:

$$2R — O\bullet \quad \overset{H}{\underset{H}{C}} = \overset{H}{\underset{H}{C}} \longrightarrow R — O — \overset{\overset{H}{|}}{\underset{\underset{H}{|}}{C}} — \overset{\overset{H}{|}}{\underset{\underset{H}{|}}{C}}\bullet + R — O\bullet$$

Termination takes place when two free radicals combine to make a molecule:

$$R — O — (CH_2)_p — CH_2 — CH_2\bullet \quad \bullet CH_2 — CH_2 — (CH_2)_q — O — R$$

$$\downarrow$$

$$R — O — (CH_2)_p — CH_2 — CH_2 — CH_2 — CH_2 — (CH_2)_q — O — R$$

freezing point: the temperature at which a liquid turns to a solid. A pure liquid has a sharp freezing point, which is the same as its ***melting point.***

Dissolving a solute in a liquid lowers its freezing point. Adding antifreeze to the water in an engine lowers the freezing point and so prevents the coolant freezing in winter. When there is a threat of ice, the highway authorities scatter salt on roads because a mixture of salt and water freezes at temperatures well below 0°C.

frequency of electromagnetic radiation is the number of complete waves passing any point per second. The SI unit of frequency, ν, is the hertz (Hz). Frequencies of ***electromagnetic radiation*** vary from about 10^5 Hz for radio waves up to 10^{20} Hz for gamma rays.

All electromagnetic radiation travels at the same speed, c, in a vacuum. The ***frequency*** (ν), ***wavelength*** (λ) and speed (c) are simply related by $c = \nu\lambda$.

Quantum theory shows that the higher the frequency the higher the energy of the quanta (photons) of radiation.

Friedel–Crafts reaction: a method for forming C—C bonds to build up the carbon skeleton by adding a side chain to an ***arene*** such as benzene. In a Friedel–Crafts reaction a *halogenoalkane* or an *acyl chloride* undergoes an *electrophilic substitution* reaction with an arene. The reaction takes place in the presence of a catalyst such as aluminium chloride (a Lewis acid).

One example is the substitution of a group with three carbon atoms in place of a hydrogen atom in the benzene ring, using 2-chloropropane.

Friedel–Crafts alkylation with a chloroalkane – benzene reacting with 2-chloropropane

The reaction with an acyl chloride produces a *ketone*.

Friedel–Crafts acylation with an acyl chloride – benzene reacting with ethanoyl chloride

This important type of reaction was discovered and developed jointly by the French organic chemist Charles Friedel (1832–1899) and the American James Crafts (1839–1917). The reaction is used both on a laboratory and on an industrial scale. A Friedel–Crafts reaction to make ethylbenzene is the first step in the production of the *addition polymer* poly(phenylethene), more often called polystyrene.

The purpose of the of the aluminium chloride is to produce an *electrophile* which attacks the benzene ring.

Formation of the electrophile

$$R-C(=O)-Cl + AlCl_3 \longrightarrow R-C^+(=O) + AlCl_4^-$$

Electrophilic attack and substitution

The mechanism of Friedel–Crafts acylation

froth flotation is a process for separating the valuable mineral from a rock made up of several minerals. The rock is finely crushed to separate the minerals and then stirred with a solution of *surfactants* as a stream of air bubbles rises up through the mixture. The surfactants are selected to make sure that the particles of metal *ore* are caught up by the bubbles and rise to the surface where they can be skimmed off. The unwanted minerals (or gangue) sink and flow away as a stream of waste.

fuel cell: an electrochemical cell which is continuously supplied with fuel and oxidising agent. A fuel cell produces electric power from a fuel, directly, without having to burn it. This avoids the production of *atmospheric pollutants*.

The ideal fuel for a fuel cell is pure hydrogen, H_2, but other hydrogen-rich fuels can also be used, such as methane (CH_4) and methanol (CH_3OH). When hydrogen is used, no CO_2 is produced and the only by-product is water, H_2O.

At the negative electrode of a hydrogen fuel cell: $2H_2(g) \rightarrow 4H^+(aq) + 4e^-$

At the positive electrode, which is supplied with oxygen: $O_2(g) + 4H^+(aq) + 4e^- \rightarrow 2H_2O(l)$

fuels burn in air or oxygen to release energy. Many power stations use the *combustion* of *fossil fuels* to raise steam and generate electricity. Most of the fuels for transport are made from fractions produced by the *fractional distillation of oil*. There is growing interest in the use of *biofuels*.

Petrol is a blend of *hydrocarbons* based on the gasoline fraction (hydrocarbons with 5–10 carbon atoms). Jet fuel is produced from the kerosene, or paraffin, fraction (hydrocarbons with 10–16 carbon atoms). Fuel for diesel engines is made from diesel oil (hydrocarbons with 13–25 carbon atoms).

Fuels have to be refined to remove components which would harm engines or cause air pollution when they burn.

fullerenes are molecular allotropes of carbon. The best-known example is the 'bucky ball', buckminsterfullerene, C_{60}. Other spherical fullerenes have formulae C_{28}, C_{32}, C_{50}, and C_{70}. There are also cylindrical fullerenes called carbon *nanotubes*.

functional group: the atoms and bonding which give a series of organic compounds its characteristic properties and is responsible for most of the reactions. The hydrocarbon chain which makes up the rest of any organic molecule is generally inert to most common reagents such as acids and alkalis.

Examples of functional groups include:
- $C = C$ in *alkenes*
- —Cl in *chloroalkanes*
- —OH in *alcohols*
- —NH_2 in *amines*
- $\diagdown C = O$ in *aldehydes* and *ketones*
- —COOH in *carboxylic acids*.

Functional groups can be identified by their reactions (see *organic analysis*) and by using instrumental techniques such as *infrared spectroscopy* and *NMR spectroscopy*.

functional group level: a classification of *functional groups* based on the number of bonds from the carbon atom to an electronegative element.

Level	Name	Examples	Structures with bonds to electronegative atoms shown bold
4	carbon dioxide	CO_2	O $=$ C $=$ O
3	carboxylic acid	RCOOH, RCOOR, $RCONH_2$, RCN	H $-$ C $-$ C $-$ OH (with H, H on C and O double bond on second C)
2	aldehyde / ketone	RCHO, RCOR	H $-$ C $-$ C $-$ C $-$ H and H $-$ C $-$ C $-$ H
1	alcohol	ROH, RNH_2, RCl, RBr, RI	H $-$ C $-$ C $-$ OH
0	alkane	R $-$ R	H $-$ C $-$ H

Oxidation of a functional group moves the carbon atom in a compound up a level.
Reduction of a functional group moves the carbon atom in a compound down a level.
Hydrolysis reactions involve changes within a level.

fundamental particles are the particles that make up an *atom*. In chemistry the fundamental particles are protons, neutrons and electrons. Physicists with high-energy particle accelerators have shown that protons and neutrons are not fundamental but in turn made up of quarks.

	Relative mass	Charge
proton	1	+1
neutron	1	0
electron	5×10^{-4} (negligible)	−1

fusion means either melting or joining. (See *enthalpy change of melting* and *nuclear fusion*.)

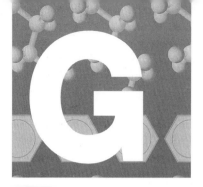

gamma radiation (γ) is high-energy *electromagnetic radiation* given off during *radioactive decay*. The new nucleus formed when a radioactive atom emits an alpha or beta particle is often in an excited state; it gives off gamma rays as it loses energy. Emission of gamma rays does not cause a change in the structure of the nucleus.

The wavelengths of gamma rays are shorter than those of *X-rays*. Gamma radiation is *ionising radiation* which is more penetrating than alpha or beta radiation and is stopped only by several centimetres of lead.

gas: see *kinetic theory of gases* and *gas volume calculations*.

gas constant: the constant R in the *ideal gas* equation $pV = nRT$. The value of the constant depends on the units used for pressure and volume. If all quantities are in SI units with the pressure in N m^{-2} (pascals) and volume in cubic metres (m^3), then $R = 8.314$ J K^{-1} mol^{-1}.

gas laws: the laws which describe the behaviour of *ideal gases* and are summarised by the ideal gas equation. The gas laws include *Boyle's law*, *Charles' law* and *Avogadro's law*.

gas chromatography (GC) is a sensitive technique for analysing complex mixtures. The technique is used for compounds which vaporise on heating without decomposing. This type of chromatography not only separates the chemicals from a sample, but also gives a measure of how much of each is present.

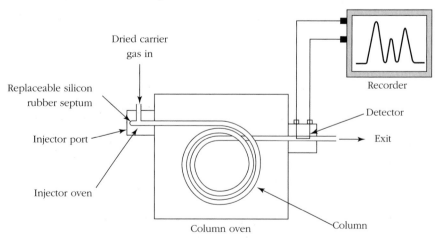

The main features of gas chromatography. The carrier gas takes the mixture of chemicals through the column where they separate. As the compounds leave the column they are detected and measured.

In gas chromatography the mobile phase is a gas, such as helium, which carries the mixture of volatile chemicals through a long tube containing the stationary phase. The column is coiled inside an oven. Heating the column makes it possible to analyse any chemicals that turn to vapour at the temperature of the oven.

The analyst injects a small sample into the column where it enters the oven. Volatile solids are dissolved in a solvent before injection. The chemicals in the sample turn to gases and mix with the carrier gas. The gases then pass through the column.

The components in the mixture separate as they pass through the column. After a time the chemicals emerge one by one. They pass into a detector which sends a signal to a recorder as each compound appears. A series of peaks, one for each compound in the mixture, make up the chromatogram. The position of a peak on a GC print out is a record of how long it took for a compound to pass through the column. This is the retention time.

The area under each peak gives an indication of the relative amounts of the compounds in the mixture. If the peaks are narrow it is good enough to measure the peak heights.

Some GC instruments have capillary columns. These are 20–60 metres long with a very small internal diameter. Capillary columns are often made of silica with an outer coating of a polymer. The stationary phase is the inner surface of the column which adsorbs chemicals to a greater or lesser extent. The inner surface may be coated with a solid adsorbent or a thin film of a liquid.

Other GC columns are steel or glass tubes that are packed with a powder. The powder is an inert solid coated with a thin film of liquid. In these columns the stationary phase is the liquid coating. There is an equilibrium between molecules dissolved in a liquid coating on the surface of the stationary phase and molecules in the carrier gas. Chemicals separate because they differ in their relative solubility in the stationary liquid. When this type of column is used the technique is called gas–liquid chromatography.

(See also *GC–MS*)

Applications of GC include:
- tracking down the source of oil pollution from the pattern of peaks, which acts like a fingerprint for any batch of oil
- measuring the level of alcohol in urine or blood samples from drivers
- detecting and measuring pesticides in river water.

gas tests are used to identify gases produced during *qualitative analysis*.

Gas	Test	Observations
Hydrogen	Burning splint	Burns with a 'pop'
Oxygen	Glowing splint	Splint bursts into flame (relights)
Carbon dioxide	Limewater (aqueous calcium hydroxide)	Turns milky white
Hydrogen chloride (hydrogen bromide and iodide react in the same way)	Smell	Pungent
	Blue litmus	Turns red
	Ammonia vapour (from a drop of concentrated ammonia solution on a glass rod)	Thick white smoke

Gas	Test	Observations
Chlorine	Colour	Greenish-yellow
	Smell	Pungent – bleach-like
	Effect on moist blue litmus paper	Turns the paper red and then bleaches it
	Moist starch-iodide paper	Turns blue-black
Sulfur dioxide	Smell	Pungent
	Blue litmus	Turns red
	Acid dichromate(VI)	Turns green
Hydrogen sulfide	Smell	'Bad eggs'
	Burning splint	Gas burns – yellow deposit of sulfur
	Lead(II) ethanoate	Turns brown-black
Ammonia	Smell	Pungent
	Red litmus	Turns blue
Nitrogen dioxide	Colour	Orange-brown
	Blue litmus	Turns red
Water vapour	Appearance	'Steams' in the air
	Anhydrous cobalt(II) chloride paper	Turns from blue to pink

gas volume calculations use the gas laws and equations to calculate the volumes of gases involved in reactions. Gas volume calculations are straightforward when the reactants and products are all gases. Equal volumes of gases contain equal amounts in moles when measured under the same conditions – this is *Avogadro's law*. As a result the ratio of the gas volumes in a reaction must be the same as the ratio of the numbers of moles in the equation (see *Gay-Lussac's law of combining volumes*).

Worked example

What volume of oxygen reacts with 60 cm³ methane and what volume of carbon dioxide forms if all gas volumes are measured under the same conditions?

Notes on the method

Write the balanced equation. Note that below 100°C the water formed condenses to an insignificant volume of liquid.

Answer

The equation for the reaction is:

$$CH_4(g) + 2O_2(g) \rightarrow CO_2(g) + 2H_2O(l)$$
$$\text{1 mol} \quad \text{2 mol} \qquad \text{1 mol}$$

So 60 cm³ methane reacts with 120 cm³ oxygen to form 60 cm³ carbon dioxide.

The other approach to gas volume calculations is based on the fact that the volume of a gas, under given conditions, depends only on the amount of gas in moles. It does not matter which gas is involved so long as the gas is behaving (at least approximately) like an *ideal gas*.

volume of gas/cm³ = amount of gas/mol × molar volume/cm³mol⁻¹

The molar volume of a gas can be taken to be 24 000 cm³ at room temperature and pressure. So under laboratory conditions the volume of gas formed in a reaction can be estimated from this relationship:

volume of gas/cm³ = amount of gas/mol × 24 000 cm³mol⁻¹

Worked example

What volume of hydrogen is produced under laboratory conditions when 1.95 g zinc reacts with excess acid?

Notes on the method

Start by writing the equation for the reaction. Convert the quantity of zinc to an amount in moles.

Answer

The equation for the reaction is:

$$Zn(s) + 2HCl(aq) \rightarrow ZnCl_2(aq) + H_2(g)$$

The amount of zinc $= \dfrac{1.95 \text{ g}}{65 \text{ g mol}^{-1}} = 0.03$ mol

1 mol zinc produces 1 mol hydrogen.

Volume of hydrogen = 0.03 mol × 24 000 cm³ mol⁻¹ = 720 cm⁻³

Gay-Lussac's law of combining volumes states that when gases are involved in reactions, the ratios of the volumes of gases are simple whole numbers so long as all measurements are taken at the same temperature and pressure. For example 50 cm³ hydrogen react with 50 cm³ chlorine to form 100 cm³ of hydrogen chloride. The ratios are 1:1:2. The law is named after the French chemist Joseph Gay-Lussac (1778–1850).

Avogadro's law accounts for Gay-Lussac's observations. If equal volumes of gases contains equal numbers of molecules, under the same conditions, it follows that the ratios of the volumes are also the ratios of the number of molecules as the equation shows.

$$H_2(g) + Cl_2(g) \rightarrow 2HCl(g)$$

GC–MS is a combined technique that couples *gas chromatography* (GC) with *mass spectrometry* (MS) to give a very powerful system for separating, identifying and measuring complex mixtures of chemicals. GC–MS is widely used in drug detection, investigation of fires, environmental monitoring, as well as in airport security. Some space probes carry tiny GC–MS systems for analysing samples collected in space or on the surface of planets such as Mars.

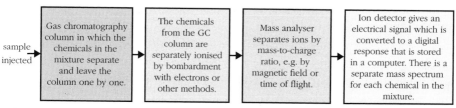

| sample injected → | Gas chromatography column in which the chemicals in the mixture separate and leave the column one by one. | → | The chemicals from the GC column are separately ionised by bombardment with electrons or other methods. | → | Mass analyser separates ions by mass-to-charge ratio, e.g. by magnetic field or time of flight. | → | Ion detector gives an electrical signal which is converted to a digital response that is stored in a computer. There is a separate mass spectrum for each chemical in the mixture. |

Schematic diagram showing the key features of a GC–MS system

GC–MS overcomes some of the limitations of gas chromatography. Similar compounds often have similar retention times in GC which means that they cannot be identified by chromatography alone even if the conditions are carefully standardised. Also GC alone cannot identify any novel chemicals because there are no standards that can be used to determine retention times under given conditions.

GC–MS produces a mass spectrum for each of the chemicals separated on the GC column. These spectra can be used like fingerprints to identify the compounds because every chemical has a unique mass spectrum.

A computer connected to a GC–MS system stores all the data. The computer can be linked to a library of spectra on known compounds. The computer compares the mass spectrum of each chemical in a mixture to mass spectra in the library. It automatically reports a list of likely identifications along with the probability that the matches are correct.

gel: a *colloid* in which a liquid is finely dispersed in a solid. Table jelly is a typical gel with water loosely held in a network of large gelatine molecules. Warming or even shaking can break up the solid network so that the gel liquefies.

general formula: see *homologous series*.

geometrical isomerism: a term that is now discouraged by *IUPAC* (see *cis–trans isomers* and *E–Z isomers*).

germanium (Ge) is an element with some metallic and some non-metallic features. It is a metalloid – shiny like a metal but hard and brittle like a non-metal. It is the element below silicon but above tin in *group 4* of the periodic table with the *electron configuration* [Ar]$3d^{10}4s^24p^2$. Germanium is a *semiconductor*.

giant structures: crystal structures in which all the atoms or ions are strongly linked by a network of bonds extending throughout the crystal. Substances with giant structures generally have high melting and boiling points.

All *metals* consist of giant structures of atoms held together by *metallic bonding*. Metal giant structures are good conductors of electricity because of the delocalised bonding electrons. The layers of atoms in a metal can slide over each other so that metals are *malleable* and *ductile*.

Ionic compounds have giant structures held together by *ionic bonding*. Ionic compounds do not conduct electricity when solid but they do conduct when the ions are free to move on melting the compound or dissolving it in water.

A few *non-metal* elements consist of giant structures of atoms held together by *covalent bonds*. Examples are carbon (as *graphite* and *diamond*) and silicon. Typically these elements do not conduct electricity because the electrons in covalent bonds are localised between pairs of atoms. Graphite is the exception with *delocalised electrons* over the layers of carbon atoms.

Some compounds of non-metals with non-metals, such as *silicon(iv) oxide* and *boron nitride*, have giant structures with covalent bonding. These compounds are non-conductors because each bonding pair of electrons is localised between two atoms.

Many silicates are based on networks of silicon and oxygen atoms, as in the clay minerals. Chains, sheets and three-dimensional arrays of silicate ions combined with metal ions give rise to many different minerals.

Gibbs free energy: see *free energy change*. The concept is named after the US physical chemist, Josiah Gibbs (1839–1903). Hence the symbol ΔG.

glasses are *ceramic* materials which are rigid like solids but which are not crystalline. Glass is made by melting one or more oxides in a furnace. The liquid glass is cooled until thick enough to mould and then shaped and cooled further until it sets solid.

It is possible to make glass from pure *silicon(IV) oxide* (SiO_2) but it melts at 1700°C. Molten silicon dioxide is a thick, sticky liquid (*viscous*). As the liquid cools back to its melting point, the atoms cannot move freely enough to return to the ordered arrangement of crystalline silica. They become 'frozen' into a disordered state.

Most glass is made from silica mixed with other oxides which melt at a lower temperature than pure silica. Windows, bottles and drinking glasses are made of soda lime glass. Lead glass is used for decorative cut glassware. *Borosilicate glass* is used to make ovenware and laboratory glassware.

global warming is the gradual warming of the Earth's atmosphere which is leading to climate change. The theory is that global warming is the result of an increased concentration of *greenhouse gases*, such as carbon dioxide, in the air which is enhancing the *greenhouse effect*.

graphic formulae: an alternative name for *displayed formulae*.

graphite is one of the *allotropes* of carbon, consisting of a giant structure of atoms. Carbon has by far the highest melting point (over 3800 K) of any element. Graphite has an almost ideal combination of physical properties for use as a high temperature *ceramic* except that it has to be used in a non-oxidising environment. Graphite bonded with clay is used to make crucibles for melting metals. Large blocks of graphite are used as *refractories* to line furnaces.

gravimetric analysis is a method of quantitative analysis for finding the composition and formulae of compounds based on accurate weighing of reactants and products. A familiar example, in schools, is to find the formula of magnesium oxide by weighing a piece of magnesium in a crucible and then weighing the magnesium oxide produced after strong heating in air. Chemists have developed sophisticated and precise methods of gravimetric analysis which include *combustion analysis* to find the formulae of organic compounds.

green chemistry is the design of chemical processes and products that reduce or eliminate the production and use of hazardous chemicals.

The key principles of green chemistry cover five broad areas of the production and the use of chemicals:

- Changing to *renewable resources* – this means choosing raw materials or chemical feedstocks that can be renewed instead of relying on crude oil and natural gas.
- Finding alternatives to hazardous chemicals – this includes producing chemical products that are effective but not harmful and developing manufacturing processes that avoid toxic intermediates and solvents that are hazardous to people and the environment (see *hazard*).
- Developing new *catalysts* – this includes the development of catalysts to make possible processes with high *atom economies* and catalysts which are highly selective so that only the desired product is formed, thus cutting down wastes (see *ethanoic acid manufacture* and *ibuprofen*).

- Improving energy *efficiency* – this involves devising processes that run at low temperatures and pressures, and making good use of the energy released by chemical changes (for example with the help of *heat exchangers*).
- Reducing *waste* and preventing pollution – this includes minimising waste so that there is less to treat or dispose of, *recycling* materials, creating *biodegradable* products which do not persist in the environment but break down to harmless chemicals, and the increasing use of sensitive methods of analysis to detect *pollution*.

Many of the changes brought about by applying these principles not only make the manufacture of chemicals safer but also promotes *sustainable development* of the chemical industry.

greenhouse effect: the effect of some gases in the air which keeps the surface of the Earth about 30°C warmer that it would be if there were no *atmosphere*. Without the greenhouse effect there would be no life on Earth.

When radiation from the Sun reaches the Earth's atmosphere, about 30% is reflected into space, 20% is absorbed by gases in the air and about half reaches the surface of the Earth. The warm surface radiates energy back into space but at longer, infrared wavelengths. Some of the infrared radiation is absorbed by gases in the air and warms up the atmosphere. This is the greenhouse effect.

The gases in the air which absorb infrared radiation are called *greenhouse gases*. The concentrations of greenhouse gases in the air are rising because of human activity such as burning *fossil fuels* and agriculture. This is enhancing the greenhouse effect. There is growing evidence that this is responsible for *global warming*.

greenhouse gases are the gases in the air which absorb infrared radiation. The main natural greenhouse gases are carbon dioxide, methane, dinitrogen oxide and water vapour. *CFCs*, *HCFCs* and *HFCs* are also greenhouse gases. The *Kyoto protocol* is an international agreement to control emissions of greenhouse gases.

Nitrogen and oxygen make up most of the air but they do not absorb infrared radiation. This is because they are not polar. They are never polar even when they vibrate and so they cannot absorb infrared radiation (see *infrared spectroscopy*).

Grignard reagents are reactive organic compounds containing magnesium atoms which are used to form C—C bonds in organic synthesis. They are examples of *organometallic compounds*. The reagents were discovered by the French organic chemist Victor Grignard (1871–1935).

Grignard showed that *halogenoalkanes* react with magnesium in dry ether (ethoxyethane):

$$RBr + Mg \rightarrow RMgBr \qquad \text{where R represents an alkyl group.}$$

The C—Mg bond in a Grignard reagent is polarised with the metal atom at the positive end of the dipole and the carbon atom at the negative end. As a result, the carbon atom attached to magnesium is nucleophilic and C—C bonds form when Grignard regents take part in *nucleophilic substitution* or *nucleophilic addition* reactions. This is what makes the reagents so useful in synthesis.

group 1

R — H
hydrocarbon

$H_2O/H^+(aq)$

CO_2 (dry ice)

R — MgBr
Grignard reagent
(R = alkyl)

$$R - C \overset{O}{\underset{O - MgCl}{\diagdown}}$$

$H_2O/H^+(aq)$

$$R - C \overset{O}{\underset{OH}{\diagdown}}$$
carboxylic acid

aldehyde
or ketone

$$\overset{R'}{\underset{R''}{\diagup}} C = O$$

$$R - \overset{R'}{\underset{R''}{\overset{|}{\underset{|}{C}}}} - OMgBr$$

$H_2O/H^+(aq)$

$$R - \overset{R'}{\underset{R''}{\overset{|}{\underset{|}{C}}}} - OH$$

R' and R'' = alkyl
or hydrogen

alcohol (primary,
secondary or tertiary)

Summary of the reactions of Grignard reagents

ground state: the stable state of an atom or molecule with the electrons in the lowest available energy levels.

group: a vertical column of elements in the *periodic table*. Elements in the same group have similar chemical properties because they have the same outer *electron configuration*.

There are trends in properties down a group as the number of full inner electron shells increases and the atoms get larger. Generally the metallic characteristics of elements increase down a group. In *group 1* and *group 2* the metals get more reactive down the group.

Non-metal characteristics decrease down a group. Down *group 7* the halogens get less reactive. *Group 4* shows a trend from non-metals at the top to metals at the bottom with germanium, a *metalloid*, in the middle.

IUPAC now recommends that the groups should be numbered from 1 to 18. Groups 1 and 2 are the same as before. Groups 3 to 12 are the vertical families of *d-block elements*, the groups traditionally numbered 3 to 8 then become groups 13 to 18.

group 1 elements belong to the family of *alkali metals* with similar chemical properties because they all have one electron in an outer *s orbital*. These elements are more similar to each other than the elements in any other group. Even so, because of the increasing number of full, inner shells there are trends in properties down the group from lithium to caesium. The element in period 7, francium, is very rare and all its isotopes are radioactive.

lithium, Li	[He]2s^1	rubidium, Rb	[Kr]5s^1
sodium, Na	[Ne]3s^1	caesium, Cs	[Xe]6s^1
potassium, K	[Ar]4s^1		

The metals are powerful *reducing agents*. They react by losing the outer s electron to form M^+ ions. The first ionisation energies decrease down the group as the increasing number of full shells means that the outer electrons get further away from the same effective nuclear charge (see *shielding*).

Atomic and ionic radii increase down the group. For each element the 1+ ion is smaller than the atom because of the loss of the outer shell of electrons. The tendency to react and form ions increases down the group.

The small size of the lithium ion means that it has a relatively high *polarising power* and is heavily *hydrated* in solution. As a result lithium is in some ways not typical of the group as a whole.

The metals are soft and easily cut with a knife. They are shiny when freshly cut but quickly dull in air as they react with moisture and oxygen. The metals are stored under oil.

All the metals burn brightly in oxygen on heating to form ionic oxides. Lithium forms a simple oxide, Li_2O. Sodium forms mainly the *peroxide*, Na_2O_2. Potassium forms a *superoxide*, KO_2. The oxides are basic; they react with acids to form salts:

$$Li_2O(s) + 2HCl(aq) \rightarrow 2LiCl(aq) + H_2O(l)$$

All the metals react with water to form hydroxides and hydrogen. The rate and violence of the reaction increases down the group. Lithium reacts steadily with cold water. Caesium reacts explosively.

All the metals react vigorously with chlorine to form colourless, ionic chlorides, M^+Cl^-, which are soluble in water. The crystal structures depend on the size of the metal ion (see *caesium chloride structure* and *sodium chloride structure*).

The hydroxides are:
- similar in that they all have the formula MOH, and are soluble in water forming alkaline solutions (they are *strong bases*)
- different in that their solubility increases down the group.

The carbonates are similar in that they all have the formula M_2CO_3 and, with the exception of lithium, do not decompose on heating (see *thermal stability of carbonates*). The carbonates of sodium and potassium are soluble, forming alkaline solutions because the carbonate ion is a *base*.

The nitrates are:
- similar in that they all have the formula MNO_3, are colourless crystalline solids, are very soluble in water and decompose on heating
- different in that they become more difficult to decompose down the group. Lithium nitrate, like magnesium nitrate, decomposes on heating to the oxide, nitrogen dioxide and oxygen. The nitrates of sodium and potassium need strong heating to decompose and they form the nitrite:

$$2KNO_3(s) \rightarrow 2KNO_2(s) + O_2(g)$$

group 2 elements belong to the family of *alkaline earth metals* with similar chemical properties because they all have two electrons in an outer *s orbital*:

beryllium, Be	$[He]2s^2$	strontium, Sr	$[Kr]5s^2$
magnesium, Mg	$[Ne]3s^2$	*barium*, Ba	$[Xe]6s^2$
calcium, Ca	$[Ar]4s^2$		

The first member of the group, beryllium, is not a typical member of group 2. This is partly because of the small size of the Be^{2+} ion.

The following similarities and differences refer to the elements Mg, Ca, Sr and Ba. The symbol M is used to represent any one of these elements.

The metals are *reducing agents* which react by losing their two s electrons to form M^{2+} ions. The first and second *ionisation energies* decrease down the group as the increasing number of full shells means that the outer electrons are further away from the same effective nuclear charge (see *shielding*).

Atomic radii and *ionic radii* increase down the group. For each element the 2+ ion is smaller than the atom because of the loss of the outer shell of electrons. The tendency to react and form ions increases down the group.

The metals are harder and denser than group 1 metals and have higher melting points. In air the surface of the metals is covered with a layer of oxide

All the metals burn brightly in oxygen on heating to form white, ionic oxide $M^{2+}O^{2-}$. These are *basic oxides*. They react with acids forming salts.

$$BaO(s) + 2HNO_3(aq) \rightarrow Ba(NO_3)_2(aq) + H_2O(l)$$

All the metals react with water to form hydroxides, and hydrogen or acids to form salts and hydrogen. Magnesium reacts only slowly with hot water. Barium reacts quite fast even with cold water. Barium is so reactive with air and moisture that it is generally stored under oil like the alkali metals.

All the metals react vigorously with chlorine to form colourless, ionic chlorides MCl_2. Unlike the group 1 chlorides, the chlorides of this group are usually *hydrated*. They are soluble in water.

The hydroxides are:
- similar in that they all have the formula $M(OH)_2$ and are to some degree soluble in water, forming alkaline solutions
- different in that their solubility increases down the group.

The carbonates are:
- similar in that they all have the formula MCO_3, are insoluble in water and decompose on heating: e.g. $CaCO_3(s) \rightarrow CaO(s) + CO_2(g)$
- different in that they become more difficult to decompose down the group (see the *thermal stability of carbonates* and *calcium carbonate*).

The nitrates are:
- similar in that they all have the formula $M(NO_3)_2$, are colourless crystalline solids, are very soluble in water and decompose to the oxide on heating:
$$2Mg(NO_3)_2(s) \rightarrow 2MgO(s) + 4NO_2(g) + O_2(g)$$
- different in that they become more difficult to decompose down the group.

The sulfates are:
- similar in that they are all colourless solids with the formula MSO_4
- different in that they become less soluble down the group.

group 4 (IUPAC group 14) is a group of elements which shows a trend from *non-metals* at the top to *metals* at the bottom. All these elements have four electrons in their outer shell. The characteristic oxidation states of the group are +4 and +2. The +2 state becomes more important down the group and is the more stable state of lead. This is an example of the *inert pair effect*.

carbon, C $[He]2s^22p^2$ *tin*, Sn $[Kr]4d^{10}5s^23p^2$

silicon, Si $[Ne]3s^23p^2$ *lead*, Pb $[Xe]4d^{14}6s^23p^2$

germanium, Ge $[Ar]3d^{10}4s^23p^2$

The two common *allotropes* of carbon (diamond and graphite) consist of covalent giant structures. Silicon and germanium have diamond-like giant structures and are *semiconductors*.

The room temperature allotrope of tin has a metallic structure. The low temperature allotrope, grey tin, has the non-metallic diamond structure. Lead has a metallic structure.

The two metals, tin and lead, have lower melting points than the non-metals with giant structures.

The group 4 compounds in the +4 oxidation state generally behave more like the compounds of non-metals:

- the +4 oxides (CO_2, SiO_2, SnO_2 and PbO_2) are acidic or, in the case of tin and lead, amphoteric with a bias towards *acidic oxide* behaviour
- the +4 chlorides are molecular liquids which (with the exception of CCl_4) are rapidly hydrolysed by water (see *hydrolysis of non-metal chlorides*).

The compounds of tin and lead in the +2 oxidation state are more typical of metallic compounds:

- the +2 oxides (SnO and PbO) are *amphoteric compounds* but with a bias towards *basic oxide* properties
- the +2 chlorides are white solids; $PbCl_2$ is an ionic solid.

The +4 oxidation state is more stable for tin, so that tin(II) compounds are reducing agents. Tin(II) chloride reduces iron(III) to iron(II) for example.

The +2 oxidation state is the more stable state for lead. So PbO_2 is a strong oxidising agent. Lead(IV) oxide will oxidise chloride ions to chlorine gas.

group 7 (IUPAC group 17) elements belong to the family of *halogens* with similar chemical properties because they all have seven electrons in the outer shell – one fewer than the next noble gas in group 8. The element *astatine* in group 7 is very rare and highly radioactive. The halogens are all toxic.

fluorine, F $[He]2s^22p^5$ *bromine*, Br $[Ar]3d^{10}4s^24p^5$

chlorine, Cl $[Ne]3s^23p^5$ *iodine*, I $[Kr]4d^{10}5s^25p^5$

The symbol X is often used to represent a halogen.

The halogens all consist of diatomic molecules, X_2, with the atoms linked by a single *covalent bond*. The *bond enthalpies* get smaller down the group from chlorine to iodine, but fluorine is out of line with this trend.

All the halogens are volatile. *Intermolecular forces* increase down the group as the numbers of electrons in the atoms increase, so melting points and boiling points rise down the group. Iodine molecules are the most polarisable. Fluorine and chlorine are gases at room temperature, bromine is a liquid and iodine a solid.

Halogen atoms are highly electronegative. They form ionic compounds or compounds with *polar covalent bonds*. *Electronegativity* decreases down the group. The bonding in aluminium fluoride is ionic, but anhydrous aluminium chloride is a covalent solid.

The halogens are powerful *oxidising agents*. Fluorine is the strongest oxidising agent and iodine the weakest in the group. (See *displacement reactions*.)

Hot iron burns in chlorine forming iron(III) chloride. The metal reacts much less vigorously with iodine forming iron(II) iodide. (See also *halide ions* and *hydrogen halides*.)

group 8 (IUPAC group 18) is a family of colourless, unreactive elements called the noble gases. The gases exist as single atoms. The *intermolecular forces* between the atoms are very weak, so at room temperature they behave very much like *ideal gases.* They have very low melting and boiling points. The strength of the forces between the atoms increases as the number of electron shells increases. The larger atoms or more *polarisable*, so melting points and boiling points rise down the group.

First *ionisation energies* decrease down the group as the outer electrons get further away from the nucleus. The ionisation energies of the elements near the bottom of the group are low enough for them to be able to take part in chemical reactions and form compounds.

Helium, neon and *argon* do not form compounds with other elements. *Krypton* combines with fluorine. *Xenon* forms compounds with fluorine and oxygen.

Do you need revision help and advice?

Go to pages 388–407 for a range of revision appendices that include plenty of exam advice and tips.

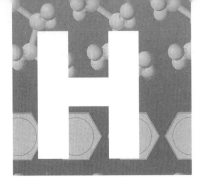

Haber process: the process for *ammonia manufacture* developed by the German physical chemist Fritz Haber (1868–1934).

haemoglobin is the red globular *protein* in blood which carries oxygen from the lungs to the cells in body tissues. A haemoglobin molecule consists of four poly*peptide* chains, each with a nitrogen atom forming a dative bond to an Fe^{2+} ion in a haem group. So there are four haem groups in each haemoglobin molecule.

Nitrogen atoms in the porphyrin ring form four more dative bonds to the Fe^{2+} ion. So the porphyrin ring acts as a *polydentate ligand*. This leaves one remaining site on the metal ion which can accept a pair of electrons from an oxygen molecule (in oxyhaemoglobin which is bright red) or a water molecule (in deoxyhaemoglobin which is dull red).

A haem group

The reactions between haem groups and oxygen or water are reversible *ligand substitution reactions*, allowing haemoglobin to pick up and release oxygen. The reaction with *carbon monoxide* is irreversible, which explains why the gas is dangerously toxic.

half-cell: an electrode dipping into a solution of ions. Two half-cells connected by a salt bridge make up an *electrochemical cell*.

half-equation: an *ionic equation* used to describe either the gain or the loss of electrons during a redox process. Half-equations help to show what is happening during a *redox reaction*. Two half-equations combine to give an overall balanced equation for a redox reaction.

Iron(III) ions can oxidise iodide ions to iodine. This can be shown as two half-equations:

electron gain (reduction): $Fe^{3+}(aq) + e^- \rightarrow Fe^{2+}(aq)$

electron loss (oxidation): $2I^-(aq) \rightarrow I_2(s) + 2e^-$

The number of electrons gained must equal the number lost. So the first of these half-equations must be doubled to arrive at the overall balanced equation:

$$2Fe^{3+}(aq) + 2e^- \rightarrow 2Fe^{2+}(aq)$$

$$2I^-(aq) \rightarrow I_2(s) + 2e^-$$

$$2Fe^{3+}(aq) + 2I^-(aq) \rightarrow 2Fe^{2+}(aq) + I_2(s)$$

half-life (chemical): chemists measure the half-lives of chemical reactions. The half-life of a reaction is the time for the concentration of one of the reactants to fall by half. Half-lives help to identify first order reactions. At a constant temperature, the half-life of a *first order reaction* is the same wherever it is measured on a concentration/time graph. So the half-life is independent of the initial concentration.

half-life (radioactive): the time for half the atoms in a sample of a radioactive *isotope* to decay. It is also the time for the count rate for *alpha decay* or *beta decay* from a sample to fall by half. Half-lives for radioactive isotopes can be as short as a fraction of a second or as long as millions of years.

Half-lives for *radioactive decay* are unaffected by changes in temperature, pressure or chemical state. This is the basis of the technique used to find the age of the remains of living things or of rocks. Knowing the half-lives it is possible to estimate ages by measuring the proportions of different *nuclides* in a sample.

Radiocarbon dating is based on the radioisotope carbon-14. It can be used to measure the age of materials containing carbon that are up to about 60 000 years old. Archaeologists use the technique to estimate the age of organic samples recovered from digs, such as bones. During photosynthesis, plants take in carbon dioxide from the air to make starch and other plant chemicals. These chemicals contain the same proportion of carbon-14 as is found in the atmospheric carbon dioxide at the time. Humans and other animals eat plants. The carbon from the plants gets into their bones. In life, and after death, the carbon-14 in the organic material decays. By measuring the proportion of carbon-14 in a bone, or other remains, it is possible to calculate the age of the material knowing the half-life of the radioactive isotope.

The radioactive isotopes of uranium make it possible to measure the ages of rocks on geological timescales ranging from about 1 million years to over 4.5 billion years old. There are two radioactive isotopes of uranium which break down through a series of alpha and beta decays to lead. Uranium-238 decays to lead-206, with a half-life of 4.47 billion years, while uranium-235 decays to lead-207 with a half-life of 704 million years. It is possible to determine the age of a rock by measuring these two ratios: $^{206}Pb/^{238}U$ and $^{207}Pb/^{235}U$.

halide ions are the ions of the halogen elements. They include the fluoride, F^-, chloride, Cl^-, bromide, Br^- and iodide, I^-, ions. (See also *silver halides*.)

Warming sodium chloride with concentrated sulfuric acid produces clouds of hydrogen chloride gas. This *acid–base reaction* can be used to make hydrogen chloride.

$$NaCl(s) + H_2SO_4(l) \rightarrow HCl(g) + NaHSO_4(s)$$

Both sulfuric acid and hydrogen chloride are strong acids. The reaction goes from left to right because the hydrogen chloride is a gas and escapes from the reaction mixture, so the reverse reaction cannot happen.

This type of reaction cannot be used to make hydrogen bromide or hydrogen iodide, because bromide and iodide ions are strong enough reducing agents to reduce sulfur from the +6 state to lower *oxidation states*.

The reactions of halide ions with sulfuric acid show that there is a trend in the strength of the halide ions as *reducing agents*:

- chloride ions do not reduce sulfuric acid at all, so the only gaseous product is hydrogen chloride gas
- bromide ions turn to orange bromine molecules as they reduce H_2SO_4 to SO_2 (mixed with some hydrogen bromide gas)
- iodide ions are the strongest reducing agents, they turn into iodine molecules as they reduce H_2SO_4 to S and H_2S; scarcely any hydrogen iodide forms.

So the trend as reducing agents is $I^- > Br^- > Cl^-$. In *group* 7, iodine is the weakest oxidising agent so it has the least tendency to form negative ions. Conversely iodide ions are the ones that most readily give up electrons and turn back into iodine molecules. (See also *displacement reactions*.)

hallucinogens: drugs which affect the brain powerfully and cause people to experience hallucinations. Some hallucinogens occur naturally in plants such as the cactus, peyote, which produces mescaline. Other hallucinogens are synthetic chemicals such as LSD and MDMA (the *amphetamine* called ecstasy).

halogenation: see *chlorination, bromination, iodination*.

halogens is the family name of the *group* 7 elements. The halogens are *fluorine, chlorine, bromine, iodine* and *astatine*. The name halogen means 'salt-former' and is based on the fact that the elements combine with most metals to form salts (containing *halide ions*).

halogen carriers polarise chlorine and bromine molecules so that they can take part in *electrophilic substitution* reactions with benzene and other arenes. Halogen carriers include $AlCl_3$ and $FeCl_3$ for chlorination and $AlBr_3$ and $FeBr_3$ for bromination.

halogenoalkanes (or haloalkanes) are compounds formed by replacing one or more of the hydrogen atoms in *alkanes* with halogen atoms.

1-iodobutane
(primary)

2-chloro-2-methylpropane
(tertiary)

2-bromobutane
(secondary)

Names and structures of some halogenoalkanes

A primary halogenoalkane has the halogen atom at the end of the chain. A secondary compound has the halogen atom somewhere along the chain but not at the ends. A tertiary halogenoalkane has the halogen atom at a branch in the chain.

Chloromethane, bromomethane and chloroethane are gases at room temperature. Most other halogenoalkanes are colourless liquids which do not mix with water.

Halogen atoms are more electronegative than carbon atoms so a carbon–halogen bond is polar. The characteristic reactions of haloalkanes are *nucleophilic substitution* reactions.

$$CH_3CH_2CH_2CH_2Br \xrightarrow[\text{with NaOH(aq)}]{\text{heat}} CH_3CH_2CH_2CH_2OH$$
alcohol

$$\xrightarrow[\text{in ethanol}]{\text{heat with KCN}} CH_3CH_2CH_2CH_2CN$$
nitrile

$$\xrightarrow{\text{heat with } NH_3} CH_3CH_2CH_2CH_2NH_2$$
primary amine
(secondary and tertiary
amines also form)

Nucleophilic substitution reactions of 1-bromobutane

The *hydrolysis* of halogenoalkanes makes it possible to distinguish chloro-, bromo- and iodo- compounds. Heating the compound with alkali releases halide ions. Acidifying with nitric acid and then adding silver nitrate produces a precipitate of the *silver halide*.

The rates of reaction of halogenoalkanes are in the order RI > RBr > RCl, where R represents an alkyl group.

The C—I bond is the longest and the weakest (as measured by the mean *bond energy*). The C—Cl bond is the shortest and the strongest. Bond polarity does not appear to be a factor in determining the rates because chlorine is the most electronegative of the elements so the C—Cl bond is the most polar.

An *elimination reaction* happens on warming a halogenoalkane with a solution of potassium hydroxide in ethanol. The reaction goes more readily in compounds with a halogen atom at a branch in the carbon chain.

Halogenoalkanes are used as:
* solvents (e.g. dichloromethane)
* refrigerants (e.g. the *HFC*, CF_3CH_2F which has replaced the *CFC*, CF_2Cl_2)
* pesticides (e.g. bromomethane which is being phased out because of its powerful ozone-depleting properties)
* fire retardants (e.g. halons, such as CBr_2ClF which are also being phased out).

There are growing restrictions on the uses of many halogenoalkanes because of concern about their hazards to health, their persistence in the environment and their effect on the *ozone in the stratosphere*.

halogenoarenes are compounds formed by replacing one or more of the hydrogen atoms in an *arene* with halogen atoms. An example is chlorobenzene which is a colourless liquid at room temperature.

Structure of chlorobenzene

hardness is a property of materials which shows how easy they are to dent or scratch (see the *Mohs scale of hardness*). The term is also used in the expression *hard water*.

hard water does not easily lather with *soap* but instead forms a greasy scum. Water is hard if it contains calcium or magnesium ions. Scum is a *precipitate* formed when soap mixes with water containing these ions.

$$Ca^{2+}(aq) \quad + \quad 2C_{17}H_{35}COO^-(aq) \quad \rightarrow \quad Ca(C_{17}H_{35}COO)_2(s)$$

octadecanoate (stearate)   insoluble precipitate
ions in soap   (scum)

Ion exchange resins and other methods of *water treatment* soften water by removing calcium and magnesium ions.

hazard: anything that can cause harm if things go wrong is a hazard. The chance of harm actually being done is the risk. *Risk* assessments are carried out to determine the likelihood of anything going wrong during a procedure and on the seriousness of any possible injury.

Hazard warning signs identify chemicals which can harm people. The signs have been agreed internationally and they have very specific meanings. Words beside or below the signs indicate the degree of hazard. Very toxic and *toxic chemicals*, for example, carry the same sign but are distinguished by the words alongside.

 Toxic chemicals may involve serious acute or chronic health risks and even cause death if breathed in, swallowed or absorbed through the skin.

 Corrosive chemicals destroy living tissues on contact.

 Harmful chemicals may involve limited health risks if breathed in, swallowed or absorbed through the skin.

 Irritant chemicals are non-corrosive but may cause inflammation or lesions through immediate, prolonged or repeated contact with skin or eyes.

 Highly or extremely flammable chemicals are liquids which are graded by the temperature at which they catch fire in air (flash point) and their volatility (boiling point), and solids which may catch fire after brief contact with a flame or which give off flammable gases in contact with water.

 Oxidising chemicals which in contact with other substances produce a reaction that has a large heating effect and may ignite flammable substances.

Hazard warning signs for chemicals

HCFCs (hydrochlorofluorocarbons) are a group of synthetic compounds containing hydrogen, chlorine, fluorine and carbon. They do not occur naturally. HCFCs were developed to replace *CFCs* in the 1980s when they were found to be destroying *ozone in the stratosphere*. HCFCs can replace CFCs for refrigeration, *aerosol* propellants, *foam* manufacture and air conditioning. Unlike the CFCs, most HCFCs are broken down in the lowest part of the atmosphere so that they are very much less damaging to the ozone layer. Unfortunately HCFCs are also very powerful *greenhouse gases*, despite their very low atmospheric concentrations.

heat exchangers are widely used in the chemical industry to transfer energy from a hot liquid or gas to a cooler liquid or gas.

A laboratory water condenser is a simple example of a heat exchanger. The water flowing in the jacket round the inner tube cools the hot vapour from a *distillation* column. The water flows in the opposite direction to the vapour.

An industrial heat exchanger is normally made of steel rather than glass and has many tubes carrying the hotter *fluid* through the surrounding container of cooler fluid.

Heat exchangers are important to the economics of industry. In *sulfuric acid manufacture* by the Contact process, the reaction of sulfur dioxide with oxygen to make sulfur trioxide is highly exothermic. The gases must be cooled after each pass through the *catalyst* to maximise the *yield*. Instead of letting the energy go to waste, the hot gases are cooled in heat exchangers in which water turns to steam. The steam is then used to generate electricity. Typically a sulfuric acid plant has no fuel bills because it can generate all the electricity it needs. This helps to make the process commercially viable.

heavy metals are metals such as cadmium, *mercury* and lead which have relatively high relative atomic masses. The term does not have a precise chemical meaning but crops up in descriptions of pollution caused by *toxic* heavy metal ions.

heavy water is the common name for deuterium oxide, D_2O. This is water in which both the atoms linked to oxygen are the *deuterium* isotope of hydrogen.

helium (He) is the first member of the group of noble gases, *group 8* (IUPAC group 18). Helium is the second most common element in the universe after hydrogen. The energy of stars such as the Sun comes from *nuclear fusion* which turns hydrogen nuclei into helium nuclei. Helium was detected on the Sun by *spectroscopy* during a total eclipse in 1868 over 20 years before it was isolated and identified on Earth.

On Earth one of the main sources of helium is *natural gas*. Some natural gas wells produce up to 7% helium. The gas boils at 4.2 K and is the coldest liquid available to study the properties of materials at low temperatures.

Helium is used to provide an inert atmosphere for welding and for growing crystals of pure *semiconductors* such as germanium and silicon.

Divers breathe a mixture of helium with oxygen. Helium is less soluble in blood than nitrogen and so this mixture reduces the risk of divers suffering from the 'bends' when they come to the surface.

Helium is less dense than air under the same conditions of temperature and pressure. It is a much safer gas than hydrogen for filling airships and weather balloons.

Henderson–Hasselbalch equation: an equation which helps chemists to explain the behaviour of *buffer solutions* and *acid–base indicators*. The equation is derived by rearranging the expression for K_a of a weak acid and then putting both sides of the equation into logarithmic form. The advantage of taking logs is to produce an equation which can be easily used to calculate *pH* values, which are themselves logarithmic.

For a weak acid HA:

$$HA(aq) \rightleftharpoons H^+(aq) + A^-(aq)$$

$$K_a = \frac{[H^+(aq)][A^-(aq)]}{[HA(aq)]}$$

This rearranges to give:

$$[H^+(aq)] = \frac{K_a[HA(aq)]}{[A^-(aq)]}$$

Taking logs and substituting pH for $-\lg[H^+(aq)]$ and pK_a for $-\lg K_a$, gives:

$$pH = pK_a + \lg\frac{[A^-(aq)]}{[HA(aq)]} \quad \text{because} \quad -\lg\frac{[HA(aq)]}{[A^-(aq)]} = +\lg\frac{[A^-(aq)]}{[HA(aq)]}$$

In a mixture of a weak acid and one of its salts, the weak acid is only slightly ionised while the salt is fully ionised, so it is often accurate enough to assume that all the anions come from the salt present and all the unionised molecules from the acid.

Hence: $pH = pK_a + \lg\dfrac{[salt]}{[acid]}$, which is the Henderson–Hasselbalch equation.

So in a buffer solution with [acid] = [salt], $pH = pK_a + \lg 1 = pK_a$, since $\lg 1 = 0$.

Diluting a buffer solution does not change [salt]/[acid] so the pH does not change.

The equation also shows why it takes a big change in the ratio [salt]:[acid] in a buffer solution to make a significant change in pH. Since $\lg 1 = 0$, $\lg 10 = 1$ and $\lg 100 = 2$, the ratio has to change by a factor of ten to make the pH change by one unit.

The equation also explains why acid–base indicators change colour roughly in the range $pH = pK_{in} \pm 1$. When $pH = pK_a$, [HIn] = [In$^-$] and the two different colours of the indicator are present in equal amounts, so the indicator is mid-way through its colour change.

Add a few drops of acid and the pH falls. The characteristic acid colour of the indicator is distinct when [HIn] ≈ 10 × [In$^-$]. At this point $pH = pK_a + \lg 0.1 = pK_a - 1$, since $\lg 0.1 = -1$.

Add a few drops of alkali and the pH rises. The characteristic alkaline colour of the indicator is distinct when [In$^-$] ≈ 10 × [HIn]. At this point $pH = pK_a + \lg 10 = pK_a + 1$, since $\lg 10 = +1$.

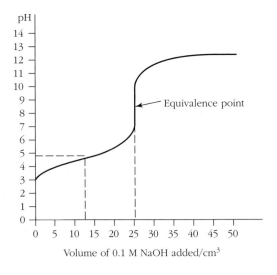

Curve showing the change of pH when 25 cm³ 0.1 mol dm⁻³ ethanoic acid, a weak acid, is titrated with a strong base. Half way to the end-point, half the acid has been converted to its salt, so [acid] = [salt] and pH = pK$_a$ = 4.8.

herbicides: see *pesticides.*

Hess's law makes it possible to calculated enthalpy changes which cannot be measured. The law states that the enthalpy change for a reaction is the same whether it takes place in one step or a series of steps. So long as the reactants and the products are the same, the overall enthalpy change will be the same whether the reactants are converted to products directly or through two or more intermediate reactions.

Hess's law is used to calculate:
- *enthalpy changes of formation* from enthalpy changes of combustion
- *enthalpy changes of reaction* from enthalpy changes of formation
- lattice energies in *Born–Haber cycles.*

heterocyclic compounds are compounds with rings made of carbon atoms with atoms of at least one other element such as nitrogen, sulfur or oxygen. The bases in *DNA* are heterocyclic compounds. Adenine is an example. Many *alkaloids* are also heterocyclic compounds.

The structure of the heterocyclic compound nicotine or 3-(1-methyl-2-pyrrolidyl)pyridine. Nicotine is present in the leaves of tobacco plants. Pure nicotine is a colourless liquid. It is highly toxic. Nicotine from tobacco plants is used as an insecticide but the compound is better known as the addictive drug in tobacco smoke.

nicotine

heterogeneous catalyst: a *catalyst* which is in a different *phase* from the reactants. Generally a heterogeneous catalyst is a solid while the reactants are gases or in solution in a solvent.

The advantage of a heterogeneous catalyst is that it can be held in a reaction vessel as the reactants flow in and the products flow out. There no difficulty in separating the products from the catalyst.

Heterogeneous catalysts are used in large-scale continuous processes such as *catalytic cracking*. This type of catalyst is also used in the Haber process for *ammonia manufacture*, in the Contact process for *sulfuric acid manufacture* and in the production of *methanol*.

Platinum metal alloyed with other metals such as rhodium is an important catalyst. It is used to oxidise ammonia during *nitric acid manufacture*. It is also used in *catalytic converters*. In a catalytic converter the expensive catalyst is finely divided and supported on the surface of a ceramic to increase the surface area in contact with the exhaust gases.

Heterogeneous catalysts work by adsorbing reactants at *active sites* on the surface of the solid. *Catalyst poisons* can stop catalysts working by being adsorbed onto the active sites.

Nickel acts as a catalyst for the addition of hydrogen to $C=C$ in unsaturated compounds by adsorbing hydrogen molecules which probably split up into single atoms held on the surface of the metal crystals.

If a metal is to be a good catalyst for a *hydrogenation* reaction it must not adsorb the hydrogen so strongly that the hydrogen atoms become unreactive. This happens with tungsten. Equally, if adsorption is too weak there will not be enough adsorbed atoms for the reaction to go at a useful rate, as is the case with silver. The strength of adsorption must have a suitable intermediate value. Suitable metals for catalytic hydrogenation are nickel, platinum and palladium.

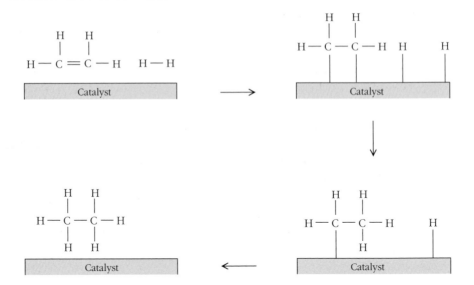

Possible mechanism for the hydrogenation of an alkene in the presence of a nickel catalyst

heterogeneous equilibrium: an equilibrium system in which the substances involved are not all in the same *phase*. An example is the equilibrium state involving two solids and a gas, formed on heating calcium carbonate in a closed container.

$$CaCO_3(s) \rightleftharpoons CaO(s) + CO_2(g)$$

The concentrations of solids do not appear in the expression for the equilibrium constant. Pure solids have a constant 'concentration'.

$$K_c = [CO_2(g)] \text{ or } K_p = p_{CO_2}$$

The same applies to the equilibrium between a solid and its saturated solution in water (see *solubility product constant*).

heterolytic bond breaking (fission) is a type of bond breaking which produces ionic *intermediates* in reactions. In heterolytic fission, a *covalent bond* breaks so that one of the atoms joined by the bond takes both of the shared pair of electrons while the other is left with none. This type of bond breaking happens during *electrophilic substitution, nucleophilic substitution, electrophilic addition* and *nucleophilic addition* reactions and is favoured when reactions take place in *polar solvents* such as water.

Heterolytic bond breaking

$$H{\overset{\times\times}{\underset{\times\times}{\ast}}}Br{\overset{\times\times}{\underset{\times\times}{\times}}} \longrightarrow H^+ + {\overset{\times\times}{\underset{\times\times}{\ast}}}Br{\times}^-$$

$$H\overset{\frown}{-}Br \longrightarrow H^+ + Br^-$$

hexagonal close-packed structure (hcp): see *close-packed structures*.

HFCs (hydrofluorocarbons) are composed only of carbon, hydrogen and fluorine. They are colourless, odourless and chemically unreactive gases They contain no chlorine so that, unlike *CFCs*, they have no known effects on the *ozone in the stratosphere*. However, HFCs are powerful *greenhouse gases* so they do contribute to *global warming*. HFCs have been used to replace CFCs as refrigerants and *aerosol* propellants.

high-performance liquid chromatography (HPLC) is a sophisticated version of liquid chromatography. The technique is able to separate components in a mixture which are very similar to each other.

High performance is achieved by packing very small particles of a solid such a silica packed into a steel column. A typical column is about 10–30 cm long and has an internal diameter of about 4 mm.

The use of fine particles increases the surface area of the stationary phase. This makes the separation efficient but it means that a high-pressure pump is necessary to force the solvent through the tightly packed column. As a result the techniques is sometimes called high-pressure liquid chromatography.

An advantage of HPLC is that it is carried out at room temperature and so can analyse mixtures that decompose before they vaporise, such as many biological molecules. Organic molecules that break down on heating cannot be studied using *gas chromatography*. This makes the method suitable for separating biological molecules such as proteins. One important application of HPLC is to study what happens to drugs as they are metabolised in the body.

homogeneous catalyst: a *catalyst* which is in the same *phase* as the reactants. Typically the reactants and the catalyst are dissolved in the same solution.

Transition element ions can be effective homogeneous catalysts because they gain and lose electrons as they change from one *oxidation state* to another. The oxidation of iodide ions by persulfate ions, for example, is very slow.

$$S_2O_8^{2-}(aq) + 2I^-(aq) \rightarrow 2SO_4^{2-}(aq) + I_2(aq)$$

The reaction is catalysed by iron(III) ions in the solution. A possible mechanism is that iron(III) is reduced to iron(II) as it oxidises iodide ions to iodine. The $S_2O_8^{2-}(aq)$ ions oxidise the iron(II) back to iron(III) ready to oxidise some more of the iodide ions and so on.

Acid catalysis, **base catalysis** and *autocatalysis* are other examples of homogeneous catalysis.

homogeneous equilibrium: an equilibrium in which all the substances involved are in the same *phase*.

homologous series: a series of closely related organic compounds. The compounds in a homologous series have the same functional group and can be described by a general formula. The formula of one member of the series differs from the next member by CH_2. Primary *alcohols* are an example of a homologous series with the general formula $C_nH_{(2n + 2)}O$. Physical properties, such as the boiling point, show a steady trend in values along a homologous series.

homolytic bond breaking (fission) is a type of bond breaking which produces *free radicals* with unpaired electrons. In homolytic fission, a *covalent bond* breaks so that the atoms joined by the bond separate, each taking one of the shared pair of electrons:

Homolytic bond breaking

189

hormones are chemical messengers carried round the body in the bloodstream. Hormones produce a response from particular target cells. Hormones are produced by glands and pass directly into the blood. Examples of hormones are the *steroid* sex hormones and *insulin* from the islet cells in the pancreas, adrenaline from the adrenal glands near the kidneys and sex hormones produced by the testes in men and the ovaries or other organs in women.

Hund's rule is one of the rules which help to predict electron configurations of atoms (see *aufbau principle*). The rule states that the most stable arrangement of electrons in an atoms is the one with as many unpaired electrons as possible, all with parallel *spins*. For example, in a nitrogen atom the three p-electrons have parallel spins and each occupies a separate p energy level.

$$\text{N} \quad \text{1s } \boxed{\text{↑↓}} \quad \text{2s } \boxed{\text{↑↓}} \quad \text{2p } \boxed{\text{↑}}\boxed{\text{↑}}\boxed{\text{↑}}$$

Electron-in-box diagram to show the configuration of a nitrogen atom

hybridisation: see *valency theory.*

hydration takes places when water molecules bond to ions or add to molecules. Water molecules are *polar* and so they are attracted to both positive and negative ions.

It is hard to see why the charged ions in a crystal of sodium chloride separate and go into solution in water with only a small energy change. Where does the energy come from to overcome the attraction between the ions? The explanation is that the ions are so strongly hydrated by the polar water molecules that the sum of the **enthalpy changes of hydration** nearly balances the **lattice energy** giving rise to a small **enthalpy change of solution**.

Some metal ions are hydrated in crystals as well as solution. In many hydrated salts the metal is present as a complex ion. This is true of copper(II) ions in blue copper sulfate. Cobalt(II) ions also bond to water molecules when blue (anhydrous) cobalt(II) chloride turns pink with water.

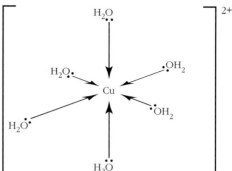

Hydrated copper(II) ion. Here coordinate (dative covalent) bonds hold the water molecules to the metal ions.

Hydrogen ions are hydrated in solution. They do not float around freely, but bond to water molecules forming *oxonium ions.*

The term hydration is also used to describe the addition of water to molecules such as *alkenes*. Ethene, for example, is hydrated when mixed with steam and passed over a phosphoric acid catalyst:

$$CH_2 = CH_2(g) + H_2O(g) \xrightarrow[\text{heat under pressure}]{\text{phosphoric acid catalyst}} CH_3CH_2OH(l)$$

hydrazine (N_2H_4) is a colourless, fuming liquid which is manufactured by oxidising ammonia with sodium chlorate(I) in the presence of gelatin. Hydrazine, like ammonia, is a base and a reducing agent. It has a *lone pair of electrons* on each nitrogen atom:

Structure and shape of a hydrazine molecule

hydrides are compounds of elements with hydrogen. The hydrides of the group 1 and 2 metals are ionic and contain the hydride ion, H^-.

$$2Na(s) + H_2(g) \rightarrow 2Na^+H^-(s)$$

Ionic hydrides are ionic crystals. They are rapidly hydrolysed by water because the hydride ion is a strong base.

$$H^-(s) + H_2O(l) \rightarrow H_2(g) + OH^-(aq)$$

The hydrides of non-metals are covalently bonded molecular liquids and gases. They include the *hydrogen halides*, *water*, hydrogen sulfide and *ammonia*. There are very many hydrides of carbon including the *alkanes*, *alkenes* and *arenes*.

The properties of the hydrides of the three highly *electronegative* elements nitrogen, oxygen and fluorine are affected by *hydrogen bonding*.

Hydrogen also forms *interstitial hydrides* with some *d-block elements*.

hydrocarbons are compounds which consist of just carbon and hydrogen. Important classes of hydrocarbons are the *alkanes*, *alkenes*, *alkynes* and *arenes* (see also *cyclic hydrocarbons*).

hydrochloric acid is a solution of hydrogen chloride gas in water, HCl(aq). Hydrogen chloride is a strong acid so the solution in water is fully ionised into aqueous hydrogen ions and chloride ions. The concentration of commercial concentrated hydrochloric acid is about 12 mol dm^{-3}. Dilute hydrochloric acid is commonly used as a laboratory reagent. Hydrochloric acid has the advantage of being cheaper and safer to dilute than sulfuric acid. It is a non-oxidising acid unlike dilute nitric acid.

hydrogen is the commonest element in the universe. Stars, like the Sun, consist mainly of hydrogen and get their energy from *nuclear fusion* which turns hydrogen into *helium*.

Hydrogen atoms are the smallest of all atoms consisting, normally, of one proton and one electron in the 1s *orbital*. There are two other *isotopes*: *deuterium* and *tritium*.

Life on Earth depends on hydrogen. Most organic compounds contain hydrogen and life requires water – hydrogen oxide.

Hydrogen is an important industrial chemical. It is produced from steam and natural gas for *ammonia manufacture*. It is also one of the products of the *electrolysis of brine*. Hydrogen is used for the *hydrogenation* of vegetable oils for margarine. The element is used in *fuel cells* and could be the basis of a *hydrogen economy*.

Hydrogen forms a wide range of *hydrides* with other elements.

hydrogen bonding is a type of attraction between molecules which is much stronger than other types of *intermolecular force*, but much weaker than covalent bonding.

Hydrogen bonding affects molecules in which hydrogen is covalently bonded to one of the three highly *electronegative* elements nitrogen, oxygen and fluorine.

The hydrogen atoms lie between two highly electronegative atoms. They are hydrogen bonded to one of them and covalently bonded to the other. The *covalent bond* is highly polar. The small hydrogen atom ($\delta+$) can get close to the other electronegative atom ($\delta-$) to which it is strongly attracted.

The three atoms associated with a hydrogen bond are always in a straight line.

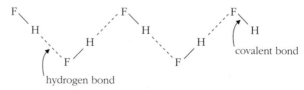

Hydrogen bonding in hydrogen fluoride

Hydrogen bonding accounts for:

- the relatively high boiling points of ammonia, water and hydrogen fluoride which are out of line for the trends in the properties of the other hydrides in groups 5, 6 and 7
- the open structure of *ice*
- the tertiary structure of *proteins*
- the pairing of bases in a *DNA* double helix
- the pairing up (*dimerisation*) of carboxylic acids in a non-aqueous solvent.

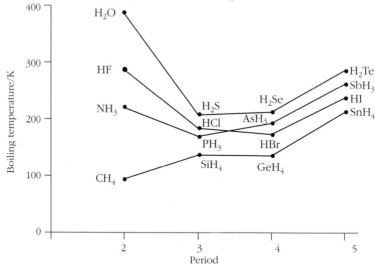

Boiling points for the hydrides of the elements in groups 4, 5, 6 and 7

hydrogen economy is an economy in which hydrogen replaces petrol, diesel or natural gas as the fuel for transport. There are no natural sources of hydrogen gas on Earth so the gas is a secondary energy source that has to be made by *steam reforming* natural gas and steam, or by *electrolysis* of water. The attraction of using hydrogen as a fuel is that it is non-toxic and forms only water when it burns. The idea is to use the hydrogen in *fuel cells* to drive electric vehicles. In order to reduce *greenhouse gas* emissions the hydrogen has to be made using electricity from nuclear power or from renewables such as wind or solar power.

One approach being considered for storing hydrogen as a fuel is to absorb the gas into a solid lattice to form a compound such as an *interstitial hydride*.

Critics of the hydrogen economy argue that using electricity to make hydrogen for use in vehicles fitted with fuel cells is inherently inefficient. They argue that the *efficiency* of an economy based on supplying electricity direct to electric vehicles is much higher.

hydrogen electrode: a *half-cell* of great theoretical importance but of limited practical significance. By definition the *standard electrode potential* of the hydrogen electrode is zero, $E^{\ominus}\,[\frac{1}{2}H_2(g)|H^+(aq)] = 0.00$ V.

A standard hydrogen electrode sets up an equilibrium between hydrogen ions in solution (1 mol dm^{-3}) and hydrogen gas (at 1 bar pressure) all at 298 K on the surface of a platinum electrode coated with platinum black.

By convention, when a standard hydrogen electrode is the left-hand electrode in an electrochemical cell, the cell voltage is the electrode potential of the right-hand electrode.

Hydrogen electrodes are inconvenient and so most measurements are taken with alternative *reference electrodes* acting as secondary standards.

$$Pt[H_2(g)] \mid 2H^+(aq) \mathbin{\vdots\vdots} Zn^{2+}(aq) \mid Zn(s)$$

Conventional diagram to represent the cell which defines the standard electrode potential of the Zn^{2+}(aq)|Zn(s) electrode

hydrogen emission spectrum: the spectrum of radiation from excited hydrogen gas. There is a bluish glow when a high voltage is applied across the electrodes at each ends of a glass tube containing hydrogen at low pressure. A spectroscope produces a line spectrum which can be recorded photographically.

Each line in the *atomic spectrum* corresponds to electrons dropping from a higher energy level to a lower level. The series of lines are named after the people who first discovered or studied them.

The pattern of energy levels in a hydrogen atom, with just one electron, is simpler than in other atoms with more than one electron. In a hydrogen atom all the orbitals in the same shell have the same energy. So the energy levels correspond to the main shells.

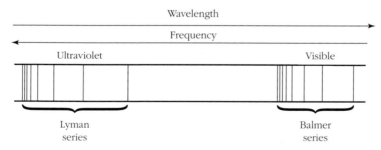

The series of lines in the hydrogen emission spectrum

The lines of the Balmer series are in the visible region. The lines of the Lyman series are in the ultraviolet. The bigger the energy jumps, the higher the *frequency* of *electromagnetic radiation* emitted.

The energy gaps between energy levels get smaller as electrons recede from the nucleus. As a result, in each series the differences between the energy jumps get smaller and smaller until they converge. They converge at the high frequency (big jump) end. The biggest jump in the Lyman series is for an electron dropping back from the very edge of an atom to the lowest energy level. The size of this jump is the ionisation energy for a hydrogen atom. So the *ionisation energy* of hydrogen can be calculated from the frequency of the convergence limit of the Lyman series since (according to *quantum theory*) $\Delta E = h\nu$.

hydrogen halides are compounds of hydrogen with the *halogens*. They are all colourless, molecular compounds with the formula HX where X stands for F, Cl, Br or I.

The bonds between hydrogen and the halogens are *polar*. The H—F bond is so polar that the properties of the compound are affected by *hydrogen bonding*.

Hydrogen chloride, hydrogen bromide and hydrogen iodide are similar in that they are:
- colourless gases at room temperature and fume in moist air
- very soluble in water, forming acid solutions (hydrochloric, hydrobromic and hydriodic acids)
- *strong acids* so they ionise completely in water and react with *ammonia* to form *ammonium ions*.

Hydrogen chloride, hydrogen bromide and hydrogen iodide show some trends down *group 7* in that they:
- have boiling points which rise down the group from HCl to HI
- become less thermally stable – heating does not decompose hydrogen chloride but a hot wire will decompose hydrogen iodide into hydrogen and iodine
- become easier to oxidise to the halogen – hydrogen iodide is a strong *reducing agent*.

Hydrogen fluoride is rather different from the other hydrogen halides:
- it is a liquid at room temperature (boiling point 19.9°C) because of hydrogen bonding
- it is a *weak acid*
- its solution in water (hydrofluoric acid) reacts with glass and can be used for etching.

hydrogen ions (H^+) are hydrogen atoms which have lost an electron. Since a hydrogen atom consists of one proton and one electron this means that a hydrogen ion is just a *proton*. In water, hydrogen ions do not float around freely, they become attached to water molecules forming oxonium ions, H_3O^+. A lone pair on the oxygen atom forms a *dative covalent bond* with the hydrogen ion.

Bonding in an oxonium ion

Hydrogen ions (protons) transfer from an acid to a base during an *acid–base reaction*. This gives rise to the *Brønsted–Lowry theory* definition of an acid as a proton donor and a base as a proton acceptor.

hydrogen peroxide (H_2O_2) is a pale blue liquid when pure. Like water it is a *hydride* of oxygen and it is a *hydrogen bonded*, molecular liquid.

Bonding of a molecule of hydrogen peroxide

Hydrogen peroxide is usually supplied in solution in water. Solutions are sometimes labelled with the 'volume strength'. For example, '20 volume' hydrogen peroxide decomposes to give 20 cm³ oxygen from 1 cm³ of solution (measured under standard conditions). The concentration of 20 volume hydrogen peroxide is 1.67 mol dm⁻³.

Only in peroxides does oxygen have the oxidation state −1, in between 0 and −2. So hydrogen peroxide can act as both an *oxidising agent* and as a *reducing agent*.

Hydrogen peroxide decomposes slowly on standing in the dark, but much more rapidly in the presence of a catalyst such as powdered manganese(IV) oxide or the *enzyme* catalase. This is a *disproportionation reaction* with oxygen starting in the −1 state and ending up in the −2 and 0 states. Hydrogen peroxide, as it were, oxidises and reduces itself.

$$2H_2O_2(aq) \rightarrow 2H_2O(l) + O_2(g)$$

0	O_2	
−1	H_2O_2	O_2^{2-}
−2	H_2O	O^{2-}

Oxidation states of oxygen

Hydrogen peroxide oxidises cobalt(II) to cobalt(III) in the presence of ammonia, and chromium(III) to chromium(VI) under alkaline conditions. As an oxidising agent hydrogen peroxide is also used to *bleach* paper and textiles. It is used to manufacture the bleaches used in washing powders. In water treatment it destroys micro-organisms. The great advantage of hydrogen peroxide as a bleach is that it turns into water when it reacts, so it does not contaminate or pollute.

$$H_2O_2(aq) + 2H^+(aq) + 2e^- \rightarrow 2H_2O(l)$$

Hydrogen peroxide is not a powerful reducing agent but it will decolorise an acid solution of manganate(VII) ions. A redox titration with *potassium manganate(VII)* is one way of determining the concentration of a solution of hydrogen peroxide.

$$H_2O_2(aq) \rightarrow O_2(g) + 2H^+(aq) + 2e^-$$

hydrogenation is a reaction which adds hydrogen to a compound. Hydrogenation converts liquid vegetable oils to the solid fats used as ingredients of *margarine*. This process is described as hardening. In these reactions hydrogen adds to double bonds in the hydrocarbon chains of unsaturated *fatty acids* in *triglycerides*. A disadvantage of hardening vegetable oils by hydrogenation is that it leads to the formation of *trans fats*.

Hydrogen adds to C=C double bonds at room temperature in the presence of a platinum or palladium catalyst, or on heating in the presence of a nickel catalyst. (See also *heterogeneous catalyst*.).

hydrolysis: a reaction in which a compound is split apart in a reaction involving water. Hydrolysis reactions are often catalysed by acids or alkalis.

$$CH_3C \underset{OC_2H_5}{\overset{O}{\diagup}} + OH^-(aq) \xrightarrow[\text{aqueous alkali}]{\text{heat with}} CH_3C \underset{O^-}{\overset{O}{\diagup}} + C_2H_5OH$$

ester   salt of ethanoic acid   ethanol

Hydrolysis of the ester ethyl ethanoate catalysed by alkali

Hydrolysis is used to make *soap* from fats and oils. This is an example of *ester* hydrolysis.

Other examples include *hydrolysis of metal aqua ions*, *hydrolysis of non-metal chlorides*, *hydrolysis of an organic chloro compound* and the *hydrolysis of salts*.

hydrolysis of metal aqua ions means that solutions of the salts containing *aluminium*(III), *iron*(III) and *chromium*(III) ions are acidic. In the hydrated iron(III) ion, for example, electrons in the water molecules are pulled towards the highly polarising iron ion, making it easier for the water molecules linked to the metal ion to give away protons. As a result the hydrated iron(III) ion is an *acid*. The solution is acidic enough to release carbon dioxide when added to sodium carbonate.

$$[Fe(H_2O)_6]^{3+} + H_2O \rightleftharpoons [Fe(H_2O)_5OH]^{2+} + H_3O^+$$

$$[Fe(H_2O)_5OH]^{2+} + H_2O \rightleftharpoons [Fe(H_2O)_4(OH)_2]^+ + H_3O^+$$

Adding a base such as hydroxide ions to a solution of hydrated iron(III) ions removes a third proton, producing an uncharged complex. The uncharged complex is much less soluble in water and precipitates as an orange-brown jelly-like precipitate – hydrated iron(III) hydroxide.

$$[Fe(H_2O)_4(OH)_2]^+(aq) + OH^-(aq) \rightarrow [Fe(H_2O)_3(OH)_3](s) + H_2O(l)$$

The smaller charge on 2$^+$ metal ions means that their polarising power is less and their aqua complexes have a much smaller tendency to ionise.

hydrolysis of non-metal chlorides splits the compounds into an *oxoacid* (or hydrated oxide of the non-metal) and hydrogen chloride. Non-metal chlorides are typically molecular liquids at room temperature. Examples are CCl_4, $SiCl_4$ and PCl_3. Most of these compounds do not mix with water but react with it. For example:

$$SiCl_4(l) + 2H_2O(l) \rightarrow SiO_2(s) + 4HCl(g)$$

The exception is CCl_4 which does not react with water. This is an example of *kinetic inertness*. The four large chlorine atoms prevent water molecules from using their *lone pairs of electrons* to attack the small central carbon atom.

hydrolysis of an organic chloro compounds brings about a *substitution reaction* that replaces the chlorine atom with an —OH group. *Acyl chlorides* are rapidly hydrolysed while *halogenoarenes*, such as chlorobenzene, are inert to hydrolysis. In comparison with an *halogenoalkanes*, R—Cl, the order of reactivity is:

$$R—COCl > R—Cl > \underset{}{\bigcirc}—Cl$$

Acyl chlorides are very reactive because the presence of the polar C=O bond means that *nucleophiles* such as water can readily attack the carbon atom leading to an *addition-elimination reaction*.

Interaction between the *delocalised electrons* in the benzene ring and the *lone pairs of electrons* on the chlorine atom in chlorobenzene strengthens the C—Cl bond, makes it less polar and means that nucleophiles cannot attack the carbon atom so that the compound is relatively inert to hydrolysis.

The rate of hydrolysis of a chloroalkane varies with the nature of the alkyl group (R). Generally tertiary chloro compounds are more reactive than the equivalent primary compounds. It is not easy to predict or account for the trend because the mechanism of *nucleophilic substitution* is S_N1 for tertiary compounds and S_N2 for primary compounds.

hydrolysis of salts changes the pH of a solution of a salt by altering the concentrations of H_3O^+ and OH^- ions. Some salts are acidic because of the *hydrolysis of metal aqua ions*. Solutions of the salts of weak bases are also acidic. This includes the salts of ammonia and amines. The NH_4^+ ion in an ammonium salt is an acid because ammonia (the weak base) does not have a strong hold on the protons it accepts.

$$NH_4^+(aq) + H_2O(l) \rightleftharpoons NH_3(aq) + H_3O^+(aq)$$

Solutions of the salts of weak acids are alkaline. This includes the salts of:
- carbonic acid (carbonates)
- sulfurous acid (sulfites)
- ethanoic acid (ethanoates)
- hydrogen sulfide (sulfides).

The negative ions from weak acids are strong bases – they take hydrogen ions from water molecules and turn back into the acid.

$$CH_3COO^-(aq) + H_2O(l) \rightleftharpoons CH_3COOH(aq) + OH^-(aq)$$

hydrosphere: the watery part of the Earth's surface. This includes the oceans, rivers, lakes, groundwater and glaciers. Over 80% of the hydrosphere is in the oceans and shallow seas. Almost all the rest of the water is trapped in sediments.

The oceans are linked to the rest of the hydrosphere by the *water cycle* – water evaporates from the sea into the atmosphere, falls as rain or snow onto land and then returns to the oceans via rivers and lakes.

hygroscopic substances absorb water from the air. The chemicals used as drying agents in *desiccators* are hygroscopic, they include anhydrous calcium chloride and calcium oxide. It is difficult to weigh hygroscopic substances accurately which makes them unsuitable as *primary standards* for titrations.

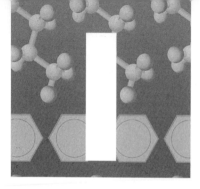

ibuprofen is a drug used in *medicines* formulated to treat pain, inflammation and fever.

Ibuprofen was developed to reduce inflammation in the joints of people suffering from rheumatoid arthritis. Testing showed that the drug was also an effective pain reliever and a rival to aspirin. The research and development programme by the Boots company took 30 years (see *drug development*). The Boots synthesis was complex. Since the original patent ran out, a new synthesis has been developed by applying the principle of **green chemistry**.

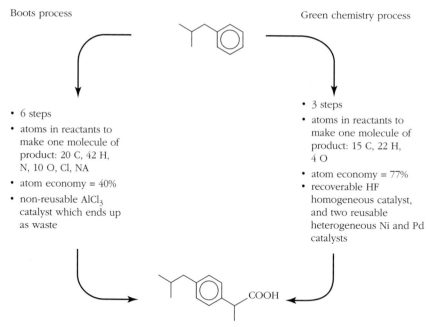

Boots process

Green chemistry process

- 6 steps
- atoms in reactants to make one molecule of product: 20 C, 42 H, N, 10 O, Cl, NA
- atom economy = 40%
- non-reusable AlCl₃ catalyst which ends up as waste

- 3 steps
- atoms in reactants to make one molecule of product: 15 C, 22 H, 4 O
- atom economy = 77%
- recoverable HF homogeneous catalyst, and two reusable heterogeneous Ni and Pd catalysts

Two routes to making ibuprofen from the same starting material. Note that the ibuprofen molecule has an asymmetric carbon atom. It is a chiral compound.

ice: the solid state of water which is remarkable because, at the freezing point, it is less dense than liquid water and so floats. Ice owes its open cage-like structure to **hydrogen bonding**. Each water molecule is hydrogen bonded to four others. The hydrogen bonds are 0.177 nm long which is much longer than the length of a covalent O—H bond (0.094 nm). The distance between the oxygen atoms is equal to the sum of the lengths of a hydrogen bond and a normal O—H bond. Note that in every instance the three atoms O—H ---- O are in a straight line.

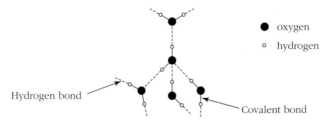

oxygen
hydrogen

Hydrogen bond

Covalent bond

Structure of ice

The view from some angles shows the oxygen atoms in ice arranged in puckered hexagons – a pattern which is reflected in the hexagonal shapes of snowflakes.

When ice melts, the open structure partly collapses as some of the hydrogen bonds break, so at 0°C the liquid is denser than the solid.

ideal gases: gases which obey the ideal gas equation $pV = nRT$, where p is the *pressure* in N m^{-2} (pascals), V is the volume in m^3, T is the temperature on the *Kelvin* scale, n the *amount* in moles and R the *gas constant*. It is important to convert all the quantities to *SI units* before substituting values in the equation.

The ideal gas equation incorporates all the *gas laws*. For example, with a fixed amount of gas n is constant, so at constant temperature $pV =$ constant, which is *Boyle's law*.

In practice *real gases* deviate from ideal gas behaviour. Under laboratory conditions the gases which are close to behaving like ideal gases are those that are well above their boiling points such as the *noble gases*, nitrogen, oxygen and hydrogen. For these gases at room temperature and pressure, the assumptions of the *kinetic theory of gases* are nearly true. The attractive forces between these non-polar molecules are small enough to be ignored and the volume of the molecules is insignificant compared to the total volume of gas.

The ideal gas equation can be used to determine the *molar mass* of gases and of other substances which evaporate easily.

ignition temperature: the lowest temperature at which a substance will spontaneously catch fire in air. There is no need for a flame to ignite a gas or *vapour* at or above its ignition point. Examples of ignition temperatures in air are hexane at 487°C and ethanol at 558°C.

immiscible liquids are liquids which do not mix. As a rule *non-polar solvents* such as hexane and other hydrocarbons do not mix with polar liquids such as water.

Oily hydrocarbon liquids are usually less dense than water so they form a separate, colourless layer on top of water. Liquid organic halogen compounds such as dichloromethane are often denser than water so they give rise to a separate liquid layer underneath water.

Liquids which are immiscible are useful in *solvent extraction*.

immobilised enzymes are *enzymes* bound to a solid so that they can be easily recovered from a reaction mixture by filtering or centrifuging. Alternatively the solid holding the enzyme can be contained in a column and the raw material (*substrate*) passed through the column turning into products in the presence of the trapped enzyme *catalyst*.

Immobilised enzymes are used to convert starch from maize into glucose and then to convert much of the glucose to fructose which is sweeter than glucose or sucrose (see *sweetness*). Using immobilised enzymes has meant a big fall in the cost of high-fructose syrup used to make colas and other soft drinks.

incineration is one of the methods available for disposing of *waste*. Burning plastic waste or domestic refuse in a modern incinerator releases energy to generate electricity. Burning all the domestic waste from homes in Europe could generate 5% of European energy needs.

Just dumping rubbish in *landfill* is expensive and a waste of resources. Incineration greatly reduces the volume of waste to be dumped in landfill sites. It also replaces *fossil fuels*, thus reducing the overall emission of greenhouse gases.

There has been strong opposition to burning waste. Old incinerators were inefficient, operated at too low a temperature and often were not designed to generate electricity. Incomplete combustion produces *carbon monoxide* which is toxic. Another worry is that if waste contains chlorine compounds such as PVC an incinerator will give off corrosive or toxic chemicals such as hydrogen chloride or *dioxins*. High temperature combustion at 800–1250°C helps to achieve complete combustion and limit the emission of toxic chemicals. Modern incinerators have elaborate gas-cleaning systems to control the emission of harmful gases and dust particles.

indicators are used to detect the end-point during titrations. *Acid–base indicators* change colour over a range of pH values. Starch is used as an indicator in *iodine–thiosulfate titrations*.

induced dipoles: see *intermolecular forces*.

inductive effect: a term used to describe the extent to which electrons are pulled away from, or pushed towards, an atom by the atoms or group to which they are bonded. The term crops up sometimes in accounts of reaction mechanisms.

More *electronegative* atoms pull electrons away from a carbon atom, so the carbon carries a slight positive charge. This means that that carbon atom is open to attack by nucleophiles, which happens when carbon is bonded to oxygen, chlorine or bromine. Electronegative atoms can also increase the acid strength of carboxylic acids (see *acid strength of organic hydroxy compounds*).

Other groups, especially *alkyl groups*, have a slight tendency to push electrons towards the carbon atom to which they are bonded. One of the effects of this kind of inductive effect is that a secondary *carbocation* is more stable than a primary carbocation. This helps to account for the products formed during *electrophilic addition* of compounds such as HBr to unsymmetrical alkenes (see *Markovnikov's rule*).

inert chemicals: a chemical is inert if it has no tendency to react under given circumstances. The lighter noble gases, helium and neon, live up to the original name for group 8 (IUPAC group 18). They are inert towards all other reagents.

Nitrogen is a relatively unreactive gas which can be used to create an 'inert atmosphere' free of oxygen which is much more reactive. Nitrogen is not inert in all circumstances. It reacts, for example, with hydrogen in the *Haber process* and with oxygen at high temperatures to form *nitrogen oxides*.

Sometimes there is no tendency for a reaction to go because the reactants are *stable*. This is so if the *free energy change* for the reaction is positive.

Sometimes there is no reaction even though thermochemistry suggests that it should go. The free energy change is negative (and the total entropy change is positive) so the change is *feasible*. A high activation energy means that the rate of reaction is very, very slow. Methane, for example, does not burn in oxygen at room temperature even though it would be energetically favourable for it to do so. This is an example of a substance being kinetically inert.

inert gases: the older name for the group of *noble gases* which was used until the discovery of compounds formed by xenon.

inert pair effect: describes the observation that, for groups 3, 4 and 5 of the periodic table, the oxidation state which is 2 below the highest state becomes increasingly stable down the group. In *group 4*, for example, the main oxidation state for carbon and silicon is +4. The +2 oxidation state becomes important in the chemistry of tin and lead. In the chemistry of lead the +2 state is even more stable relative to the +4 state than it is in the chemistry of tin. What this means is that two of the four electrons in the outer shell become less available for bonding down the group, hence the term 'inert pair effect'. This effect is best regarded as a reminder of a trend rather than an explanation.

infrared spectroscopy (IR spectroscopy) is an analytical technique used to identify functional groups in organic molecules. Most compounds absorb IR radiation. The wavelengths they absorb correspond to the natural frequencies for the *vibration of bonds* in the molecules bend and stretch (see *absorption of radiation*). It is *polar covalent bonds* such as O—H, C—O and C=O which absorb strongly as they vibrate.

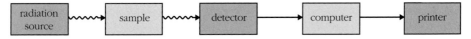

Essential features of a single-beam IR spectrometer

Bonds vibrate in particular ways and absorb radiation at specific wavelengths. Spectroscopists have found that it is possible to correlate absorptions in the region 4000–1500 cm⁻¹ with the stretching or bending vibrations of particular bonds. As a result the infrared spectrum gives valuable clues to the presence of functional groups in organic molecules. Hydrogen bonding broadens the absorption peaks of —OH groups in alcohols and even more so in carboxylic acids.

Wavenumber ranges

4000 cm⁻¹	2500 cm⁻¹	1900 cm⁻¹	1500 cm⁻¹	650 cm⁻¹
C — H	C ≡ C	C = C		
O — H	C ≡ N	C = O		
N — H			fingerprint region	
single bond stretching vibrations	triple bond stretching vibrations	double bond stretching vibrations		

The main regions of the infrared spectrum and important correlations between bonds and observed absorptions

Molecules with several atoms can vibrate in many ways because the vibrations of one bond affect others close to it. Complex pattern of vibrations can be used as a 'fingerprint' to be matched against the recorded IR spectrum in a database. Comparing the IR spectrum of a product of synthesis with the spectrum of the pure compound can be used to check that the product is pure.

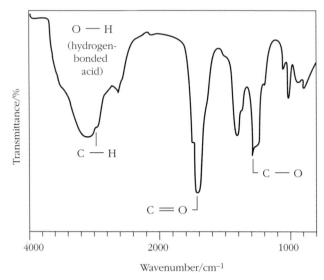

IR spectrum of ethanoic acid. The units along the bottom are wavenumbers (in cm⁻¹). IR wavenumbers range from 400 cm⁻¹ to 4500 cm⁻¹. Spectroscopists find the numbers more convenient than wavelengths. The wavenumber is the number of waves in 1 cm. Transmittance on the vertical axis measures the percentage of radiation that passes through the sample. The troughs appear at the wavenumbers corresponding to the radiation that the compound absorbs strongly.

IR is an analytical tool that can be used to monitor the progress of an organic synthesis. Comparing the spectrum of the final product with the known spectrum in a database can be used to check that the product is pure.

inhibitor: a chemical which stops or slows down a specific chemical reaction. Inhibitors are sometimes called negative *catalysts*. Liquid *monomers* such as phenyl(ethene) are often supplied with a dissolved inhibitor to prevent polymerisation in the bottle.

Enzyme inhibitors are very important in the control of metabolism. Competitive inhibition reduces the rate of a reaction catalysed by an enzyme in the presence of a molecule with a similar shape to the normal *substrate*. The inhibitor binds to the *active site* of the enzyme and stays there unchanged, stopping normal enzyme activity.

initial-rate method: a method for finding the order of a chemical reaction and the form of its *rate equation*. This method is used for reactions that have several terms in the rate equation, which makes other methods hard to use.

The method is based on finding the rate immediately after the start of a reaction. This is the one point when all the concentrations are known.

The procedure is to make up a series of mixtures in which all the initial concentrations are the same except one. A suitable method is used to measure the change of concentration with time for each mixture (see *rates of reaction*). The results are used to plot concentration–time graphs. The initial rate for each mixture is then found by drawing tangents to the curves at the start and calculating their gradients.

A *clock reaction* provides a simplified method of applying the initial-rate method.

Worked example

The initial-rate method was used to study the reaction A + 2B → C + D. The initial rate was calculated from five graphs plotted to show how the concentration of B varied with time for different initial concentrations of reactants.

Experiment	Initial concentration of A/mol dm^{-3}	Initial concentration of B/mol dm^{-3}	Initial rate of reaction/mol dm^{-3} s^{-1}
1	0.10	0.05	2×10^{-4}
2	0.20	0.05	8×10^{-4}
3	0.30	0.05	18×10^{-4}
4	0.20	0.10	8×10^{-4}
5	0.20	0.20	8×10^{-4}

What is:
- the rate equation for the reaction?
- the value of the rate constant?

Notes on the method

Recall that the rate equation cannot be worked out from the balanced equation for the reaction.

First study the experiments in which the concentration of A varies but the concentration of B stays the same. How does doubling or tripling the concentration of A affect the rate?

Next study the experiments in which the concentration of B varies but the concentration of A stays the same. How does doubling the concentration of B affect the rate?

Substitute values for any one experiment in the rate equation to find the value of the rate constant, k. Take care with the units.

Answer

From experiments 1, 2 and 3: doubling $[A]_{initial}$ increases the rate by a factor of 4 (2^2). Tripling $[A]_{initial}$ increases the rate by a factor of 9 (3^2). So rate α $[A]^2$.

From experiments 2, 4 and 5: doubling $[B]_{initial}$ does not change the rate. So rate α $[B]^0 = 1$.

The reaction is second order with respect to A and zero order with respect to B. The rate equation is:

$$\text{rate} = k[A]^2$$

Rearranging this equation, and substituting values from experiment 2:

$$k \quad \frac{\text{rate}}{[A]^2} \quad \frac{8 \times 10^{-4} \text{ mol dm}^{-3} \text{ s}^{-1}}{(0.2 \text{ mol dm}^{-3})^2}$$

$$k = 0.02 \text{ mol}^{-1} \text{ dm}^3 \text{ s}^{-1}$$

inorganic chemistry is the study of the chemical elements and their compounds. In an advanced level course the study of the elements concentrates on the reactions of selected elements and of their simpler compounds such as oxides, chlorides and hydrides. Inorganic chemistry includes the chemistry of the element carbon and a few of its compounds such as carbon monoxide, carbon dioxide, tetrachloromethane and the carbonates. The chemistry of most other carbon compounds belongs to *organic chemistry*.

Most of the reactions of inorganic chemicals belong to one of these four types:

- *acid–base reactions*
- *redox reactions*
- *ionic precipitation*
- reactions which form *complex ions* (or *ligand substitution reactions*).

insulin is the *hormone*, produced by special cells in the pancreas, which helps to control the level of glucose in the blood. Digestion of food after a meal raises the level of glucose in blood. In response the pancreas secretes insulin which circulates round the body and speeds up the rate at which cells take up glucose. Insulin is a *protein* consisting of two folded chains of amino acids which are *cross-linked* by disulfide links.

intermediate bonding is bonding which is neither purely ionic nor purely covalent. In most compounds the bonding between the atoms of different elements is intermediate between the two extremes.

The *electronegativity* values for two elements are a guide to the extent to which the bonds between them will be covalent, ionic or intermediate. Electronegativity values are particularly useful for discussing *polar covalent bonds*.

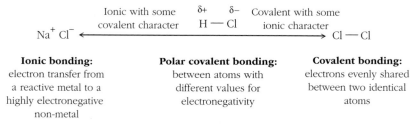

A spectrum from purely ionic to purely covalent bonding

Where the bonding is largely ionic, the degree of covalent character is determined by the *polarising power* of the metal (positive) ion which distorts neighbouring negative ions, giving rise to a degree of electron sharing.

intermediates in reactions are atoms, molecules, ions or *free radicals* which do not appear in the balanced equation but which are formed during one step of a reaction, then used up in the next step. Chemists can use spectroscopy to detect intermediates which exist for only a short time during a reaction.

Examples of reaction intermediates are the free radicals formed during a *free-radical chain reaction* and the carbocations formed during *nucleophilic substitution* by the S_N1 mechanism and during *electrophilic addition* to alkenes.

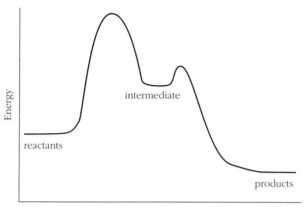

Energy changes in the course of a reaction. Note the energy dip as the intermediate forms. The dip is not deep enough for the intermediate to exist as a stable product. Contrast this with the energy profile for a reaction with just a transition state between reactants and products (see **activated complex**).

intermolecular forces are weak attractive forces between molecules. Without intermolecular forces there could be no molecular liquids or solids. Also, **real gases** would behave more like **ideal gases**.

Weak intermolecular forces arise from electrostatic attractions between dipoles, including attractions between:

- molecules with permanent dipoles such as hydrogen chloride
- a permanent dipole in one molecule and a dipole induced in a neighbouring molecule, such as the attraction between hydrogen bromide and ethene molecules
- temporary dipoles created fleetingly in non-polar atoms or molecules (**London forces**)

Examples of substances with non-polar atoms and molecules include the noble gases, the halogens and the alkanes. When non-polar atoms or molecules meet there are fleeting repulsions and attractions between the nuclei of the atoms and the surrounding clouds of electrons. Temporary displacements of the electrons lead to temporary dipoles. It is the attractions between these transient dipoles which give rise to the tendency for the molecules to cohere.

The greater the number of electrons, the greater the polarisability of the molecule and the greater the possibility for temporary, induced dipoles. This explains why the boiling points rise down **group 7** (IUPAC group 17, the halogens) and **group 8** (IUPAC group 18, the noble gases). For the same reason the boiling points in **alkanes** increase with the increasing number of carbon atoms in these hydrocarbon molecules.

In polymers the intermolecular forces become larger as molecules get longer so that there is a larger area of contact over which intermolecular forces can operate. The strength of **thermoplastic polymers** depends on the total strength of intermolecular forces between tangled, long-chain molecules.

Weak intermolecular forces due to attractions between temporary and permanent dipoles are often called **van der Waals forces**. Forces of this kind are roughly a hundred times weaker than covalent bonds.

The properties of some molecular compounds are affected by stronger intermolecular forces called *hydrogen bonding*. Hydrogen bonds are about ten times stronger than van der Waals forces.

internal combustion engine: an engine in which the fuel burns in a cylinder or combustion chamber in the engine. Expansion of the hot gases from the burning fuel drives a piston (in a *petrol* or *diesel fuel* engine) or a turbine (in a jet engine). The chemical changes and high temperatures inside internal combustion engines can lead to the formation of *atmospheric pollutants*.

interstitial hydrides are compounds formed by *d-block elements* with hydrogen. The hydrogen atoms fit into the spaces between the atoms in the metal crystal. The holes are the interstices of the crystal, hence the name.

The amounts of hydrogen in the lattice vary with the temperature and pressure. The compounds do not have a definite composition. The hydride of titanium has the variable formula TiH_x, where x is less than 2.

Palladium reacts particularly well with hydrogen and can absorb nearly a thousand times its own volume of the gas at room temperature.

Interstitial hydrides release hydrogen gas on heating so they have been investigated as a possible safe way of storing hydrogen as part of a *hydrogen economy*.

intramolecular forces are the strong *covalent bonds* between the atoms in a molecule. They are about one hundred times stronger than the weak *intermolecular forces* between molecules. Substances consisting of small molecules melt and boil at relatively low temperatures because of the weakness of the forces between the molecules. The strong intramolecular forces in small molecules do not normally break during melting and boiling.

iodination is a reaction in which an iodine atom replaces a hydrogen atom in an organic molecule. One example is the iodination of propanone – a reaction which crops up in the study of reaction rates. The reaction is *acid catalysed*.

$$CH_3COCH_3(aq) + I_2(aq) \rightarrow CH_2ICOCH_3(aq) + H^+(aq) + I^-(aq)$$

iodine (I) is a lustrous grey-black solid at room temperature, formula I_2, which *sublimes* when gently warmed to give a purple vapour. Iodine is a *halogen*, the fourth element in *group 7*, with the *electron configuration* $[Kr]4d^{10}5s^25p^5$.

Iodine consists of diatomic molecules with pairs of atoms held together by single *covalent bonds*. The molecules are non-polar so the *intermolecular forces* are relatively weak, but stronger than in bromine because the atoms are larger and have more electrons. This makes iodine molecules more *polarisable* than bromine molecules. (See *crystal structures of non-metals*.)

Iodine is only slightly soluble in water. It is much more soluble in a solution of potassium iodide because of the formation of the *triiodide ion*. Iodine solution in aqueous potassium iodide is a brownish-yellow colour. Iodine dissolves freely in non-polar solvents such as hexane forming a solution with the same colour as iodine vapour.

Iodine, like the other halogens, is an *oxidising agent* but a less powerful oxidising agent than bromine.

Iodine reacts with metals to form iodides. Because of the polarisability of the large iodide ion, the iodides formed with small cations or highly charged cations are essentially covalent (see *polarisation of ions*). Examples are lithium iodide, magnesium iodide and aluminium iodide.

Iron(III) iodide and copper(II) iodide do not exist because iodide ions reduce the metals ions to their lower oxidation state. The only iodides of these metals are iron(II) iodide and copper(I) iodide.

Iodine oxidises hydrogen on heating forming hydrogen iodide. Unlike the reactions of chlorine and bromine, this is a reversible reaction which has been studied in detail to establish the *equilibrium law*.

$$H_2(g) + I_2(g) \rightleftharpoons 2HI(g)$$

Hydrogen iodide is a fuming, acidic gas. Like the other *hydrogen halides* it is very soluble in water and a strong acid.

As a weaker oxidising agent, iodine converts thiosulfate ions to tetrathionate ions. This is a quantitative reaction used in *iodine–thiosulfate titrations*.

Iodine and its compounds are used to make pharmaceuticals and dyes. Iodine is needed in the diet so that the thyroid gland in the neck can make the *hormone* thyroxine which regulates growth and metabolism. In many regions sodium iodide is added to table salt to supplement the iodine in the diet and drinking water and so prevent goitre.

iodine chlorides: compounds formed when chlorine reacts with iodine. For example, passing a stream of chlorine gas over solid iodine produces a red-brown liquid, iodine monochloride, ICl. With excess chlorine this turns to a yellow solid, iodine trichloride, ICl_3, which is unstable and easily decomposes back to ICl.

iodine number: a measure of the degree of unsaturation of the *triglycerides* in a *fat* or *vegetable oil*. Some *fatty acids* in fats and oils are fully saturated. They have no double bonds. Other fatty acids are unsaturated. They have double bonds. The iodine number is found by measuring how much iodine will add to double bonds per 100 g oil. The higher the iodine number the more unsaturated the fat or oil. Values range from 25–30 g per 100 g for butter which is high in saturated fats to 80–140 g per 100 g for a vegetable oil high in unsaturated fats such as soya oil.

iodine–thiosulfate titration: a useful analytical method for measuring *amounts* of *oxidising agents*. The method is based on the fact that oxidising agents convert iodide ions to iodine quantitatively. Among the oxidising agents which do this are iron(III) ions, copper(II) ions, chlorine molecules and chlorate(I) ions in *bleach*, dichromate(VI) ions with acid, iodate(V) ions with acid and manganate(VII) ions with acid.

$$2I^-(aq) \rightarrow I_2(aq) + 2e^-$$

The iodine stays in solution in excess potassium iodide turning a yellow-brown (see *triiodide ion*).

The iodine produced is then titrated with a *standard solution* of sodium thiosulfate which reduces iodine molecules back to iodide ions. This too happens quantitatively exactly as in the equation:

$$I_2(aq) + 2S_2O_3^{2-}(aq) \rightarrow 2I^-(aq) + S_4O_6^{2-}(aq)$$

The greater the amount of oxidising agent added, the more iodine formed and so the more thiosulfate needed to react with it. When thiosulfate is added from a burette the colour of the iodine gets paler. Near the end-point the solution is a very pale yellow. Adding a little soluble starch solution as an indicator near the end-point gives a sharp colour change from blue-black to colourless.

Worked example

Calculate the concentration of a solution of sodium thiosulfate standardised by this method. A 0.0642 g sample of potassium iodate(v), KIO_3, was dissolved in water in a conical flask. Excess potassium iodide was dissolved in the solution which was then acidified with dilute sulfuric acid. The iodine formed was titrated with the solution of sodium thiosulfate from a burette. The volume of sodium thiosulfate solution needed to decolorise the blue iodine-starch colour at the end-point was 24.50 cm³.

Notes on the method

Write the equations and work out the amount in moles of $S_2O_3^{2-}$ equivalent to 1 mol IO_3^-.

There is then no need to consider the amounts of iodine in the calculations.

Look up the *molar mass* of potassium iodate: $M_r(KIO_3) = 214.0$ g mol⁻¹.

Use your calculator only in the last step of the calculation to avoid repeated rounding errors.

Answer

The equation for producing iodine:

$$IO_3^-(aq) + 5I^-(aq) + 6H^+(aq) \rightarrow 3I_2(aq) + 3H_2O(l)$$

The equation for the reaction during the titration:

$$I_2(aq) + 2S_2O_3^{2-}(aq) \rightarrow 2I^-(aq) + S_4O_6^{2-}(aq)$$

So 1 mol IO_3^- produces 3 mol I_2 which then reacts with 6 mol $S_2O_3^{2-}$.

$$\text{The amount of } KIO_3 \text{ at the start} = \frac{0.0642 \text{ g}}{214.0 \text{ g mol}^{-1}}$$

6 mol $S_2O_3^{2-}$ react with the iodine formed by 1 mol IO_3^-.

So the amount of thiosulfate in 24.5 cm³ (= 0.0245 dm³) solution

$$= 6 \times \frac{0.0642 \text{ g}}{214.0 \text{ g mol}^{-1}}$$

So the concentration of the sodium thiosulfate solution

$$= 6 \times \frac{0.0642 \text{ g}}{214.0 \text{ g mol}^{-1}} \times \frac{1}{0.0245 \text{ dm}^3}$$

$$= 0.0735 \text{ mol dm}^{-3}$$

iodoform reaction: see *triiodomethane reaction*.

ion exchange is a process for swapping ions in a solution and is used to soften and deionise water. An ion-exchange resin consists of small beads of a *polymer* which has been modified so that along the chains there are ionic groups. In a cation-exchange resin the polymer has negatively charged groups which attract positive ions.

A water softener contains a cation exchanger in which all the negatively charged sites are attracting sodium ions. As hard water flows over the resin beads, the calcium ions in the *hard water* change places with the sodium ions. The softened water can then be used for washing without forming insoluble precipitates with *soaps* and other *detergents*.

Ion exchange is a reversible process. Once all the sodium ions in a bed of resin have been used up it can be recharged by running a concentrated solution of sodium chloride through it to replace calcium ions with sodium ions, ready to soften more water.

It takes two ion-exchange resins to deionise water: a cation resin holding hydrogen ions and an anion resin with hydroxide ions. As water runs through the two resins, the cation exchanger holds on to positive metal ions and replaces them with hydrogen ions. The anion exchanger holds on to negative ions and replaces them with hydroxide ions.

Then the reaction $H^+(aq) + OH^-(aq) \rightarrow H_2O(l)$ occurs, so no ions are left in solution.

ionic bonding is one of three types of strong chemical bonding. Ionic bonding occurs in compounds of metals with non-metals. Ionic compounds form particularly when the reactive *s-block elements* combine with the halogens of group 7 or oxygen in group 6. The *crystal structures of ionic compounds* are three-dimensional lattices of ions.

Ionic bonding is the result of *electrostatic forces* of attraction between positive metal ions and negative non-metal ions. The larger the charges on the ions the bigger the attractive force. The smaller the ions, the closer the charges get to each other and the stronger the force of attraction.

Electron *dot-and-cross diagrams* provide a balance sheet for keeping track of the electrons when ionic compounds form. These diagrams do not describe the mechanisms of the reactions.

$$\text{Na} \bullet \quad + \quad \overset{\times\times}{\underset{\times\times}{\times\text{Cl}\,\overset{\times}{\times}}} \quad \longrightarrow \quad \text{Na}^+ \quad + \quad \overset{\times\times}{\underset{\times\times}{\overset{\bullet}{\times}\text{Cl}\,\overset{\times}{\times}}}{}^-$$

sodium atom   chlorine atom   sodium ion   chloride ion
(2.8.1)   (2.8.7)   (2.8)   (2.8.8)

$$\text{Ca} \bullet\!\bullet \quad + \quad \overset{\times\times}{\underset{\times\times}{{}^\times\text{F}\,{}^\times}} \;\; \overset{\times\times}{\underset{\times\times}{{}^\times\text{F}\,{}^\times}} \quad \longrightarrow \quad \text{Ca}^{2+} \quad + \quad \overset{\times\times}{\underset{\times\times}{\bullet\text{F}\,{}^\times}}{}^- \;\; \overset{\times\times}{\underset{\times\times}{\bullet\text{F}\,{}^\times}}{}^-$$

calcium atom   two fluorine atoms   calcium ion   two fluoride ions
(2.8.8.2)   (2.7)   (2.8.8)   (2.8)

Dot-and-cross diagrams for the formation of sodium chloride and calcium fluoride

Energy is needed to remove electrons from metal ions (see *ionisation energy*). The energy changes on adding electrons to non-metal atoms are quite small (see *electron affinity*). Ionic crystals are stable because of the large release of energy as the ions form a crystal lattice (see *lattice energy*). The *Born–Haber cycle* helps to analyse the stability of ionic crystals. This energy cycle also helps to explain why it is generally true that the atoms of s- and p-block elements gain or lose electrons to attain the electron configuration of the nearest *noble gas* in line with the *octet rule*.

ionic character of bonds: a phrase used to describe the extent to which bonds between the atoms of two different elements are polar. The greater the difference in electronegativity between the elements, the more ionic the bonding. (See also *polar covalent bonds*, *intermediate bonding* and *polarisation of ions*.)

ionic equations describe chemical changes by showing only the reacting ions in solution. These equations leave out the *spectator ions* which remain in solution unchanged. For example, an ionic equation shows that the use of a barium salt to test for sulfate ions is essentially the same reaction whether the reagent is barium chloride or barium nitrate and whatever the source of the sulfate ions.

$$Ba^{2+}(aq) + SO_4^{2-}(aq) \rightarrow BaSO_4(s)$$

Note that ionic equations are balanced both for atoms and charges.

An ionic equation also shows that the neutralisation of any fully ionised *strong acid* by a fully ionised *strong base* is essentially the same process:

$$H^+(aq) + OH^-(aq) \rightarrow H_2O(l)$$

Ionic *half-equations* are used to describe *redox reactions*.

ionic precipitation is a reaction which produces a solid precipitate on mixing two solutions. This type of reaction can be used to make an insoluble salts from two soluble salts. Silver bromide forms as a precipitate on mixing solutions of silver nitrate and potassium bromide. This is best represented by an *ionic equation* leaving out the *spectator ions*.

$$Ag^+(aq) + Br^-(aq) \rightarrow AgBr(s)$$

Ionic precipitation reactions play a big part in *anion tests* and *cation tests*.

ionic product of water: see K_w.

ionic radius is the radius of an ion in a crystal. Ionic radii are determined by X-ray crystallography. The radius of the positive ion of an element is smaller than its atomic radius. The radius of the negative ion of an element is larger than its atomic radius.

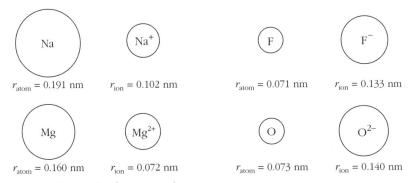

Na
$r_{atom} = 0.191$ nm

Na$^+$
$r_{ion} = 0.102$ nm

F
$r_{atom} = 0.071$ nm

F$^-$
$r_{ion} = 0.133$ nm

Mg
$r_{atom} = 0.160$ nm

Mg^{2+}
$r_{ion} = 0.072$ nm

O
$r_{atom} = 0.073$ nm

O^{2-}
$r_{ion} = 0.140$ nm

Comparison of the radii of atoms and ions

Trends in ionic radius help to account for patterns of bonding and properties in the periodic table.

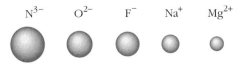

N^{3-} O^{2-} F$^-$ Na$^+$ Mg^{2+}

Ionic radii for selected elements. All the ions shown have the same electron configuration. They are isoelectronic.

Note these patterns:

- down any group the ionic radii increase as the number of inner full shells increases
- across a period the radii of ions with the same electron configuration decrease as the nuclear charge increases.

Ions in solution are **hydrated**. The radius of the ion plus the hydrating water molecules is larger than the ion in a crystal. A smaller ion is likely to be more heavily hydrated than a larger ion with the same charge.

Ion	Radius in a crystal/nm	Radius when hydrated/nm
Li^+	0.074	1.00
Na^+	0.102	0.79

ionisation energy (enthalpy) is a measure of the energy needed to remove an electron from a gaseous atom or ion. Ionisation energies give evidence for the arrangement of electrons in atoms in shells and subshells. Ionisation energies can help to explain which ions an element can form. Values for ionisation energies are used in energy cycles, such as the **Born–Haber cycle**, to investigate bonding in compounds.

Ionisation energies can be measured using a **mass spectrometer** or by studying the emission spectra of atoms (see **hydrogen emission spectrum**). In these ways it is possible to measure energy changes involving ions which do not normally appear in chemical reactions.

The first ionisation energy for an element is the energy needed to remove one mole of electrons from one mole of gaseous atoms. Successive ionisation energies for the same element measure the energy needed to remove a second, third, fourth electron and so on.

$Na(g) \rightarrow Na^+(g) + e^-$ first ionisation energy = 496 kJ mol^{-1}

$Na^+(g) \rightarrow Na^{2+}(g) + e^-$ second ionisation energy = 4563 kJ mol^{-1}

$Na^{2+}(g) \rightarrow Na^{3+}(g) + e^-$ third ionisation energy = 6913 kJ mol^{-1}

There are 11 electrons in a sodium atom so there are 11 successive ionisation energies for this element. The electron configuration of the element is $1s^2 2s^2 2p^6 3s^1$. There is 1 electron in the outer shell which is furthest from the nucleus and shielded by ten inner electrons. There are eight electrons in the second shell and these are closer to the nucleus and only have two inner **shielding** electrons. The two inner electrons feel the full attraction of the nuclear charge and are closest to the nucleus. They are hardest to remove.

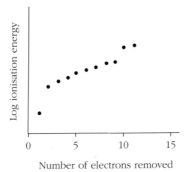

Plot of lg(ionisation energy) against number of electrons removed for sodium

So the successive ionisation energies for an element rise and there are big jumps in value each time electrons start to be removed from the next shell in towards the nucleus.

In the periodic table, ionisation energies tend to rise from left to right across a period because from one element to the next the added electron goes into the same main shell as the charge on the nucleus increases by one. The rise in values is not smooth but shows a 2–3–3 pattern, corresponding to the way that the *s orbitals* and *p orbitals* fill up.

The 2s orbital is full at Be, there is then a slight dip as the next electron goes into a p orbital with a slightly higher energy. The three p orbitals each have one electron in a nitrogen atom. The dip at oxygen happens because the next electron has to pair up with another electron. Nevertheless, the main factor is the charge on the nucleus which increases steadily across the period. This pattern repeats from Na to Ar.

Ionisation energies decrease down a group. Down a group the number of full shells increases. The increased shielding effect balances out the increasing charge on the nucleus. The outer electrons get further and further away from the same effective nuclear charge, so they are easier to remove.

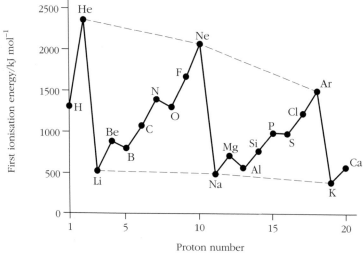

Plot of first ionisation energy against proton number for the elements H to Ca showing a periodic pattern. Note that the trend is for the values of the first ionisation energies to decrease down any group.

ionisation of water: see K_w.

ionising radiation includes all the types of *electromagnetic radiation* with enough energy per *photon* to ionise atoms. Ionising radiation includes alpha and beta particles and gamma rays from *radioactive decay*, and *X-rays*. Living cells are very sensitive to ionising radiation. High doses of ionising radiation can kill living cells and even small doses can cause mutation. Exposure to ionising radiation can, in time, give rise to cancer. The radiation dose from ionising radiation is measured in *sieverts*.

ions: electrically charged atoms or groups of atoms (see *anions, cations, ionic bonding* and *complex ions*).

H^+

						no simple ions	N^{3-}	O^{2-}	F^-	no ions formed
Li^+										
Na^+	Mg^{2+}			Al^{3+}				S^{2-}	Cl^-	
K^+	Ca^{2+}								Br^-	
Rb^+	Sr^{2+}	transition metals form more than one ion, e.g. Fe^{2+}, Fe^{3+}							I^-	
Cs^+	Ba^{2+}									
1+	2+			3+			3–	2–	1–	

metals
positive ions

non-metals
negative ions

The charges on simple ions are related to the positions of elements in the periodic table

iron (Fe) is a metal, a **d-block element** which shows the characteristic behaviour of the **transition elements**. Its **electron configuration** is: $[Ar]3d^64s^2$. The pure metal is soft and easily worked. It is strongly **ferromagnetic**. In most of its uses iron is mixed with other elements to make **steel**. Iron is so widely used that **corrosion of the metal** is a serious economic problem.

Iron can exist in more than one **oxidation state**. Iron dissolves in dilute acids to form iron(II) salts. The aqueous iron(II) ion is pale green. Oxidising agents, including oxygen in the air, chlorine and potassium manganate(VII), all oxidise iron(II) to iron(III). The aqueous iron(III) ion is yellow.

$$Fe^{2+}(aq) \rightarrow Fe^{3+}(aq) + e^-$$

Adding aqueous sodium hydroxide to solutions of iron salts helps to distinguish iron(II) from iron(III) compounds (see **cation tests**). A greenish precipitate of iron(II) hydroxide turns browny-red when exposed to the air, as it is oxidised to iron(III) hydroxide.

Iron ions form a variety of complexes in both the +2 and the +3 oxidation states. The hexaaquo complex of the iron(III) ion, $Fe(H_2O)_6^{3+}$, is acidic (see **hydrolysis of metal aqua ions**).

A very sensitive test for iron(III) is to add a dilute solution of thiocyanate ions, SCN^-. The solution turns a deep blood-red colour due to the formation of iron complexes such as $[Fe(H_2O)_5(SCN)]^{2+}$. The SCN^- **ligand** takes the place of a water molecule in the aqua complex.

An example of an iron(III) complex with a **bidentate ligand** is $[Fe(C_2O_4)_3]^{3-}$, in which the ligand is the ethanedioate ion.

Iron and its compounds act as **homogeneous catalysts** and **heterogeneous catalysts**.

iron extraction produces the metal from its oxide ores in large blast furnaces. Coke burning in air heats the furnace.

$$C(s) + O_2(g) \rightarrow CO_2(g)$$

Coke also produces the **reducing agent** by reacting with carbon dioxide further up the furnace to make carbon monoxide.

$$C(s) + CO_2(g) \rightarrow 2CO(g)$$

The carbon monoxide reduces the oxide ore to iron.

$$Fe_2O_3(s) + 3CO(g) \rightarrow 2Fe(l) + 3CO_2(g)$$

Limestone (*calcium carbonate*, $CaCO_3$) decomposes to basic calcium oxide (CaO), which combines with acidic silicon dioxide and other impurities to make a liquid *slag*. For example:

$$CaO(s) + SiO_2(s) \rightarrow CaSiO_3(l)$$

The molten metal and slag run to the bottom of the furnace, where the slag floats on the metal so that it can be tapped off separately. Most of the iron is then turned into *steel*.

irritant: see *hazard*.

isoelectric point: the pH at which the overall charge on an *amino acid* or *protein* is zero. At this pH amino acids or proteins do not move towards either of the electrodes during *electrophoresis*.

At low pH an amino acid has a positive charge because there is an extra hydrogen ion on the basic amino group. At high pH an amino acid is negatively charged because the carboxylic acid group has given away a hydrogen ion. At the pH of the isoelectric point the amino acid is present in solution as a *zwitterion* and the total charge is zero.

| in acid solution | zwitterion | in alkaline solution |

The charges on alanine at different pH values

Isoelectric points vary from one amino acid to another because of differences in their side chains.

A large protein molecule is at its least soluble in water at its isoelectric point when it is uncharged.

isoelectronic atoms and ions have the same number and arrangement of electrons. The following particles are all isoelectronic: N^{3-}, O^{2-}, F^-, Ne, Na^+, Mg^{2+} and Al^{3+}. They all have the electron configuration $1s^2 2s^2 2p^6$.

Methane and the ammonium ion are isoelectronic. Both have ten electrons – two in the first shell of the central atom and eight forming four covalent bonds with hydrogen atoms. They both have the same shape (see *shapes of molecules*). Diamond and *boron nitride* are also isoelectronic.

isomerisation is a process used in oil refining to convert straight-chain *alkanes*, such as pentane, into branched *isomers*, such as 2-methylbutane. The value of the process is that branched alkanes increase the *octane number* of *petrol*. Isomerisation happens when the hot hydrocarbon vapour passes over a *catalyst*.

isomers are compounds with the same molecular formula but different structures. Structural isomerism may arise because:

- the hydrocarbon chain is branched in different ways

$$H-\underset{\underset{H}{|}}{\overset{\overset{H}{|}}{C}}-\underset{\underset{H}{|}}{\overset{\overset{H}{|}}{C}}-\underset{\underset{H}{|}}{\overset{\overset{H}{|}}{C}}-\underset{\underset{H}{|}}{\overset{\overset{H}{|}}{C}}-H$$

butane

$$H-\underset{\underset{H}{|}}{\overset{\overset{H}{|}}{C}}-\underset{\underset{\underset{\underset{H}{|}}{H-C-H}}{|}}{\overset{\overset{H}{|}}{C}}-\underset{\underset{H}{|}}{\overset{\overset{H}{|}}{C}}-H$$

2-methylpropane

Chain isomers of C_4H_{10}

- the functional group is in a different position

$$H-\underset{\underset{H}{|}}{\overset{\overset{H}{|}}{C}}-\underset{\underset{H}{|}}{\overset{\overset{H}{|}}{C}}-\underset{\underset{H}{|}}{\overset{\overset{H}{|}}{C}}-OH$$

propan-1-ol

$$H-\underset{\underset{H}{|}}{\overset{\overset{H}{|}}{C}}-\underset{\underset{H}{|}}{\overset{\overset{OH}{|}}{C}}-\underset{\underset{H}{|}}{\overset{\overset{H}{|}}{C}}-H$$

propan-2-ol

Position isomers of C_3H_8O

- the functional groups are different.

$$H-\underset{\underset{H}{|}}{\overset{\overset{H}{|}}{C}}-\underset{\underset{H}{|}}{\overset{\overset{H}{|}}{C}}-OH$$

ethanol

$$H-\underset{\underset{H}{|}}{\overset{\overset{H}{|}}{C}}-O-\underset{\underset{H}{|}}{\overset{\overset{H}{|}}{C}}-H$$

methoxymethane

Functional group isomers of C_2H_6O

(See also *cis–trans isomers*, *E–Z isomers* and *optical isomers*.)

isomorphous compounds have the same type of chemical formula and crystallise with the same shape and structure. *Alums* are isomorphous. They form octahedral crystals.

isotactic polymer: a polymer made by *addition polymerisation* with a regular structure. In isotactic poly(propene), for example, all the methyl side chains are on the same side of the carbon chain. The molecules coil into a regular helical shape and pack together to form a highly crystalline polymer which is very strong. This is the form of the polymer which is useful for hardwearing fibres, tough mouldings in motor vehicles and containers which can hold boiling water.

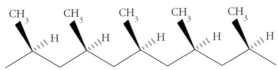

*Structure of isotactic poly(propene) with all the methyl groups on the same side of the carbon chain. Compare this with the **atactic polymer**.*

215

isotopes are *atoms* of the same element which have the same number of protons in the nucleus but a different number of neutrons. In other words the atoms of isotopes have the same *atomic number* but different *mass numbers*. The isotopes have the same chemical properties because they have the same number and arrangement of electrons (*electron configuration*). Some isotopes are radioactive (see *alpha decay*, *beta decay* and *radioactive decay*).

The isotopes of an element can be separated and detected with a *mass spectrometer*. Peaks in a mass spectrum are often multiple because of the existence of isotopes. The heights of the peaks in mass spectrum show the *isotopic abundance* of each isotope.

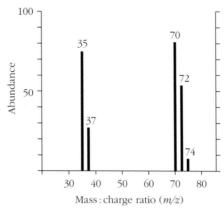

Mass spectrum of chlorine. The three peaks with larger values of m/z correspond to $^{35}Cl^{35}Cl$, $^{35}Cl^{37}Cl$ and $^{37}Cl^{37}Cl$. Fragmentation of chlorine molecules into ^{35}Cl and ^{37}Cl atoms produces the two peaks with lower m/z values.

isotopic abundance: the proportions of the different *isotopes* of an element. Chlorine, for example, has two isotopes: chlorine-35 and chlorine-37. Naturally occurring chlorine contains 75% of chlorine-35 and 25% of chlorine-37.

$$\text{So the mean relative atomic mass} = \frac{(3 \times 35) + 37}{4} = 35.5$$

Most relative atomic mass values are not whole numbers because most elements have several naturally occurring isotopes.

isotopic labelling uses *isotopes* as markers to trace the movement of chemicals or to investigate what happens to particular atoms during chemical changes (see also *tracers*). Atoms of the normally abundant isotope of an element in a molecule are replaced by a different isotope.

Isotopes have the same chemical properties so it is possible to use them to follow what happens during a change without altering the normal course of the process. Radioactive isotopes can be tracked by detecting their radiation. The fate of non-radioactive isotopes can be followed by analysing samples with a *mass spectrometer*.

Isotopic labelling has been used to investigate the mechanism of ester hydrolysis. Using water labelled with oxygen-18 (instead of the normal isotope oxygen-16) it was shown that it is the C—O bond in the ester which breaks and not the O—C_2H_5 bond. Mass spectrometry showed that the heavier oxygen atoms end up in the acid and not in the alcohol.

$$CH_3C-C\overset{O}{\underset{OC_2H_5}{\diagdown}} + H_2{}^{18}O \longrightarrow CH_3C\overset{O}{\underset{{}^{18}OH}{\diagdown}} + C_2H_5OH$$

This
bond breaks

Use of labelling to investigate bond breaking during ester hydrolysis

isotropic solids have properties which are the same in all directions. Crystals of sodium chloride are isotropic because the ions are spherical and the crystal structure is cubic. Most substances are more or less *anisotropic.*

IUPAC (International Union of Pure and Applied Chemistry) is the recognised authority for the names of chemical compounds, for chemical symbols and terms, and the values of data such as relative atomic masses. IUPAC names are systematic names based on a set of rules which make it possible to work out the chemical structure of a compound from its name. Chemists increasingly use approved IUPAC names for simpler compounds but stick to traditional names when the systematic name is complex. The systematic name 2-hydroxy-propane-1,2,3-tricarboxylic acid describes the structure but is cumbersome compared to the traditional name *citric acid* based on its occurrence in citrus fruits. (See *names of carbon compounds, names of complex ions* and *names of inorganic compounds.*)

Are you studying other subjects?

The *A–Z Handbooks (digital editions)* are available in 14 different subjects. Browse the range and order other handbooks at www.philipallan.co.uk/a-zonline.

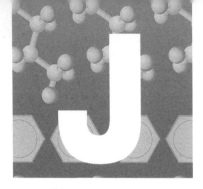

joule (symbol J) is the *SI unit* of energy. The quantities of energy transferred during chemical reactions are relatively large, so chemists generally measure energy changes in kilojoules (kJ). 1 kJ = 1000 J. The energy transfers in chemical reactions are measured in a calorimeter and calculated from the data with the help of the *specific heat capacity* of the materials being heated.

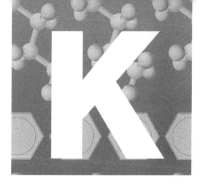

K_a is the symbol for an *acid dissociation constant* (or ionisation constant). K_a values show the extent to which acids dissociate into ions in solution. Acid dissociation constants are used to compare the strengths of relatively *weak acids* which are only partly ionised in solution.

For a weak acid represented by the formula HA:

$$HA(aq) \rightleftharpoons H^+(aq) + A^-(aq)$$

According to the *equilibrium law*, the equilibrium constant takes the form:

$$K_a = \frac{[H^+(aq)][A^-(aq)]}{[HA(aq)]}$$

where K_a is the acid dissociation constant.

K

Acid	K_a/mol dm^{-3}	
nitrous acid [nitric(III) acid], HNO_2	4.7×10^{-4}	stronger
methanoic acid, HCOOH	1.6×10^{-4}	
ethanoic acid, CH_3COOH	1.7×10^{-5}	
chloric(I) acid, HOCl	3.7×10^{-8}	
phenol, C_6H_5OH	1.3×10^{-10}	weaker

Worked example

Calculate the hydrogen ion concentration and the pH of a 0.01 mol dm^{-3} solution of ethanoic acid. K_a for the acid is 1.7×10^{-5} mol dm^{-3}.

Notes on the method

Two approximations simplify the calculation.

1   The first assumption is that $[H^+(aq)] = [A^-(aq)]$. In this example A^- is the ethanoate ion CH_3COO^-. This assumption seems obvious from the equation for the ionisation of a weak acid but it ignores the hydrogen ions from the ionisation of water. Water produces far fewer hydrogen ions than most weak acids so its ionisation can be ignored.

2   The second assumption is that so little of the ethanoic acid ionises in water that $[HA(aq)] \approx 0.01$ mol dm^{-3}. Here HA represents ethanoic acid. This is a riskier assumption which has to be checked because in very dilute solutions the degree of ionisation may become quite large relative to the amount of acid in the solution.

Answer

$$K_a = \frac{[H^+(aq)][A^-(aq)]}{[HA(aq)]} = \frac{[H^+(aq)]^2}{0.01 \text{ mol dm}^{-3}} = 1.7 \times 10^{-5} \text{ mol dm}^{-3}$$

Therefore $[H^+(aq)]^2 = 1.7 \times 10^{-7} \text{ mol}^2 \text{ dm}^{-6}$

So $[H^+(aq)] = 4.12 \times 10^{-4} \text{ mol dm}^{-3}$

$pH = -lg\,[H^+(aq)] = -lg\,(4.12 \times 10^{-4}) = 3.39$

Check the second assumption: in this case about 0.0004 mol dm^{-3} of the 0.0100 mol dm^{-3} of acid (4%) has ionised. In this instance the degree of ionisation is just about small enough to justify the assumption that [HA(aq)] ≈ the concentration of un-ionised acid.

K_b is the symbol for the base dissociation constant which measures the strength of bases by showing the extent to which they accept hydrogen ions in solution. *Base dissociation ionisation constants* are used to compare the strengths of relatively *weak bases*. For the weak base ammonia:

$$NH_3(aq) + H_2O(l) \rightleftharpoons NH_4^+(aq) + OH^-(aq)$$

The equilibrium constant, $K_b = \dfrac{[NH_4^+(aq)][OH^-(aq)]}{[NH_3(aq)][H_2O(l)]}$

In dilute solution the concentration of water is effectively constant, so the expression can be written in this form, where K_b is the base dissociation constant:

$$K_b = \frac{[NH_4^+(aq)][OH^-(aq)]}{[NH_3(aq)]}$$

K_c is the symbol for an equilibrium constant, with the concentrations measured in moles per litre (see *equilibrium law*).

Equilibrium expressions make it possible to explain the effect of changing the concentration of one of the chemicals in an equilibrium mixture. There is an equilibrium, for example, in a solution of bromine in water:

$$Br_2(aq) + H_2O(l) \rightleftharpoons HOBr(aq) + Br^-(aq) + H^+(aq)$$
orange   colourless

At equilibrium: $K_c = \dfrac{[HOBr(aq)][Br^-(aq)][H^+(aq)]}{[Br_2(aq)]}$, where these are equilibrium concentrations

In dilute solution $[H_2O(l)]$ is constant.

Adding alkali removes H$^+$(aq) from the right-hand side:

$$H^+(aq) + OH^-(aq) \rightarrow H_2O(l)$$

This briefly upsets the equilibrium. For an instant after adding alkali:

$$\frac{[HOBr(aq)][Br^-(aq)][H^+(aq)]}{[Br_2(aq)]} < K_c$$

The system restores equilibrium by bromine reacting with water to produce more of the products while lowering the concentration of Br$_2$(aq). There is soon a new equilibrium. Once again:

$$\frac{[HOBr(aq)][Br^-(aq)][H^+(aq)]}{[Br_2(aq)]} = K_c \qquad \text{but now with new values for the various concentrations.}$$

Speaking roughly, chemists say that adding alkali makes the 'position of equilibrium shift to the right'. The effect is visible because the orange colour of the solution fades. This is as *Le Chatelier's principle* predicts. The advantage of using K_c is that it makes quantitative predictions possible.

K_p is the symbol for an equilibrium constant for an equilibrium involving gases with the concentrations measured by *partial pressures*.

Equilibrium	K_p	Units of K_p (using the SI unit of pressure, Pa)
$H_2(g) + I_2(g) \rightleftharpoons 2HI(g)$	$K_p = \dfrac{(p_{NH_3})^2}{(p_{H_2})(p_{I_2})}$	no units
$N_2(g) + 3H_2(g) \rightleftharpoons 2NH_3(g)$	$K_p = \dfrac{(p_{NH_3})^2}{(p_{N_2})(p_{H_2})^3}$	Pa^{-2}
$N_2O_4(g) \rightleftharpoons 2NO_2(g)$	$K_p = \dfrac{(p_{NO_2})^2}{(p_{N_2O_4})}$	Pa

Worked example

When a mixture of 10.0 mol sulfur dioxide and 5.0 mol oxygen comes to equilibrium at 450°C and 200 kPa pressure, the amount of sulfur trioxide formed is 9.0 mol. Calculate K_p for the reaction.

Notes on the method

First write the equation for the reaction and work out the amount in moles of each gas at equilibrium. Note that 2 mol $SO_2(g)$ reacts with 1 mol $O_2(g)$ forming 2 mol $SO_3(g)$. So 9.0 mol $SO_2(g)$ must react to form 9.0 mol $SO_3(g)$.

Calculate the mole fraction of each gas. This leads to the partial pressures for the gases at equilibrium. Substitute these values in the expression for K_p and then decide on the units.

Answer

	$2SO_2(g)$	$+$	$O_2(g)$	$\rightleftharpoons$	$2SO_3(g)$
The equation:					
Initial amounts	10.0 mol		5.0 mol		0.0 mol
Equilibrium amount	1.0 mol		0.5 mol		9.0 mol
Equilibrium mole fractions	$\dfrac{1.0}{10.5}$		$\dfrac{0.5}{10.5}$		$\dfrac{9.0}{10.5}$
Equilibrium partial pressures	$\dfrac{1.0}{10.5} \times 200 \text{ kPa}$		$\dfrac{0.5}{10.5} \times 200 \text{ kPa}$		$\dfrac{9.0}{10.5} \times 200 \text{ kPa}$
	$= 19.0 \text{ kPa}$		$= 9.52 \text{ kPa}$		$= 171.4 \text{ kPa}$

Hence $K_p = \dfrac{(p_{SO_3})^2}{(p_{SO_2})^2(p_{O_2})} = \dfrac{(171.4 \text{ kPa})^2}{(19.0 \text{ kPa})^2(9.52 \text{ kPa})} = 8.55 \text{ kPa}^{-1}$

K_w is the symbol for the ionic product of water. K_w is a constant for the equilibrium produced by the ionisation of water. There are *oxonium* and hydroxide ions even in pure water because of a transfer of hydrogen ions between water molecules. This only happens to a slight extent. At equilibrium in pure water, $[H_3O^+(aq)] = 1 \times 10^{-7}$ mol dm^{-3}.

$$H_2O(l) + H_2O(l) \rightleftharpoons H_3O^+(aq) + OH^-(aq)$$

which can be written more simply as:

$$H_2O(l) \rightleftharpoons H^+(aq) + OH^-(aq)$$

The equilibrium expression is:

$$K_c = \frac{[H^+(aq)][OH^-(aq)]}{[H_2O(l)]}$$

There is such a large excess of water that $[H_2O(l)]$ is a constant, so the relationship simplifies to $K_w = [H^+(aq)][OH^-(aq)]$ where K_w is the ionic product of water.

In pure water $[H^+(aq)] = [OH^-(aq)] = 1 \times 10^{-7}$ mol dm^{-3}

Hence $K_w = 1 \times 10^{-14}$ mol^2 dm^{-6}

K_w is a constant in all aqueous solutions at 298 K. This makes it possible to calculate the pH of alkalis. (see also *pK*.)

Worked example

What is the pH of a 0.02 mol dm^{-3} solution of sodium hydroxide?

Notes on the method

Sodium hydroxide is fully ionised in solution. So in this solution

$[OH^-(aq)] = 0.02$ mol dm^{-3}

pH $= -$lg $[H^+(aq)]$

Answer

For this solution:

$$K_w = [H^+(aq)] \times 0.02 \text{ mol dm}^{-3} = 1 \times 10^{-14} \text{ mol}^2 \text{ dm}^{-6}$$

So $[H^+(aq)] = \dfrac{1 \times 10^{-14} \text{ mol}^2 \text{ dm}^{-6}}{0.02 \text{ mol dm}^{-3}} = 5 \times 10^{-13}$ mol dm^{-3}

Hence pH $= -$lg $(5 \times 10^{-13}) = 12.3$

Kekulé structure: a ring structure for benzene proposed by the German chemist Friedrich Kekulé in 1865. According to Kekulé, the idea of a ring structure came to him while he was daydreaming in front of a fire.

Representations of the Kekulé structure for benzene

The problem with this structure is that it suggests that there should be two isomers of 1,2-dichlorobenzene.

Isomers of 1,2-dichlorobenzene suggested by the Kekulé structure. Isomers with this formula have never been separated. There is only one form of the compound.

In practice it has never been possible to separate isomers of disubstituted benzenes. To get around this problem Kekulé suggested that benzene molecules rapidly alternate between the two possible structures. This is not the modern explanation. Today, chemists use the idea of **delocalised electrons** to account for the benzene structure.

Kelvin (symbol K) is the **SI unit** of temperature. Temperatures above **absolute zero** are measured in kelvins. Temperatures on the **Celsius temperature scale** use the same size units but zero is set at the freezing point of water (273.15 K). Temperature differences measured on either scale are the same and given as kelvins (K).

kerosine is a mixture of hydrocarbons produced by the **fractional distillation of oil**. Part of the kerosine fraction is refined for use as paraffin, while some is cracked to make **petrol**. Most of the kerosine fraction, however, is purified for use as the fuel for jet engines in aircraft.

ketones are **carbonyl compounds** in which the carbonyl group is attached to two **alkyl groups**. The carbonyl group is the **functional group** which gives ketones (and **aldehydes**) their characteristic properties.

Ketones are sometimes called alkanones. They are named after the alkane with the same carbon skeleton by changing the ending 'e' to 'one'. Where necessary, a number in the name shows the position of the carbonyl group.

Oxidation of secondary alcohols with acidified potassium dichromate(VI) produces ketones which, unlike aldehydes, are not easily oxidised further.

$$CH_3 - \underset{\underset{H}{|}}{\overset{\overset{OH}{|}}{C}} - CH_3 + [O] \xrightarrow[\substack{\text{sodium dichromate(VI)} \\ \text{and sulfuric acid}}]{\substack{\text{heat under reflux} \\ \text{with a mixture of}}} CH_3 - \overset{\overset{O}{\|}}{C} - CH_3 + H_2O$$

propanone

Oxidation of propan-2-ol to propanone

Fehling's solution, **Benedict's solution** and **Tollens' reagent** cannot oxidise ketones. There is no change when testing ketones with these reagents. Ketones with a methyl group next to the carbonyl group, such as propanone, give a positive result with the **triiodomethane reaction**.

Chemists can identify ketones with the help of instrumental techniques such as **mass spectrometry** and **infrared spectroscopy**. Traditionally, chemists characterised these compounds by preparing crystalline derivatives with **2,4-dinitrophenylhydrazine**.

Sodium tetrahydridoborate(III) reduces ketones to secondary alcohols, as does **lithium tetrahydridoaluminate(III)**.

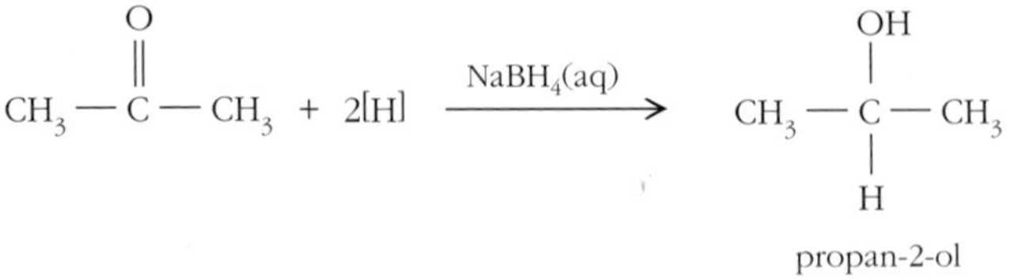

$$CH_3 - \overset{\overset{\displaystyle O}{\|}}{C} - CH_3 + 2[H] \xrightarrow{\text{NaBH}_4(aq)} CH_3 - \overset{\overset{\displaystyle OH}{|}}{\underset{\underset{\displaystyle H}{|}}{C}} - CH_3$$

propan-2-ol

Reduction of propanone to propan-2-ol

Ketones, like aldehydes, undergo addition. These are ***nucleophilic addition*** reactions. Sometimes addition is immediately followed by elimination of water in ***addition–elimination reactions***.

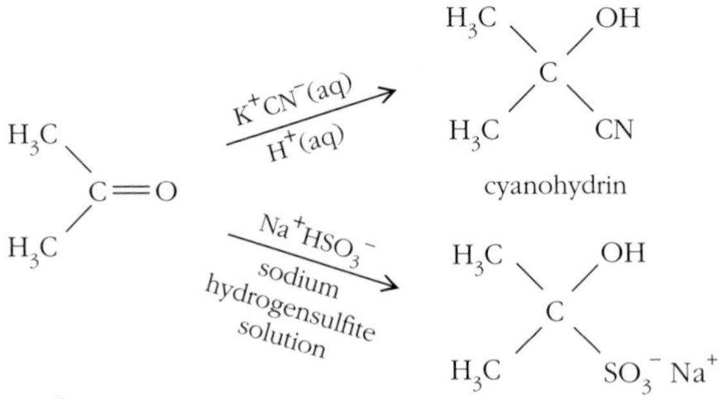

cyanohydrin

Addition reactions of propanone

Kevlar is an extremely strong, flexible, fire resistant *polymer* with a low *density*. Bulletproof vests are made of Kevlar and punctures are less likely with bicycle tyres which include the polymer. Kevlar ropes are much stronger than the same weight of steel rope. The polymer can also replace steel in motor vehicle tyres.

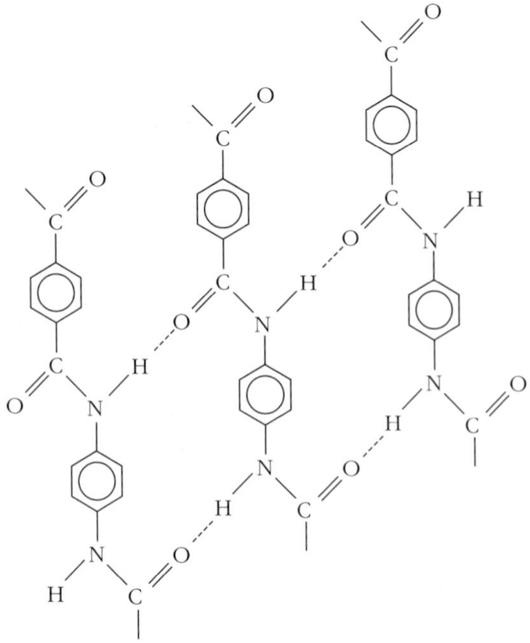

The structure of Kevlar

Kevlar is a *polyamide*. It is a polymer of an aryl amide. Polymers of this kind are called aramids. The rigid, linear polymer chains in Kevlar line up parallel to each other held together by *hydrogen bonding*.

kilogram (symbol kg) is the *SI unit* of mass. 1 kg = 1000 g.

kinetic control: a term used when the product of a reaction is the one which forms faster rather than an alternative product which is more *stable* but forms more slowly.

The production of methanal from methanol and oxygen is an example of kinetic control.

$$2CH_3OH(g) + O_2(g) \rightarrow 2HCHO(g) + 2H_2O(g)$$

This reaction is fast in the presence of a suitable catalyst at a temperature at which the alternative reaction cannot go. The alternative is burning to carbon dioxide and water. Combustion is more exothermic but has a higher *activation energy*.

$$2CH_3OH(g) + O_2(g) \rightarrow 2CO_2(g) + 4H_2O(l)$$

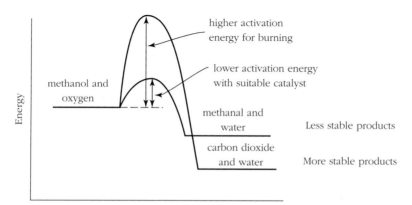

Progress of reaction from reactants to products

Reaction profiles for alternative reactions illustrating kinetic and thermodynamic control

In the absence of a catalyst, and at a higher temperature, methanol burns to produce the products which are thermodynamically more stable. In these circumstances the system is under *thermodynamic control*.

kinetic inertness: a term used when a reaction does not go, even though the reaction appears to be *feasible*. There is no change because the rate of reaction is too slow to be noticeable. There is a barrier preventing change – usually a high *activation energy*. The compound or mixture is inert.

Examples of kinetic inertness are:
- a mixture of methane (natural gas) and oxygen at room temperature
- a mixture of CCl_4 and water at room temperature (see *hydrolysis of non-metal chlorides*)
- a solution of *hydrogen peroxide* in the absence of a catalyst
- *aluminium* metal in dilute hydrochloric acid.

Nothing happens in these mixtures even though the possible reactions tend to go. In each example, thermochemistry shows that the reaction is feasible (as indicated by the equilibrium constant (see *equilibrium law*), or the *standard electrode potential*, or the total *entropy* change, or the *free energy change*) yet nothing happens.

kinetics of reactions: see *reaction kinetics*.

kinetic theory of gases: an explanation of why gases obey the *gas laws*. The theory is based on a model of an *ideal gas*. The assumptions are that:

- a gas consists of molecules in rapid, random motion
- the volume of the molecules can be neglected in comparison with the total volume of gas
- the molecules do not attract each other (no *intermolecular forces*)
- the collisions with the walls of any container are perfectly elastic.

By applying Newton's laws of motion and then assuming that the average kinetic energy of molecules is proportional to temperature, it is possible to derive the ideal gas equation from this model.

The model helps to explain why *real gases* deviate from ideal gas behaviour.

The model of gas molecules in rapid, random motion underlies the theories of *reaction kinetics* and chemical equilibrium.

knocking is a noise from a *petrol* engine heard when the mixture of fuel and air ignites too early while still being compressed by the piston. This is pre-ignition.

Compression heats up the mixture of fuel and air. Pre-ignition means that the fuel starts burning before being ignited by a spark from the spark plug. *Internal combustion engines* run powerfully if the fuel starts to burn when the piston is at the right point in the cylinder so that the expanding gases force the piston down smoothly. Knocking is a sign that a fuel with a higher *octane number* is needed. Prolonged knocking damages an engine.

krypton (Kr) is a *noble gas* which makes up less than 0.1% of *air*. The gas is separated from the other gases in the air by *fractional distillation* and is used to fill fluorescent lights and flash bulbs.

Kyoto protocol: an agreement that places a legal obligation on some countries to meet agreed reduction targets for *greenhouse gas* emissions. The agreement was negotiated at a conference held in Kyoto, Japan, in 1997 but only entered into force in February 2005 when ratification by Russia meant that there were enough signatories. The reduction targets apply to all the countries in the European Union and to specified industrialised countries. The only obligation on the other nations that have signed is that they monitor their emissions and submit regular reports. The protocol commits the industrialised nations to an average 5 per cent reduction of greenhouse gas emissions compared with 1990 levels during the first phase from 2008 to 2012.

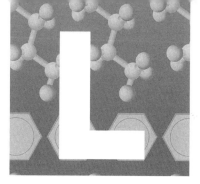

labile complexes: a term used to describe *complex ions* in which it is easy to replace the *ligands*. *Ligand substitution reactions* are difficult or impossible with inert complexes.

Chromium(III) and cobalt(III) complexes are ***kinetically inert***. Most other complexes are labile.

labile protons can quickly and easily react or move from one molecule to another. *Hydrogen bonding* affects the properties of compounds with hydrogen atoms attached to highly electronegative atoms such as oxygen or nitrogen. These molecules can rapidly exchange their labile protons as they move from one electronegative atom to another.

landfill is where local authorities, public services and industries dispose of most of their *waste*. A modern landfill site is engineered to contain waste and the products of decomposition so that *leaching* does not carry hazardous chemicals into the soil and contaminate water sources. Water percolating through rotting landfill can dissolve harmful chemicals such as the ions of toxic metals, which can be recovered from the liquids by *ion exchange* if waste disposal is managed correctly.

Micro-organisms break down the organic materials which make up more than half of household waste. At first there is oxygen present and micro-organisms convert complex carbohydrates to simple sugars. *Fermentation* begins as the conditions for *respiration* become anaerobic, turning carbohydrates, proteins and fats into organic acids, carbon dioxide and hydrogen. In time the micro-organisms start to produce the gases methane and carbon dioxide. This landfill gas can be a hazard because methane is highly flammable, but the *biogas* can also be collected from pipes buried in the waste and burnt to generate electricity.

Some landfills are for general domestic waste but others are planned to contain more hazardous materials. After dumping, heavy vehicles compact the waste and the layers of waste are then covered with soil.

lanthanide elements: the series of 14 elements between lanthanum and hafnium in the sixth period of the periodic table. Across this series of elements the electrons fill the 4f energy levels. These elements were once called the 'rare earths' but, apart from promethium, they are not rare. Cerium is five times as abundant as lead. The elements all have very similar chemical properties. They can be separated by *ion exchange* or *chromatography*. The main source of these elements is the ore monazite.

lattice: a term used to describe a network of atoms or ions in a crystal (see *crystal structures of ionic compounds*, *crystal structures of metals* and *crystal structures of non-metals*).

lattice energy (enthalpy) is the standard *enthalpy change* when one mole of an ionic compound forms from free gaseous ions.

$$Na^+(g) + Cl^-(g) \rightarrow NaCl(s) \qquad \Delta H^\ominus_{lattice} = -787 \text{ kJ mol}^{-1}$$

Note the difference between the lattice energy and the standard *enthalpy change of formation* of sodium chloride, which refers to the formation of a compound from its elements, $\Delta H^\ominus_f [NaCl(s)] = -411 \text{ kJ mol}^{-1}$.

Lattice energy is one of the thermochemical quantities which cannot be measured directly. Lattice energies are calculated indirectly with the help of the **Born–Haber cycle**.

Lattice energies give a measure of the strength of the ionic bonding in a crystal. Ionic bonding is the result of the *electrostatic forces* between the oppositely charged ions. So, for a given crystal structure, the lattice energy increases if:

• the charge on the ions increases
• the ions get smaller (bringing the charges closer together).

Lattice energies appear in thermochemical cycles to explore the factors which determine the *enthalpy change of solution* of ionic salts.

layer lattice: a crystal lattice in which there is strong bonding within layers of atoms but relatively weak bonding between the layers. It is often easy to cleave materials with a layer lattice along planes parallel to the layers. Examples of substances with layer lattices are graphite, mica and molybdenum sulfide. Graphite and molybdenum sulfide are used as *lubricants* because of the ease with which layers of atoms slide over each other.

leaching is a process for separating a material from an insoluble solid using a suitable solvent. Leaching is one of the methods used to extract metals from low-grade ores. Copper is obtained in this way from waste tips at old mines. These tips contain ores which have been oxidised by weathering. The finely divided ore is treated with sulfuric acid in a leach tank stirred by a flow of air.

$$Cu_2(OH)_2CO_3(s) + 4H^+(aq) \rightarrow 2Cu^{2+}(aq) + 3H_2O(l) + CO_2(g)$$

Leaching also describes the extraction of nitrates and other fertilisers from soil by rainwater. Water draining through the soil can carry away soluble nitrate ions into lakes and streams, causing *eutrophication*. Precautions are necessary in *landfill* to prevent hazardous chemicals leaching from the decomposing wastes into the soil and then into underground water sources.

lead (Pb) is a greyish metal often used on roofs of buildings because it resists corrosion and is malleable so that it can be bent to fit the shape and create a waterproof flashing where roof tiles meet walls. Lead is a dense metal and is an effective shield to stop X-rays and other radiations. Lead melts at a relatively low temperature for a metal and, when alloyed with tin, produces solder with an even lower melting point (see *eutectic mixture*).

Lead comes below tin in *group 4* of the periodic table. It forms compounds in two oxidation states: +2 and +4. The +2 state is more stable so that compounds in the +4 state are oxidising agents.

Lead forms two soluble lead(II) compounds: the nitrate and the ethanoate. Lead ions are colourless is solution. Adding alkali to a solution of lead(II) ions gives a white precipitate of Pb(OH)$_2$ which dissolves in excess, showing that this hydroxide to be an *amphoteric compound*.

The pigment chrome yellow consists of lead chromate which precipitates on mixing a solution of lead(II) ions with a solution of chromate(VI) ions. As well as being used by artists and printers, chrome yellow is the pigment in the yellow stripes painted along roads to control parking.

lead–acid cell: an *electrochemical cell* that is rechargeable because the chemical changes at the electrodes are reversible. The working of the cells involves lead in three *oxidation states* as Pb(0), Pb(II) and Pb(IV).

In a fully charged lead–acid cell the negative electrode consists of lead metal. The positive electrode consists of lead coated with lead(IV) oxide. The electrodes dip into a solution of sulfuric acid. These are the electrode processes as current flows:

- at the negative electrode:

$$Pb(s) + SO_4^{2-}(aq) \rightarrow PbSO_4(s) + 2e^-$$

- at the positive electrode:

$$PbO_2(s) + SO_4^{2-}(aq) + 4H^+(aq) + 2e^- \rightarrow PbSO_4(s) + 2H_2O(l)$$

The equations help to explain why the cell is rechargeable. The two changes which happen as a current flows from the cell both produce an insoluble solid, lead sulfate. This traps the lead(II) ions beside the electrode instead of dissolving in the electrolyte. So when the current is reversed to charge up the cell, the two reactions can be reversed, turning lead(II) back to lead metal at one electrode and back to PbO_2 at the other.

A 12 volt car battery usually consists of six 2 V lead–acid cells connected in series. Banks of lead–acid cells can be used to provide emergency standby power for lighting wherever a mains power cut could be serious. Lead–acid cells have the great advantage that they can be recharged many times and last for a long time.

leaving group: an atom or group of atoms which breaks away from a molecule during a *substitution reaction* or an *elimination reaction*.

Nucleophilic substitution of a hydroxide ion for a bromide ion in 1-bromobutane. The bromide ion is the leaving group.

Some groups are better at leaving than others. A water molecule is a better leaving group than a hydroxide ion. This explains why the substitution reactions of *alcohols* go better in acid conditions.

Nucleophilic substitution of a bromide ion for a hydroxide ion in propan-2-ol. The reaction goes faster under acid conditions. The —OH group in the alcohol is protonated, so the leaving group is a water molecule.

229

Le Chatelier's principle

Le Chatelier's principle is a qualitative guide to the effect of changes in concentration, pressure or temperature on a system at equilibrium. The principle was suggested as a general rule by the French physical chemist Henri Le Chatelier (1850–1936).

The principle states that when the conditions of a system at equilibrium change, the position of equilibrium shifts in the direction that tends to counteract the change.

The principle can be used to discuss the conditions chosen for **ammonia manufacture**. The equilibrium system involved is:

$$N_2(g) + 3H_2(g) \rightleftharpoons 2NH_3(g) \qquad \Delta H = -92.4 \text{ kJ mol}^{-1}$$

There are four moles of gases on the left side of the equation but only two moles on the right. Le Chatelier's principle predicts that raising the pressure will make the equilibrium shift from left to right. This reduces the number of molecules and so tends to reduce the pressure. So increasing the pressure increases the proportion of ammonia at equilibrium.

The reaction is **exothermic** from left to right and so **endothermic** from right to left. Chatelier's principle predicts that raising the temperature will make the position of equilibrium shift in the direction which takes in energy (tending to cool the mixture). So raising the temperature lowers the proportion of ammonia at equilibrium.

Le Chatelier's principle can be misleading, especially when predicting the effects of changes in concentration or pressure. It is generally better to use the **equilibrium law** and the values of K_c or K_p to predict the effect of changes in concentration or pressure.

Lewis acid–base theory: a theory which gives a very broad definition of acids and bases in terms of electron pairs. A Lewis acid is a molecule or ion which can form a bond by accepting a pair of electrons. A Lewis base is a molecule or ion which can form a bond by donating a pair of electrons.

The formation of an **oxonium ion** is a Lewis acid–base reaction between the proton (a Lewis acid) and a water molecule (a Lewis base). So, in this theory, it is the proton rather than the proton donor, which is an acid.

*Formation of an oxonium ion as the Lewis base, water, forms a **dative covalent bond** with a proton*

This much wider definition of acids and bases describes the formation of a **complex ion** as a reaction between a metal ion (a Lewis acid) and **ligands** (Lewis bases).

$$Ni^{2+} \quad + \quad 6:NH_3 \quad \longrightarrow \quad [Ni(NH_3)_6]^{2+}$$

Lewis	Lewis	Complex ion
acid	base	

Complex formation as a Lewis acid–base reaction

Aluminium chloride ($AlCl_3$) is a Lewis acid. This accounts for its use as a catalyst in **Friedel–Crafts reactions**.

Aluminium chloride acting as a Lewis acid

Chemists normally use the **Brønsted–Lowry theory**. When they use the terms acid and base they mean 'proton donor' and 'proton acceptor'. To signal that they are using the wider Lewis definition they refer to 'Lewis acids' and 'Lewis bases'.

life-cycle assessment (LCA) is a detailed process for analysing the environmental effects of a product or process 'from the cradle to the grave'. The analysis evaluates the materials used, the inputs of energy and water as well as the emissions to the environment (such as *greenhouse gases*) over the whole life of a product. The life cycle starts from extracting *raw materials*, through manufacture, distribution, retail, use, re-use and maintenance to *recycling* and *waste* management. The results of the assessment can help to show whether or not the product or process can be regarded as an example of *sustainable development*.

ligand substitution reactions involve swapping one *ligand* for another in a *complex ion*. These ligand exchange reactions are often reversible.

The ligands NH_3 and H_2O are similar in size and are uncharged. With these ligands, exchange reactions take place without change in the *coordination number* of the metal ion.

$$[Cu(H_2O)_6]^{2+}(aq) + 4NH_3(aq) \rightleftharpoons [Cu(NH_3)_4(H_2O)_2]^{2+}(aq) + 4H_2O(l)$$

Reversible ligand exchange reaction of copper(II) ions. Note that in this example substitution is incomplete. The aqua complex is pale blue but the ammine complex is a very deep blue.

The chloride ion is larger than uncharged ligands such as water, so that fewer chloride ions can fit round a central metal ion. Ligand exchange involves a change in coordination number

$$[Cu(H_2O)_6]^{2+}(aq) + 4Cl^-(aq) \rightleftharpoons [CuCl_4]^{2-}(aq) + 6H_2O(l)$$

Ligand exchange with change of coordination number. The aqua complex is pale blue but the chloro complex is yellow.

ligands are the molecules or ions bound to the central metal ion in a complex ion. Examples of molecules which can act as ligands are water and ammonia. Examples of ions which act as ligands are hydroxide ions, chloride ions and cyanide ions. Ligands have one or more *lone pairs of electrons* which can form *dative covalent bonds* to the metal ion (see *monodentate ligands*, *bidentate ligands*, *polydentate ligands* and *EDTA*).

limewater is a test reagent for detecting carbon dioxide. The reagent is a saturated solution of calcium hydroxide in water. The reagent is a colourless solution. Bubbling carbon dioxide through the solution produces a white precipitate of calcium carbonate (see *gas tests*).

$$Ca(OH)_2(aq) + CO_2(g) \rightarrow CaCO_3(s) + H_2O(l)$$

limiting reactant: the chemical in a reaction mixture which is present in an amount which limits the theoretical *yield*. Often in a chemical synthesis some of the reactants are added in excess to make sure that the most valuable chemical is converted as far as possible to the required product. The limiting reactant is the one which is not in excess and so is used up (if the reaction goes to completion).

linear relationship: two variables x and y are related linearly if a plot of y against x on a graph is a straight line. The general equation for a straight line graph is $y = mx + c$, where m is the gradient of the line and c the intercept with the y-axis. If the value of c is zero, the line passes through the origin of the graph, in which case the graph describes a proportional relationship.

lipids: a broad class of biological compounds which are soluble in organic solvents such as ethanol, but insoluble in water. Lipids are very varied and include *fatty acids*, the *triglycerides* in *fats* and *vegetable oils*, phospholipids and *steroids* such as *cholesterol*. Lipids release more energy per gram when oxidised than carbohydrates. This makes lipids important as a concentrated energy store in living organisms.

lipid/water solubility: a characteristic of *pharmaceuticals* which affects what happens to drugs in the body. Ideally drugs are taken by mouth in the form of tablets or capsules. This is only possible if the drug can survive in the stomach, which is highly acidic. The drug then has to cross from the gut into the bloodstream and reach its target without being broken down in the liver or excreted by the kidneys.

Cell membranes consist largely of *lipids*. This means that a drug has to be sufficiently fat-loving (lipophilic) if it is to cross from the gut to the bloodstream. However, fat in the body extracts any drug which is too lipophilic from the bloodstream so that it ceases to be available as an active chemical.

A drug that is too water-loving (hydrophilic) may be unable to cross the fatty barriers of cell membranes. It may also be quickly removed by the kidneys and excreted in the urine. Functional groups that make a molecule hydrophilic are —OH, —NH$_2$, —COOH and —COO$^-$. Functional groups which make a molecule lipophilic include alkyl and aryl (hydrocarbon) side chains.

liquids flow to take the shape of their container. The atoms, molecules or ions in a liquid are free to move about but with nothing like the freedom of the molecules of a *gas*. The atoms or molecules of a liquid are generally slightly more widely spaced than in a solid but the density of packing is only slightly less. Liquids lack the order of crystalline solids. Liquids, like solids, are hard to compress. Unlike gases, they have a definite volume.

liquid crystal: a state of matter which is more ordered than a liquid but less ordered than a solid. Although the liquid crystal state was first noticed as long ago as 1888, it is only since the early 1970s that such crystals have been developed for use in digital watches, calculators and some laptop computers.

Thermotropic liquid-crystals are pure chemicals which exist in the liquid crystal state over a range of temperatures between the ordered solid state and the disordered liquid state.

Some liquid crystals consist of rod shaped molecules and can exist in a nematic liquid phase in which there is a general tendency for the molecules to be aligned in the same direction. The use of these materials in display screens depends on the way that the transmission of *plane polarised light* by a liquid crystal varies with the orientation of the molecules. The orientation can be controlled with a small applied voltage if the molecules are polar.

lithium (Li) is a soft, shiny metal which turns dark grey in air. It is the first member of *group 1* with the electron configuration [He]2s^1.

Like other group 1 metals, lithium:
- is stored in oil
- floats on water and reacts, but quite slowly, forming hydrogen and LiOH which is soluble and strongly alkaline
- burns in air with a coloured flame (bright red) forming an oxide (Li_2O)
- forms an ionic, crystalline chloride.

The small size of the lithium ion means that some features of lithium chemistry are not typical of the group as a whole:
- hydration – the Li$^+$ ion is heavily hydrated in water and lithium has the most negative hydration energy in group 1; as a result the standard electrode potential for Li$^+$(aq)|Li(s) is the most negative for the group
- solubilities – lithium carbonate and lithium fluoride are only slightly soluble while most other simple compounds of group 1 metals are soluble
- thermal stability – lithium carbonate decomposes on heating while all the other group 1 carbonates are stable; lithium nitrate decomposes to the oxide on heating unlike the nitrates of sodium and potassium
- covalent bonding – lithium forms a wide range of covalent molecular compounds such as ethyl lithium, C_2H_5Li.

lithium polymer cells have been developed from earlier lithium ion cells to provide the power in many mobile phones and notebook computers. These rechargeable cells are relatively light because lithium has a low density. The reactivity of lithium means that these *electrochemical cells* have a relatively large voltage. The anode consists of lithium atoms trapped within a graphite layer lattice. The electrolyte is a polymeric material which can conduct lithium ions. The cathode consists of a compound of lithium, cobalt and oxygen or of lithium, manganese and oxygen.

lithium tetrahydridoaluminate(III) ($LiAlH_4$) is a powerful reducing agent used to reduce *aldehydes*, *ketones*, *esters* and *carboxylic acids* to alcohols. Some chemists still use the older name, lithium aluminium hydride.

In its reactions the tetrahydridoaluminate(III) ion can be regarded as a source of hydride ions (H$^-$). The reagent is rapidly hydrolysed by water so it has to be used in an anhydrous solvent such as dry ether. Where possible *sodium tetrahydridoborate(III)* is preferred because it is easier and safer to use.

lithosphere: the rigid outer layer of the Earth which consists of the crust and the upper part of the mantle. The lithosphere is divided into tectonic plates. The crust is the upper layer of the lithosphere with a distinct chemical composition. The crust makes up less than 0.0001% of the volume of the planet. The crust consists largely of oxygen (46.6% by mass) mainly combined with silicon (27.7%) in *silicate* minerals. Only a few other elements are abundant: aluminium (8.1%), iron (5.0%), calcium (3.6%), sodium (2.8%), potassium (2.6%) and magnesium (2.1%). All other elements together make up the remaining 1.5%.

litre (symbol l) is a unit of *volume* equal to 1000 cm^3. A litre is the same as a decimetre cubed (dm^3). 1 dm^3 = 10 cm × 10 cm × 10 cm = 1000 cm^3. Chemists use dm^3 when calculating because the units can then be worked through consistently. While m^3 is the *SI unit* of volume, dm^3 is accepted as convenient, and is preferred to the litre.

localised electrons: electron pairs forming covalent bonds between two atoms. Localised electrons are not free to move through a structure, so giant structures with normal covalent bonds, such as diamond and silicon dioxide, do not conduct electricity, unlike giant structures with *delocalised electrons* such as metals and graphite.

locating agent: a reagent used in *thin-layer chromatography* or *paper chromatography* to find the position of spots of colourless chemicals on a chromatogram. The locating agent reacts with the colourless chemical to form a coloured spot. Iodine vapour can be used as a general locating agent for organic compounds. Ninhydrin is used as the locating agent for *amino acids*.

logarithms in chemistry are of two kinds, logarithms to base 10 (lg) and natural logarithms to base e (ln).

Chemists use logarithms to base 10 to handle values which range over several orders of magnitude. Logs to base 10 are defined such that:

- $\lg 1000 = 3$
- $\lg 100 = 2$
- $\lg 10 = 1$
- $\lg 1 = 0$
- $\lg 0.1 = -1$
- $\lg 0.01 = -2$

In general $\lg 10^x = x$

The hydrogen ion concentration in aqueous solutions typically ranges from 1 to 10^{-14} mol dm^{-3}. The definition of pH ($-\lg [H^+(aq)]$) covers this range in a scale running from 0 to 14.

Also by definition: $\lg xy = \lg x + \lg y$ and $\lg x^n = n \lg x$

so $\lg 1/x = \lg x^{-1} = -\lg x$.

Natural logarithms appear in relationships in thermochemistry and chemical kinetics. They follow similar mathematical rules as logarithms to base 10. In general $\ln e^x = x$.

London forces are the weakest *intermolecular forces*. The forces are sometimes called induced dipole–induced dipole attractions because they arise from the temporary dipoles that form in non-polar molecules as a result of the fleeting repulsions and attractions between the nuclei of the atoms and the surrounding clouds of electrons. London forces are the attractive forces that cause non-polar molecules to condense to liquids and to freeze into solids when the temperature is lowered.

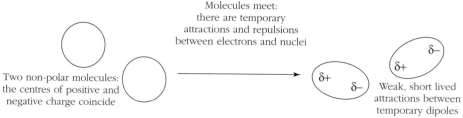

An illustration of the origin and effect of London forces as a result of the formation of temporary, induced dipoles in molecules

lone pair of electrons: a pair of electrons in the outer shell of one of the atoms in a molecule or ion which is not involved in bonding.

$$\text{H}-\overset{\displaystyle \text{H}}{\underset{\displaystyle \bullet\bullet}{\overset{\displaystyle |}{\text{O}}}}: \qquad \text{H}-\overset{\displaystyle \text{H}}{\underset{\displaystyle \bullet\bullet}{\overset{\displaystyle |}{\text{N}}}}-\text{H} \qquad \text{H}-\overset{\bullet\bullet}{\underset{\bullet\bullet}{\text{O}}}:^{-} \qquad :\overset{\bullet\bullet}{\underset{\bullet\bullet}{\text{Br}}}:^{-}$$

Examples of molecules and ions with lone pairs of electrons

Lone pairs of electrons:
- affect the *shapes of molecules*
- form *dative covalent bonds*
- are important in the chemical reactions of compounds such as *water* and *ammonia*
- are important in *nucleophilic addition* and *nucleophilic substitution* reactions.

(See also *Lewis acid–base theory*.)

lubricants are used to reduce friction wherever one surface slides over another. Lubricating oils for engines are usually based on oil fractions with molecules in the size range C_{20} to C_{34}. Synthetic lubricants for specialist purposes include *silicones*.

luminescence is a term which includes all the examples of substances emitting light at low temperatures, in contrast to incandescence, which is the emission of light by objects when they are hot, such as the filament of an electric light bulb.

Examples of luminescence are:
- fluorescence – the immediate emission of light by a material exposed to other radiation such as UV radiation, X-rays or an electron beam
- phosphorescence – delayed and slow emission of light from a material after exposure to other forms of radiation
- chemiluminescence – light given out during chemical changes (such as from light sticks, fireflies or glow-worms)
- triboluminescence – light given out when crushing or breaking a material (such as the flashes of light seen in a darkened room when crushing sugar or opening a self-adhesive envelope).

Aiming for a grade A*?

Don't forget to log on to **wwwphilipallan.co.uk/a-zonline** for advice.

235

magnesium (Mg) is a reactive metallic element in *group 2* of the periodic table. Its electron configuration is [Ne]3s². Samples of the silvery-white metal usually look grey because they are covered with a layer of magnesium oxide.

Magnesium is an essential element in biological systems where it is found as complexes of the Mg^{2+} ion. Chlorophyll, the green pigment required for photosynthesis in plants, has structural similarities to the haem group in haemoglobin but the ion at the centre of the molecule is a magnesium ion.

The commercial source of magnesium metal is *electrolysis* of molten magnesium chloride obtained either from seawater or from salt deposits. The low density of the metal helps to make light *alloys*, especially with aluminium. These alloys, which are strong for their weight, are especially useful for cars and aircraft. The reactivity of magnesium makes it suitable as the 'sacrificial' metal in anodes used to protect pipelines and ships from *corrosion*.

Magnesium burns very brightly in air with an intense white flame forming the white solid magnesium oxide (MgO). It is used in fireworks and flares.

Magnesium reacts very slowly with cold water but much more rapidly on heating in steam:

$$Mg(s) + H_2O(g) \rightarrow MgO(s) + H_2(g)$$

Magnesium forms ionic compounds with non-metals in which the metal is in the +2 oxidation state as Mg^{2+}.

Magnesium oxide is a white solid made by heating magnesium carbonate. It is a *basic oxide*. In water it turns to magnesium hydroxide which is slightly soluble. The *antacid* milk of magnesia is a suspension of magnesium hydroxide in water. Magnesium oxide has a high melting point and is used as a refractory ceramic to line furnaces (see *refractories*).

The reactions of the metal, its oxide or its carbonate with dilute acids provide a direct method for making soluble magnesium salts.

$$H_2SO_4(aq) \begin{array}{c} \xrightarrow{+ \ Mg(s)} MgSO_4(aq) \ + \ H_2(g) \\ \xrightarrow{+ \ MgO(s)} MgSO_4(aq) \ + \ H_2O(l) \\ \xrightarrow{+ \ MgCO_3(s)} MgSO_4(aq) \ + \ CO_2(g) \ + \ H_2O(l) \end{array}$$

Three reactions which produce magnesium sulfate

Epsom salts consist of hydrated magnesium sulfate, $MgSO_4.7H_2O$, which is a laxative.

magnetic resonance imaging (MRI) applies the principles of nuclear magnetic resonance (see *NMR spectroscopy*) to provide images of the structure and function of the body for medical diagnosis and research. MRI provides much better images of soft tissues than methods based on X-rays. This makes the technique especially useful for studying the brain and cardiovascular system.

The person to be investigated lies inside a very powerful magnet. The magnetic field tends to align the protons in water molecules in the body. A radiofrequency signal is used to detect energy jumps as protons line up with or against the magnetic field. A computer analyses the signals to produce images.

malleable: metals are malleable if they can be hammered into shape without breaking. It is possible to make very thin sheets of a malleable metal such as gold by hammering.

manganese (Mn) is a hard, grey brittle metal, a *d-block element* with the electron configuration $[Ar]3d^5 4s^2$.

The main oxidation states of manganese are:

- +7 in MnO_4^- – the purple manganate(VII) ion, which is a strong oxidising agent especially in acid solution
- +4 in MnO_2 – an insoluble, black compound which is an oxidising agent in acid solution
- +2 in Mn^{2+} – the pink manganese(II) ion in salts such as manganese(II) sulfate.

Under special conditions it is also possible to produce solutions with red manganese(III) ions or green manganate(VI) ions. Manganate(VI) is stable in alkaline solution but *disproportionates* to manganate(VII) and manganese(IV) oxide on adding acid. (See also *potassium manganate(VII)*.)

M

margarine is a solid emulsion consisting of water finely dispersed in *vegetable oils*, with a high enough proportion of hardened oils to make the product a spreadable solid. *Hydrogenation* is used industrially to add hydrogen to double bonds in oils. This produces saturated fats which are solid at room temperature – hence the description of the process as 'hardening'. Manufacturers bubble hydrogen though the oil in the presence of finely divided nickel, which is the *catalyst*. The formation of *trans fats* during hardening means that manufacturers now generally use interesterification to raise the melting points of *triglycerides* for use in spreads.

Markovnikov's rule: a rule which predicts the main product when a compound HX (such as H—Br, H—OSO$_3$H or H—OH) adds to an unsymmetrical *alkene* (such as propene). The rule is that the hydrogen atom adds to the carbon atom that already has the most hydrogen atoms attached to it. This pattern was first reported by the Russian chemist Vladimir Markovnikov who studied a great many alkene *addition reactions* during the 1860s.

Markovnikov's rule

The mechanism for electrophilic addition helps to account for this rule. When HBr is added to propene there are two possible intermediate *carbocations*. The secondary carbocation is preferred because it is slightly more stable than the primary carbocation. The secondary ion has two alkyl groups pushing electrons towards the positively charge carbon atom (see the *inductive effect*). This helps to stabilise the ion by 'spreading' the charge over the ion.

Markovnikov's rule: an explanation

mass (symbol m) is a fundamental physical quantity. The *SI unit* of mass is the kilogram (kg). 1 kg = 1000 g.

Mass can be determined by measuring the pull of gravity on a specimen. This is how chemists usually measure the masses of chemicals with a chemical balance.

Alternatively masses can be found by seeing how much an object accelerates when a force acts on it. This is how chemists measure the masses of ionised atoms and molecules in *mass spectrometry*.

mass number (A) is the total number of protons and neutrons in the nucleus of an atom. The *isotopes* of an element have the same atomic number but different mass numbers. The alternative term is nucleon number.

mass spectrometry: an accurate instrumental technique for determining *relative atomic masses* and *relative molecular masses*. Mass spectrometry can also help to determine molecular structures and to identify unknown compounds.

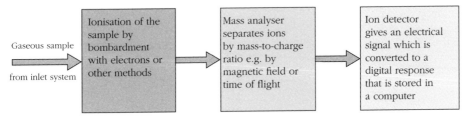

Schematic diagram to show the key features of a mass spectrometer

Inside a mass spectrometer there is a high vacuum. This allows ionised atoms and molecules from the chemical being tested to be studied without interference from atoms and molecules in the air.

In the instrument, a beam of high-energy electrons bombards the molecules of the sample. This turns them into ions by knocking out one or more electrons.

There are five main types of mass spectrometer. They differ in the method used to separate ions with different mass-to-charge ratios. One type uses an electric field to accelerate ions into a magnetic field which then deflects the ions onto the detector. A second type accelerates the ions and then separates them by their flight time through a field-free region. A third type, the so-called transmission quadrupole instrument, is now much the most common because it is very reliable, compact and easy-to-use. It varies the fields in the instrument in a subtle way to allow ions with a particular mass-to-charge ratio to pass through to the detector one at a time.

The mass spectrum for an element shows the relative abundance of different *isotopes* of the element. This makes it possible to calculate the relative atomic mass for the element.

Bombarding molecules with high-energy electrons not only ionises them but usually splits them into fragments. As a result the mass spectrum generally consists of a 'fragmentation pattern'. Molecules break up more readily at weak bonds or at bonds which give rise to stable fragments. The highest peaks correspond to positive ions which are relatively more stable, such as tertiary *carbocations* or ions such as RCO^+ (the acylium ion) or the fragment $C_6H_5^+$ from aromatic compounds.

When molecular compounds are being analysed, the peak of the ion with the highest mass is usually the whole molecule, ionised. So the mass of this 'parent ion' is the relative molecular mass of the compound. High-resolution mass spectrometry can be used to the determine *molecular formula* of a compound.

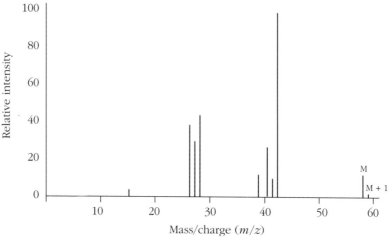

Mass spectrum of butane, C_4H_{10}. The mass to charge ratio (m/z) is the ratio of the relative mass of an ion to its charge where z is the number of charges. Usually z = 1. The peak at m/z = 58 corresponds to the parent ion. The pattern of fragments is characteristic of this compound. The very small peak at 59 is the 'M + 1' peak, which is present because of the presence of carbon-13 atoms. Carbon-13 makes up 1.1% of natural carbon so in a molecule with four carbon atoms the 'M + 1' peak is 4.4% of the molecular ion peak.

The presence of isotopes shows up in spectra of organic compounds containing chlorine or bromine atoms. Chlorine has two isotopes ^{35}C and ^{37}C. Chlorine-35 is three times more abundant than chlorine-37. If a molecule contains one carbon atom its molecular ion appears as two peaks separated by two mass units. The peak with the lower value of m/z is three times higher than the peak with the higher value.

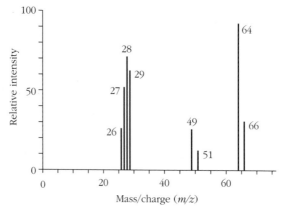

Mass spectrum of chloroethane

Bromine consists largely of two isotopes, ^{79}Br and ^{81}Br, which are roughly equally abundant. If a molecule contains one bromine atom the molecular ion shows up as two peaks separated by two mass units with roughly equal intensity.

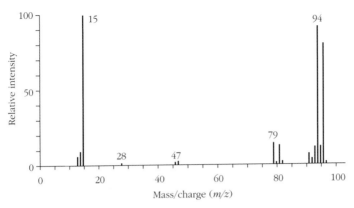

Mass spectrum of bromomethane

Chemists who synthesise new compounds can study their fragmentation patterns. They identify the fragments from their masses and then piece together likely structures with the help of evidence from other methods of analysis such as *infrared spectroscopy* and *NMR spectroscopy*.

The combination of *gas chromatography* (GC) with mass spectrometry is of great importance in modern chemical analysis. First, GC separates the chemicals in an unknown mixture, such as a sample of polluted water; then mass spectrometry detects and identifies the components (see *GC–MS*).

Maxwell–Boltzmann distribution: the distribution of molecular kinetic energies for a gas at a particular temperature.

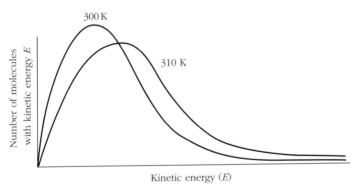

Maxwell–Boltzmann distribution of molecular kinetic energies in a gas at two temperatures

The modal speed gets higher as the temperature rises. The area under the curve gives the total number of molecules. This does not change as the temperature rises so the peak height falls as the curve widens.

The Boltzmann distribution is important in the *collision theory* of reaction rates and helps to account for the effects of temperature changes and *catalysts* on *rates of reaction*.

measurement uncertainty arises from *errors of measurement* which mean that there are unavoidable differences between measured values and the true values. Measurement uncertainty in chemical measurements can arise from a range of sources such as:

• the purity of the chemicals used as standards – suppliers of chemicals provide a range of grades of chemicals, including analytical grades which are specially pure for quantitative work

- features of the measuring devices – such as the precision of balances or thermometers, the slight variations in the calibration of glassware, and the difficulty of reading from scales
- aspects of procedure – such as changes in the laboratory temperature which affect the volume of glassware which is calibrated at a particular temperature
- the behaviour of the person carrying out the procedure – such as making the judgement of when the indicator colour change corresponds to the end-point.

Examples of measurement uncertainty for laboratory glassware:
- 50 cm³ grade B burette; ±0.08 cm³ when used to deliver 25 cm³
- 25 cm³ grade B pipette; ±0.04 cm³
- 50 cm³ measuring cylinder; ±0.6 cm³.

Uncertainties in other measurements such as mass, temperature, time and pH depend on the quality and precision of the instruments available. The overall uncertainty of a balance, for example, combines uncertainties arising from its precision, calibration and the readability of the scale. For a 2-place balance a reasonable estimate of the total uncertainty when used correctly in school or college is ± 0.01 g while for a 3-place balance the uncertainty can be taken as ± 0.001 g.

The uncertainty in measuring a temperature with a common 0–100°C thermometer may be as much as ± 1°C. This may not lead to significant error if the quantity used in calculating a result is the absolute temperature in kelvin. This would only introduce a percentage uncertainty of about ± 0.03% in a value of 300 K. However, in a thermochemistry experiment to measure a temperature rise of 5 K the same uncertainty gives rise to a percentage error of 20%, which is too high. In thermochemistry it is important to use a thermometer graduated in fifths or tenths of a degree. The measurement uncertainty with a thermometer of this kind can be taken to be ±0.1°C.

In most quantitative experiments the final results are calculated from a number of measurements. The total uncertainty is determined by combining the individual uncertainties. In an advanced chemistry course, when adding or subtracting measurements it is acceptable to add the measurement uncertainties.

In relationship where quantities are multiplied or divided, the values generally have different units. Here the first step is to calculate the percentage uncertainties.

$$\text{percentage uncertainty} = \frac{\text{uncertainty in the value}}{\text{value}} \times 100\%$$

The percentage uncertainties are ratios. They do not have units and can be added to arrive at a total percentage uncertainty for a calculated result.

mechanism of a reaction: a description of how a reaction takes place showing step by step the bonds which break and the new bonds which form.

Some mechanisms involve *homolytic bond breaking* with *free radical* intermediates. Other mechanisms involve *heterolytic bond breaking* and ionic intermediates.

Mechanisms with ionic intermediates can be classified according to the type of attacking molecules or ions involved. *Nucleophiles* attack atoms at the δ+ end of *polar covalent bonds*. *Electrophiles* attack regions which are electron rich, especially double bonds and the delocalised electrons in arenes.

Evidence used to support a proposed mechanism include:

- the *rate equation* for the reaction which indicates which species are involved in the *rate-determining step*.
- identification of *intermediates* using *spectroscopy* or other methods
- use of *isotopic labelling* to track what happens to particular atoms during a reaction.

medicine: a substance or mixture of substances used to treat a disease or to give relief from the symptoms of disease. A medicine normally contains one or more active *drugs* dissolved in water or mixed with an inert solid. Mixing the active ingredient with an inert material makes it easier to give an accurate dose.

Doctors must prescribe some medicines – these are 'prescription only medicines' (POMs). New medicines may start in this category but later be made available 'over the counter' once they have been proved safe.

Medicines which people can buy from a pharmacist without a prescription are called over-the-counter (OTC) drugs. The sale of some of these medicines has to be supervised by a pharmacist – these are P medicines. People can buy other medicines such as mild painkillers with no supervision – these are GSL medicines, as they are on the 'general sale list'.

For a medicine taken by mouth, added flavouring may make the medicine more palatable and colouring can help to identify a drug. Other drugs are taken by injection or drawn into the lungs by a deep breath.

The pharmaceutical industry employs many chemists in *drug development* for medicines.

melting point: melting is a change of state from a solid to a liquid. Another word for melting is fusion. The melting point for a pure substance is the temperature at which the solid and liquid are in equilibrium. It is the same temperature as the *freezing point*. Melting points vary with pressure but only very slightly.

Pure compounds have sharp melting points. Measuring melting points is a way of checking the purity and identity of compounds. This technique is especially important in *organic chemistry*.

Molecular substances melt at relatively low temperatures because the intermolecular forces between molecules are weak. Giant structures, with strong bonding between the atoms, have high melting points. Energy is needed to overcome the bonds between particles as a substance melts (See *enthalpy change of melting*.)

meniscus: the curved surface of a liquid in a tube. Water forms a concave meniscus in a glass tube because water molecules can form *hydrogen bonds* with oxygen atoms at a clean glass surface. This allows water to *wet* the glass. Mercury atoms bond strongly with themselves but not with atoms in glass, so mercury has no tendency to wet glass and forms a convex meniscus.

Chemists have to allow for the shape of the meniscus when calibrating and reading graduated glassware such as pipettes and graduated flasks. The correct procedure is to adjust the level until the bottom of the meniscus just touches the mark as seen from eyes on a level with the mark.

mercury (Hg) is the only liquid metal at room temperature. Mercury vapour and mercury compounds are very *toxic chemicals*. The danger from metallic mercury comes from breathing in its vapour. This can damage the nervous system. Mercury salts damage the kidneys if they enter the body. However, the main harm for people, worldwide, comes from organic mercury. Mercury and mercury compounds in industrial wastes can get into the sea where the pollutants are taken up by shellfish and fish to be converted to organic mercury in living tissues. The mercury compounds accumulate up the food chain. Organic mercury damages the nervous system and can also cause birth defects.

metal extraction involves two main types of process to reduce metal compounds to metals:

- pyrometallurgy – reactions at high temperature often above 1000°C;
    - electrolysis of a molten compound (for example *aluminium extraction*)
    - chemical reduction by coke in a blast furnace (for example *iron extraction*)
    - chemical reduction by a more reactive metal (for example *titanium extraction*)
- hydrometallurgy – reactions at low temperature in solution in water;
    - electrowinning using *electrolysis* of an aqueous solution (for example *zinc extraction*)
    - cementation using a *displacement reaction* (for example using iron to displace copper from a solution of copper(II) sulfate).

(See also *extraction of metals.*)

metallic bonding: the strong bonding between atoms in *metal* crystals. Each metal atom in a metal crystal contributes electrons from its outer shell to a 'sea' of *delocalised electrons*. Only elements with relatively low first *ionisation energies* form metallic crystals.

By losing electrons the metal atoms become positively charged. So a metal crystal consists of positively charged metal atoms held together by a 'sea' of shared electrons.

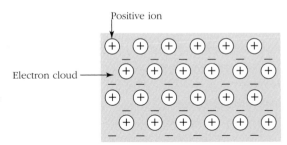

Simplified picture of metallic bonding – a giant lattice of metal ions in a sea of delocalised electrons

Metals conduct because the shared bonding electrons can drift through the crystal structure from atom to atom when there is an electric potential difference. The strength of metallic bonding means that most metals melt at relatively high temperatures.

Metals can bend without breaking because metallic bonding is not highly directional. Lines or layers of metal atoms can shift their position in a crystal without the bonds breaking.

metalloid: an element with properties which are in between those of metals and non-metals. One example of such intermediate behaviour is that metalloids are *semiconductors*. Their electrical conductivities are between those of metal conductors and non-metal insulators.

In *group 4*, the metalloid germanium comes between two non-metals at the top of the group and two metals at the bottom. Other examples of metalloids are *arsenic*, antimony and tellurium. Some chemists also classify silicon and boron as metalloids.

metals are elements on the left-hand side of the *periodic table* with one, two or three electrons in the outer shell which take part in bonding and chemical reactions. The atoms in the *crystal structures of metals* are held together by *metallic bonding*.

Physically, metals are:
- shiny when freshly polished and free of corrosion
- good conductors of electricity and thermal energy
- *malleable* and *ductile*
- (usually) solids with high melting and boiling points (only six metals melt below 100°C – mercury, which is a liquid, gallium and four group 1 metals sodium, potassium, rubidium and caesium).

Chemically, metals tend to:
- lose electrons forming positive ions (they are *reducing agents*)
- form *basic* or *amphoteric oxides* and hydroxides (which are *alkalis* if soluble in water)
- form solid, ionic chlorides.

The more an element shows these properties the greater its 'metallic character'.

methanal ($H_2C=O$) is the simplest *aldehyde*. It is also called formaldehyde. It is manufactured from *methanol* by reaction with oxygen in the presence of a silver catalyst at about 500°C. Methanal is used to make *thermosetting polymers* by reaction with melamine, urea or phenol. Formalin is an aqueous solution of methanal which can be used to preserve animal specimens.

methane hydrate: see *clathrate compounds*.

methanol (CH_3OH) is a liquid alcohol which is important as a feedstock for the chemical industry. Industry can manufacture methanol from the mixture of *carbon monoxide* and hydrogen produced by *steam reforming*:

$$CO + 2H_2 \rightleftharpoons CH_3OH \qquad\qquad \Delta H = -91 \text{ kJ mol}^{-1}$$

The yield of methanol at equilibrium is favoured by high pressures and low temperatures. A catalyst made from copper and zinc oxides on aluminium oxide pellets converts 5–10% of the reactants to methanol at 250°C and a pressure in the range 0–100 atmospheres.

Oxidation of methanol produces the *methanal* required to make thermosetting resins. Methanol is also used for *ethanoic acid manufacture* and for the *transesterification* of triglycerides to make *biodiesel*.

Methanol is very toxic. Drinking relatively small volumes can cause blindness and death. Methanol is mixed with *ethanol* to make it undrinkable.

methylbenzene is an *arene* with a methyl group substituted for hydrogen in the benzene ring. Methylbenzene is more reactive than benzene towards *electrophilic substitution* because of the *inductive effect* of the methyl group. Chlorination of the benzene ring takes place in the presence of a *halogen carrier* at 20°C to give a mixture of 1-methyl-2-chlorobenzene and 1-methyl-4-chlorobenzene. Bromination happens in a similar way and gives similar products.

Bubbling chlorine through boiling methylbenzene in ultraviolet light leads to chlorination of the methyl side chain by a *free radical* mechanism. The result is a succession of substitution products starting with (chloromethyl)benzene. Bromination, under similar conditions, also leads to side-chain substitution.

Heating a mixture of an arene carrying a hydrocarbon side chain with alkaline potassium manganate(VII) oxidises the side chain to a carboxylic acid group.

methylbenzene → benzoic acid

(1) $KMnO_4/OH^-$ and heat
(2) acidify

Oxidation of the hydrocarbon side chain of methylbenzene

micelle: a cluster of *surfactant* molecules in a solution. A micelle has the 'water-hating' (hydrophobic) hydrocarbon chains tangled together inside with the 'water-loving' (hydrophilic) ends of the molecules on the outside. The inside of a micelle is non-polar. Unlike water, a solution of a surfactant in water can dissolve oil and grease. The molecules of oily dirt diffuse to the insides of the micelles.

microwaves: *electromagnetic radiation* with wavelengths in the range 1 mm to 30 cm and frequencies in the range 10^8 to 10^9 Hz. Polar molecules absorb microwave radiation. The quanta of microwave radiation correspond to the energy jumps when molecules gain energy and rotate faster (see *quantum theory*).

In chemistry, microwaves are taking over in many contexts where controlled heating is important. This is especially true in research and development laboratories of pharmaceutical and biotechnology industries. Microwave heating can contribute to *green chemistry* by helping chemists to achieve cleaner and more efficient chemical reactions with higher yields, compared to traditional heating methods. Heating with microwaves is very efficient because they heat only the chemicals and not the apparatus or its surroundings.

Polar molecules in a material absorb microwave energy as they rotate backwards and forwards. Internal friction resulting from intermolecular forces spreads the energy and the molecules get hotter as they rotate and vibrate faster. This type of heating can speed up some reactions to a remarkable degree because the energy is absorbed directly by the molecules. Microwave heating during the synthesis of aspirin completes the process 100 times faster than conventional heating. In catalytic reactions, it is thought that microwaves selectively heat the catalyst surface to a high temperature greatly increasing the catalytic effect.

Because microwaves heat the chemicals directly, it is often possible to cut down the quantity of solvent needed for a reaction in solution. Sometimes no solvent is needed at all. This can be achieved by absorbing the reactants onto a porous form of an inert material such as alumina. Working in such ways can help to cut down the materials and energy needed to purify the product.

millilitre (symbol ml) is a unit of *volume* often used on chemical glassware, in medicine and domestically. A millilitre is one thousandth of a *litre*.

$$1000 \text{ ml} = 1000 \text{ cm}^3 = 1 \text{ dm}^3 = 1 \text{ litre}$$

minerals are the naturally occurring elements and compounds in the crust of the Earth. Most rocks are mixtures of minerals. There are about 3000 minerals of which 300 or so are important to the mineral and chemical industries.

Many *ores* contain only a small percentage of the useful mineral. The purpose of mineral processing is to separate the valuable mineral ('values') from the waste rock ('gangue').

Processing generally starts by crushing and grinding the ore to break it up and separate the mineral grains. The methods used to separate the 'values' include:
- *mechanical* separation, often based on differences in *density*
- *froth flotation*
- chemical extraction by *leaching*.

mirror image molecules: see *chiral compounds*.

miscible liquids are liquids which mix with each other. The general rule is that 'like dissolves like'. This means that:
- non-polar liquids mix freely with other non-polar liquids (for example, *hydrocarbons* mix freely, as they do in *petrol*)
- polar liquids mix with polar liquids (for example ethanol mixes with water)
- but polar and non-polar liquids do not mix (petrol floats on water).

mixed oxides are oxides which consist of a lattice of oxide ions (O^{2-}) with two or more types of metal ions, M^{n+}, in the spaces between the negative ions. So they are not mixtures of oxides but distinct compounds with a mixture of metal ions. An example of a mixed oxide is red lead (Pb_3O_4), which has a crystal lattice made up of ions in the proportions $2Pb^{2+}$: Pb^{4+}: $4O^{2-}$. Another example is magnetite (Fe_3O_4), the magnetic ore of iron which contains $2Fe^{3+}$ ions for each Fe^{2+} ion in the lattice,

mixtures are many and varied but all consist of two or more substances mixed up together. Unlike compounds, which have a definite formula, the composition of mixtures can vary widely.

The study of mixtures is very important because chemists have to know how to:
- separate mixtures – by *chromatography*, *distillation*, *filtration*, *froth flotation*, and *solvent extraction*
- formulate new mixtures for special purposes – when making *medicines*, *paints*, *detergents*, *dyes*, *pesticides*, *fertilisers* and many other products.

mobile phase: the liquid or gas which moves over the stationary *phase* during *chromatography*.

models are very important to the theory and practice of chemistry. Theoretical models are used in chemical explanations including:
- the *kinetic theory of gases*
- *collision theory*
- models of bonding (see *valency theory*).

Chemists also use physical *molecular models* and computer-generated models to represent atoms and molecules when exploring *stereochemistry* and *mechanism of the reaction* (see also *computational chemistry*).

Mohs scale of hardness: a scale for comparing the *hardness* of materials and minerals. The scale is based on scratching one substance with another. A harder material scratches a less hard material. The scale ranges from 1 for talc which is very soft, to 10 for diamond which is very hard.

molar concentration: see *concentration*. In some books you will see the term 'molarity' used for the molar concentration. Chemists still sometimes write, for example, 2.0 M to describe a solution with a molar concentration of 2.0 mol dm^{-3}.

molarity: see *molar concentration*.

molar mass: the mass per mole of a chemical (see *amount of substance*). The symbol for the molar mass is M. The unit is g mol⁻¹. As always with molar amounts the chemical *entity* must be specified. The molar mass of hydrogen atoms, $M(H) = 1$ g mol⁻¹. The molar mass of hydrogen molecules $M(H_2) = 2$ g mol⁻¹.

The molar mass for the atoms of an element is numerically equal to the *relative atomic mass* of the element. The relative atomic mass of carbon is 12. The molar mass of carbon atoms = 12 g mol⁻¹.

The molar mass of a substance is calculated by adding up the molar masses of the atoms in the formula.

$$M_r(H_2O) = [(2 \times 1 \text{ g mol}^{-1}) + 16 \text{ g mol}^{-1}] = 18 \text{ g mol}^{-1}$$
$$M_r(CuSO_4) = [63.5 \text{ g mol}^{-1} + 32 \text{ g mol}^{-1} + (4 \times 16 \text{ g mol}^{-1})] = 159.5 \text{ g mol}^{-1}$$
$$M_r(CuSO_4.5H_2O) = [159.5 \text{ g mol}^{-1} + (5 \times 18 \text{ g mol}^{-1})] = 249.5 \text{ g mol}^{-1}$$

Molar masses can be measured experimentally by *mass spectrometry*.

Another method for measuring molar masses applies to gases and to liquids that are easy to vaporise. One procedure is to inject a weighed sample of liquid into a syringe heated in an oven.

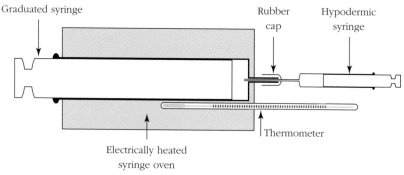

Syringe method for determining molar masses of volatile liquids

Measurements taken include the *volume* of vapour, the temperature of the vapour and its *pressure* (the *atmospheric pressure*). Measurements are converted to *SI units* and then substituted in the ideal gas equation to find the amount in moles, n.

Worked example

A 0.124 g sample of a liquid evaporated to give 45 cm³ of vapour at 100°C and a pressure of 1 atmosphere. What is the molar mass of the liquid?

Notes on the method

Convert all units to SI units. Substitute in the equation $pV = nRT$ to find n (the amount in moles).

Answer

Pressure = 101.3×10^3 N m⁻²

Volume = 45×10^{-6} m³

Temperature = 373 K

The gas constant = 8.31 J K⁻¹ mol⁻¹

$$n = \frac{pV}{RT} = \frac{101.3 \times 10^3 \text{ Nm}^{-2} \times 45 \times 10^{-6} \text{ m}^3}{8.31 \text{ J K}^{-1} \text{ mol}^{-1} \times 373 \text{ K}} = 1.47 \times 10^{-3} \text{ mol}$$

$$\text{molar mass} = \frac{\text{mass of sample}}{\text{amount in moles}} = \frac{0.124 \text{ g}}{1.47 \times 10^{-3} \text{ mol}} = 84 \text{ g mol}^{-1}$$

molar volume: the volume of one mole of a gas. The molar volume of any gas is the same under given conditions of temperature and pressure. This follows from *Avogadro's law* or from the *ideal gas* equation. Rearranging the ideal gas equation, $pV = nRT$ gives:

$$V = \frac{nRT}{p}$$

This shows that, for one mole of gas ($n = 1$), the volume (the molar volume) depends only on the temperature and pressure since R is a constant. The molar volume of an ideal gas at *stp* (273 K and 101.3×10^{-3} N m^{-2}) = 2.24 dm^{-3} mol^{-1} (= 2.24×10^{-2} m^3 mol^{-1}).

mole: the chemist's term for an amount of substance containing a particular number of atoms, molecules, ions or any other type of particles. The word entered the language of chemistry at the end of the nineteenth century and is based on the Latin word for a heap or pile.

The mole (unit: mol) is the *SI unit* for *amount of substance*. One mole is the amount of substance that contains as many specified *entities* (atoms, molecules, ions, electrons and so on) as there are atoms in exactly 12 g of the carbon-12 *isotope*. The number of entities per mole is the *Avogadro constant*.

mole fraction: a way of measuring the proportion of a substance in a mixture where it is the amounts in moles which is important and not the chemical nature of the components.

In a mixture of n_A moles of A with n_B moles of B and n_C moles of C, the mole fractions (symbol X) are given by the following:

$$X_A = \frac{n_A}{n_A + n_B + n_C}; \quad X_B = \frac{n_B}{n_A + n_B + n_C}; \quad X_C = \frac{n_C}{n_A + n_B + n_C}$$

So the mole fraction of B is the fraction of all the moles which are moles of B.

The sum of all the mole fractions is 1, so $X_A + X_B + X_C = 1$.

Mole fractions are used by chemists when studying gas mixtures (see *partial pressures* and K_p).

mole of electrons: see *Faraday constant*.

molecular formula: a formula which shows the number of atoms of each element in a molecule. The molecular formula of chlorine is Cl_2, of ammonia is NH_3 and of ethanol is C_2H_6O. The term molecular formula applies only to substances which consist of molecules.

A molecular formula is always a simple multiple of the *empirical formula*. Analysis shows that the empirical formula of hexene is C_3H_6. Data from low-resolution *mass spectrometry* shows that the *relative molecular mass* of hexane is 84. The relative mass of the empirical formula, $M_r(C_3H_6) = (3 \times 12) + (6 \times 1) = 42$. So the molecular formula is twice the empirical formula. The molecular formula of hexene is C_6H_{12}.

M

High-resolution mass spectrometry can be used to determine molecular formulae. The very strong forces that bind the protons and neutrons together in an atomic nucleus mean that the relative mass of an isotope is very slightly different from its whole number value. The mass of the carbon-12 nucleus is 12.0000 by definition. On this scale the accurate *relative isotopic mass* of oxygen-16 is 15.9949 (not 16) and of nitrogen-14 is 14.0031 (not 14). There are mass spectrometers that can determine m/z values accurately to four decimal places, making it possible to distinguish different formulae which would normally be considered to have the same relative molecular mass.

Formula	C_6H_{12}	C_5H_8O	$C_4H_8N_2$
Relative molecular mass	84.0939	84.0575	84.0688

molecular models: these are physical and computer models that are used to show the atoms in molecules, the bonding between them and the shapes of molecules in three dimensions. In many physical models there is an agreed colour code: C – black, H – white, O – red, N – blue and Cl – green. Computer-generated models often use different colours. (See *ball-and-stick models* and *space-filling models*.)

molecular orbital: a region in space in a molecule where there is a high probability of finding bonding electrons. Molecular orbitals, like *atomic orbitals*, are solutions to the Schrödinger wave equation. At a simplified, descriptive level, a molecular orbital forms by the combination of atomic orbitals of the atoms linked by a covalent bond.

Two *s orbitals*, for example, can overlap so that their wave functions interfere constructively to give rise to a bonding orbital. Electrons in the bonding orbital lie between the two nuclei, attract both nuclei and have a lower energy than they do when confined to atomic orbitals.

The two orbitals can also interfere destructively to give rise to an antibonding orbital. Electrons in an antibonding orbital are excluded from the region between the nuclei and have a higher energy than they do when in atomic orbitals.

Molecular orbitals fill with electrons according to the same rules as atomic orbitals.

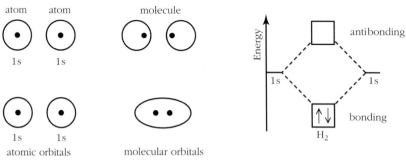

The molecular orbitals in a hydrogen molecule formed from two s orbitals in hydrogen atoms. In a hydrogen molecule both electrons are in the bonding orbital.

molecular sieve: a solid with very fine pores which can separate molecules according to size. Small molecules can fit into the holes or channels in the crystal structure while larger molecules cannot. (See *zeolites*.)

molecularity of reaction steps: the number of reacting particles taking part in one step in the *mechanism of a reaction*. Often the term refers specifically to the rate-determining step in the mechanism. The S_N1 mechanism for *nucleophilic substitution* is unimolecular. The S_N2 mechanism is bimolecular.

molecule: a group of atoms held together by covalent bonding. Most non-metals are molecular, such as H_2, N_2, P_4, S_8, Cl_2, Br_2 and I_2. Most compounds of non-metals with other non-metals are also molecular, such as H_2O, CO_2, CH_4, NH_3, HBr, $SiCl_4$, PCl_3.

The covalent bonds in molecules (*intramolecular forces*) are strong. The forces between molecules (*intermolecular forces*) are weak. Molecular substances have low melting and boiling points because of the weakness of intermolecular forces.

monobasic acid: the traditional term for a *monoprotic acid*.

monochromatic radiation is radiation of a single wavelength. Spectrometers contain a diffraction grating or prism to provide monochromatic radiation. Rotating the grating or prism allows the instrument to scan across the range of wavelengths provided by the source of radiation.

A filter in a *colorimeter* is a cheap way of selecting a narrow range of wavelengths typically with a band width of 20 to 50 nm. The disadvantage of a filter is that the operator has to change the filter to switch from one band of wavelengths to another.

Monochromatic light is also important when measuring the rotation of *plane polarised light* by optically active compounds.

monodentate ligands (unidentate ligands) are *ligands* which use a single *lone pair of electrons* to form one bond with the central metal atom in a *complex ion*. Examples of monodentate ligands are ammonia molecules, water molecules, chloride ions and cyanide ions.

monomers are the small molecules which join together in long chains to make *polymers*.

monoprotic acid: an *acid* which can give away (donate) only one *proton* per molecule. Examples of monoprotic acids are: hydrogen chloride, HCl; ethanoic acid, CH_3COOH; and nitric acid, HNO_3.

monosaccharide: see *carbohydrates*.

mordant: a chemical which fixes a dye to the fibres of a textile. Alums and other aluminium salts are used as mordants. *Dative covalent bonds* form both between —OH groups of cotton and aluminium ions and between oxygen atoms on the dye and aluminium ions.

multiple bonds are *double bonds* or *triple bonds* between atoms in molecules. In a double bond there are two shared pairs of electrons. In a triple bond there are three shared pairs.

M

names of carbon compounds: modern *IUPAC* names make it possible to work out the name from the formula and the formula from the name.

The name of an organic compound is based on the longest straight chain or main ring of carbon atoms in the skeleton of carbon atoms. If the main part is a straight chain the name is based on the corresponding *alkane*. For ring compounds the name is based on the corresponding *cyclic hydrocarbon* compound or *arene*.

Numbering the carbon atoms identifies the positions of side chains and *functional groups*, with the number repeated if there are two side groups on the same carbon atom. At all times chains are numbered from the end that gives the lowest possible numbers in the name. Numbers are omitted when there is no doubt about the position of the side chains or functional groups, as in ethanol.

Prefixes (in front) and suffixes (following) the hydrocarbon name identify the side chains and functional groups.

Where there are two or more of the same side chain or functional group the number is stated as di, tri, tetra and so on.

Prefixes are used for:
- *alkyl groups*, such as 2,3-dimethylbutane or 1,3,5-trimethylbenzene
- *halogenoalkanes*, such as 2-iodopropane or tetrachloromethane.

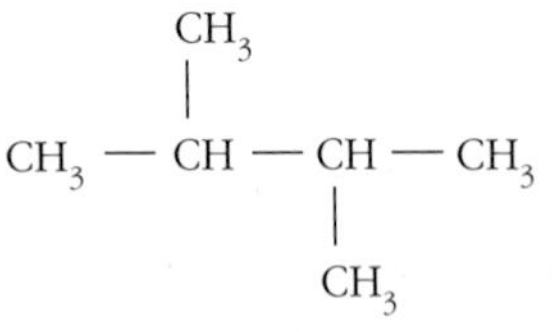

2,3-dimethylbutane

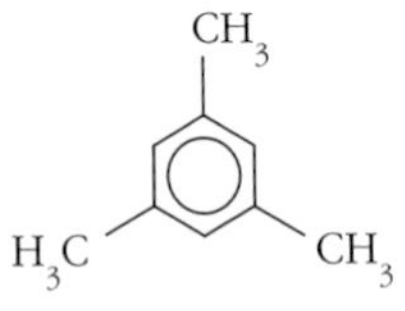

1,3,5-trimethylbenzene

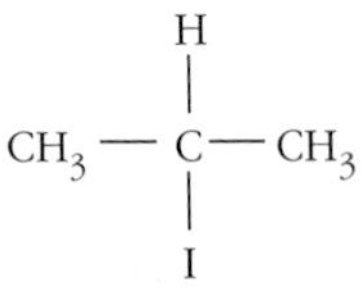

2-iodopropane

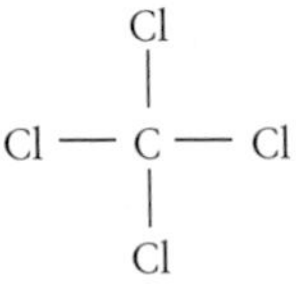

tetrachloromethane

Structures of examples named with prefixes

Suffixes are used for:
- double bonds in **alkenes**, such as but-2-ene (with a double bond between the second and third carbon atoms)
- **alcohols**, such as propane-1,2,3-triol
- **aldehydes**, such as propanal
- **ketones**, such as pentan-3-one
- **carboxylic acids**, such as hexanoic acid and ethanedioic acid
- **amines**, such as ethylamine and **phenylamine**.

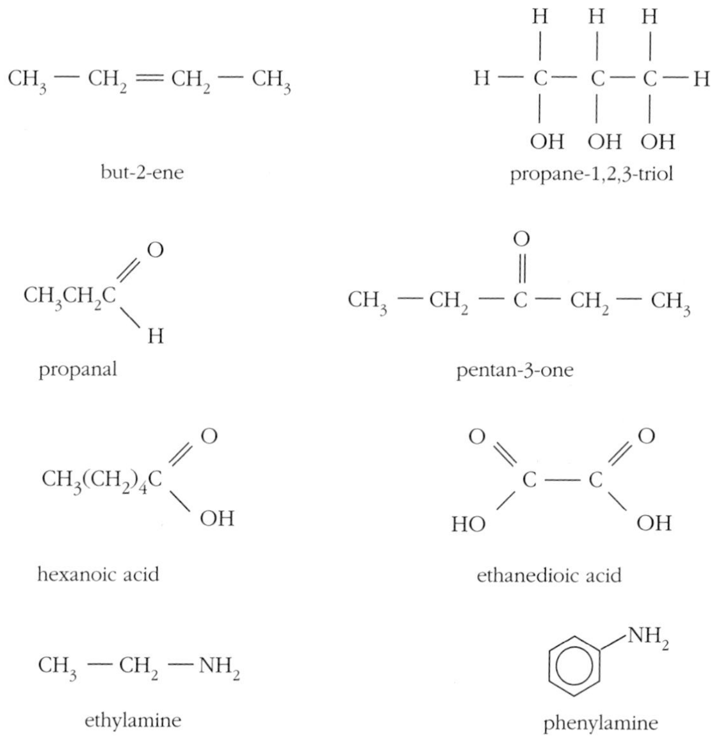

$$CH_3 - CH_2 = CH_2 - CH_3$$

but-2-ene

propane-1,2,3-triol

propanal

pentan-3-one

hexanoic acid

ethanedioic acid

$$CH_3 - CH_2 - NH_2$$

ethylamine

phenylamine

Structures of examples named with suffixes

For other examples see entries for the various series of organic compounds.

The systematic names of complex molecules can be hard to pronounce and cumbersome. Chemists still often use older, traditional names which were often based on the Latin name for the source of the compound or where it was discovered. The organic acid 2,3-dihydroxybutanedioic acid, for example, occurs in grape juice. The traditional name is tartaric acid from 'tartar' the name of the recrystallised deposits of tartaric acid salts collected from inside wine containers.

names of complex ions: the systematic names of complex ions show:
- first the number of **ligands** – 'di', 'tri', 'tetra', 'penta', 'hexa'
- then the type of ligands (in alphabetical order if there is more than one type of ligand), such as 'aqua' for water molecules, 'ammine' for ammonia, 'chloro' for chloride ions and 'cyano' for cyanide ions

- next the identity of the central metal atom in a form which shows whether or not the ion is a cation or an anion:
  - for cations (and uncharged complexes) the metal name is normal, such as silver, iron or copper
  - for anions the metal name ends in 'ate' and often has an old-fashioned style such as argentate for silver, ferrate for iron and cuprate for copper
- finally the oxidation number of the metal.

Examples:

- $[Ag(NH_3)_2]^+$ is the diamminesilver(I) ion
- $[Cu(H_2O)_6]^{2+}$ is the hexaaquacopper(II) ion
- $[CuCl_4]^{2-}$ is the tetrachlorocuprate(II) ion
- $[Fe(CN)_6]^{4-}$ is the hexacyanoferrate(II) ion
- $[Fe(CN)_6]^{3-}$ is the hexacyanoferrate(III) ion.

Note that the overall charge on the complex ion is the sum of the charges on the metal ion and the ligands.

names of inorganic compounds are becoming increasingly systematic but chemists still use a mixture of names. Most chemists prefer to call $CuSO_4.5H_2O$ hydrated copper(II) sulfate, or perhaps copper(II) sulfate-5-water but not the fully systematic name tetraaquocopper(II) tetraoxosulfate(VI)-1-water. The systematic name has much more to say about the arrangement of atoms, molecules and ions in the blue crystals but it is too cumbersome for normal use.

These are some of the basic rules for common inorganic names:

- the ending 'ide' shows that a compound contains just the two elements mentioned in the name. The more *electronegative* element comes second – for example, sodium sulfide (Na_2S) carbon dioxide (CO_2) and phosphorus trichloride (PCl_3)
- the small Roman numerals in names are the *oxidation numbers* of the elements, for example iron(II) sulfate ($FeSO_4$) and iron(III) sulfate, $Fe_2(SO_4)_3$
- the traditional names of *oxoacids* end 'ic' or 'ous' as in sulfuric (H_2SO_4) and sulfurous (H_2SO_3) acids and nitric (HNO_3) and nitrous (HNO_2) acids, where the 'ic' ending is for the acid in which the central atom has the higher oxidation number
- the corresponding traditional endings for the salts of oxoacids are 'ate' and 'ite' as in sulfate (SO_4^{2-}) and sulfite (SO_3^{2-}) and in nitrate (NO_3^-) and nitrite (NO_2^-)
- the simplified, systematic names of oxoacids and their salts are based on oxidation numbers, such that sulfurous acid is called sulfuric(IV) acid and nitrous acid is called nitric(III) acid with salts such as sodium nitrate(III).

To avoid misunderstandings, chemists give the name and formula. If necessary they may also give two names: the systematic name and the traditional name.

nanometre: the unit of length used for the sizes of atoms, molecules and ions. One nanometre (1 nm) is one thousand millionth of a metre (10^{-9} m). So there are one million nanometres in a millimetre (1000 nm = 1 mm). One nanometre is roughly five times the diameter of a hydrogen atom.

nanotechnology has developed from two key discoveries:

- the invention of the scanning tunnelling electron microscope which made it possible to observe and manoeuvre atoms and molecules
- the synthesis of nanoparticles such as *fullerenes* and carbon *nanotubes*.

The size of nanoparticles lies in the range 1 *nanometre* to 100 nm. Nanoparticles have novel properties because of their very small size. The two main reasons for changes in behaviour of materials on a nanoscale are the increase in surface area, and the dominance of quantum effects.

Traditional examples of nano-scale particles include the use of gold colloids to colour glass and ceramic glazes. Today nanoparticles are being used in sunscreens and other cosmetic products. There is concern about possible harmful effects of the use of nanoparticles in consumer products and in medicines because of their high surface reactivity and ability to cross cell membranes. The fact that nanoparticles are on the same scale as parts of living cells and larger proteins could mean that they can evade the immune system and damage cells.

nanotubes are *fullerenes* which take the form of cylinders of carbon atoms in hexagons, created as graphite-like sheets of atoms curled up into tubes. The structure may be closed at the ends with hexagons and pentagons of carbon atoms as in *buckminsterfullerene*. Nanotubes are a few *nanometres* in diameter but may be several millimetres long.

Research into the properties of nanotubes is exploring the possibility that they could be adapted to make minute diodes, transistors and other electronic devices on an atomic scale – much smaller than existing devices on silicon chips. Nanotubes are also being incorporated into plastics to make them conduct electricity. Nanotube fibres are very strong and tough. In time they may become key ingredients of new composite materials. Research is investigating the use of nanotubes to carry drugs into the body. The drug molecule can be attached to the side of the nanotube or even be placed inside it.

naphtha is a mixture of hydrocarbons from the *fractional distillation of oil*. The naphtha fraction is an important feedstock for the *petrochemical industry*. Naphtha contains *hydrocarbons* with 6 to 10 carbon atoms in their molecules.

narcotics are powerful painkillers (*analgesics*) which also lessen anxiety and give a feeling of well-being. Regular use of narcotic analgesics can lead to drug dependence. Morphine is the oldest and most well-known narcotic which is obtained from the dried juice of opium poppies (so it is an opiate). Morphine is used medically to treat patients in severe pain.

Diamorphine, otherwise known as heroin, is made from morphine. It is an even more powerful analgesic than morphine. It is also highly addictive so its medical use is limited mainly to the treatment of pain in patients who are going to die.

Codeine is another opiate. It is a much less powerful painkiller than morphine but it does not cause dependence. Unlike morphine, codeine can be taken by mouth. Codeine is an ingredient of some cough medicines because it helps to stop people coughing.

natural gas is a *fossil fuel* used for domestic heating, for raising steam in power stations and as a feedstock for the *petrochemical industry*. The composition of natural gas varies from one gas field to the next. It consists mainly of methane mixed with other *hydrocarbons* such as ethane, propane and butane together with variable amounts of carbon dioxide, nitrogen, helium and hydrogen sulfide.

Natural gas is processed before being supplied to homes. Hydrocarbons other than methane are separated for use in the petrochemical industry. Sulfur compounds are removed and a trace of a smelly chemical is added so that people notice gas leaks.

neutralisation reaction: a reaction in which an acid reacts with a base to form a *salt*:

$$HCl(aq) + NaOH(aq) \rightarrow NaCl(aq) + H_2O(l)$$

Mixing equal amounts (in moles) of hydrochloric acid with sodium hydroxide produces a *neutral solution* of sodium chloride.

Strong acids, such as hydrochloric acid, and *strong bases*, such as sodium hydroxide, are fully ionised in solution, as is the salt formed, sodium chloride. Writing ionic equations for these examples shows that neutralisation is essentially a reaction between aqueous hydrogen ions and hydroxide ions. This is supported by the values for the *enthalpy changes of neutralisation*.

$$H_3O^+(aq) + OH^-(aq) \rightarrow 2H_2O(l)$$

The surprise is that neutralisation reactions do not always produce neutral solutions. Neutralising a weak acid such as ethanoic acid with an equal amount in moles of the strong base sodium hydroxide produces a solution of sodium ethanoate, which is alkaline.

Neutralising a weak base, such as ammonia, with an equal amount of the strong acid hydrochloric acid produces a solution of ammonium chloride, which is acidic.

Where a salt has either a 'parent acid' or a 'parent base' which is weak, it dissolves to give a solution which is not neutral (see *hydrolysis of salts*). The 'strong parent' in the partnership 'wins':
- weak acid/strong base – the salt is alkaline in solution
- strong acid/weak base – the salt is acidic in solution.

neutral oxides are non-metal oxides which do not react with water to form acids. The common examples are carbon monoxide (CO), nitrogen monoxide (NO) and dinitrogen oxide (N_2O). All three of these oxides are insoluble in water.

neutral solution: a solution in which $[H^+] = [OH^-]$. At 298 K the pH of a neutral solution is 7. The pH of a neutral solution varies with temperature because the value of K_w is temperature dependent.

neutrons are the uncharged particles in the nuclei of *atoms* (see *fundamental particles*). The number of neutrons in an atom can vary without change to its chemical properties. This accounts for the existence of *isotopes*.

nickel (Ni) is a hard, greyish but shiny metal, a *d-block element* metal with the electron configuration $[Ar]3d^84s^2$. Nickel is relatively unreactive so it is used to make spatulas and crucibles. Nickel is a constituent of many alloys including some alloy *steels* and the ferromagnetic alloy alnico in permanent magnets.

The common oxidation state of nickel is +2. Nickel(II) salts, such as the sulfate, $NiSO_4$, are green.

Finely divided nickel metal is a good catalyst for *hydrogenation* reactions. It is a *heterogeneous catalyst*. It is used to 'harden' unsaturated *vegetable oils* by adding hydrogen across the double bonds.

nickel–cadmium cells are used as rechargeable 'NiCad' cells. As in a *lead–acid cells*, when a current flows the products formed at the electrodes are insoluble solids which stay put instead of dissolving in the electrolyte. This means that the electrode processes can be reversed in these electrochemical cells as the cell is recharged. The electrode processes as a NiCad cell supplies a current are:

positive electrode (reduction): $NiO(OH)(s) + H_2O(l) + e^- \rightarrow Ni(OH)_2(s) + OH^-(aq)$

negative electrode (oxidation): $Cd(s) + 2OH^-(aq) \rightarrow Cd(OH)_2(s) + 2e^-$

Unlike lead–acid cells, 'NiCad' cells must be regularly discharged fully and then recharged if they are to retain their full capacity.

Electric vehicles are using a variant of this type of cell. These nickel-metal hydride cells contain a negative electrode made of a metal alloy (M) that can absorb hydrogen instead of the cadmium.

negative electrode (oxidation): $MH(s) + OH^-(aq) \rightarrow H_2O(l) + M(s) + e^-$

nitrates are *salts* of nitric(v) acid (HNO_3) which all contain the nitrate(v) ion (NO_3^-). Nitrates are common in laboratory chemistry because they are all soluble in water. Since nitric acid is a *strong acid*, the nitrate ion is a very weak base and does not change the pH when dissolved in water.

The reason that the nitrate ion is a weak base is that the negative charge is *delocalised* over the planar ion. Delocalisation stabilises the ion (see *oxoacids*).

Heating decomposes nitrates. Hydrated nitrates first give off steam, then the usual products are the metal oxide, oxygen and the brown gas nitrogen dioxide.

$$2Mg(NO_3)_2(s) \rightarrow 2MgO(s) + O_2(g) + NO_2(g)$$

The exceptions are the nitrates of sodium and potassium which are hard to decompose and give only the nitrite and oxygen.

$$2NaNO_3(s) \rightarrow 2NaNO_2(s) + O_2(g)$$

Nitrates are *oxidising agents*. In alkaline conditions they are reduced to ammonia by aluminium metal on heating. This reaction is used as a test for nitrates since the basic ammonia gas can be detected by turning litmus paper blue (see *anion tests*). More effective than aluminium is an alloy of aluminium, zinc and copper called Devarda alloy:

$$NO_3^-(aq) + 6H_2O(l) + 8e^- \rightarrow NH_3(g) + 9OH^-(aq)$$

Plants take up nitrogen from the soil in the form of nitrate ions. So a diet rich in vegetables is rich in nitrate too. Nitrates have had a bad press because they are associated with the environmental problem *eutrophication*.

nitration of benzene is an *electrophilic substitution* reaction which takes place in the presence of a nitrating mixture of concentrated nitric and sulfuric acids. The product is nitrobenzene.

The purpose of the concentrated sulfuric acid is to produce an *electrophile* which is the nitronium ion, NO_2^+. The concentrated sulfuric acid gives a proton to the —OH group in nitric acid, which then loses a molecule of water forming the nitronium ion.

$$HO-NO_2 + H_2SO_4 \longrightarrow HO^+-NO_2 + HSO_4^-$$
$$\overset{|}{H}$$

$$\downarrow$$

$$H_2O + NO_2^+$$

Mechanism for the nitration of benzene

Nitration of arenes is important because it produces a range of useful products including the explosive TNT (trinitrotoluene, or 1-methyl-2,3,5-trinitrobenzene). Nitro compounds are also intermediates in the synthesis of chemicals used to make **polyurethanes**. Nitro groups are easily reduced to amine groups. Aryl amines, such as **phenylamine**, are important intermediates in the production of dyes, such as **azo dyes**.

nitric acid (HNO_3) is a colourless, fuming liquid in its pure state but it gradually turns yellow as it decomposes forming nitrogen dioxide. It is a highly corrosive but important chemical reagent because it can act as a:

- strong acid – nitric acid ionises fully as it dissolves in water:

$$HNO_3(l) + H_2O(l) \rightarrow H_3O^+(aq) + NO_3^-(aq)$$

  – dilute nitric acid neutralises bases producing soluble salts called **nitrates**
- powerful oxidising agent – in this role nitric acid oxidises:
  – most metals including metals such as copper which do not react with non-oxidising acids
  – non-metals, forming oxides in the highest oxidation state of the element, giving I_2O_5 with iodine, for example
  – ions in solution, such as iodide ions to iodine and iron(II) ions to iron(III)
- nitrating agent – in this role it is used for the **nitration of benzene** and other arenes.

nitric acid manufacture: a process for converting ammonia to nitric acid occurs in two stages:

- oxidation of ammonia by oxygen on the surface of a catalyst gauze made of an alloy of platinum and rhodium:

$$4NH_3(g) + 5O_2(g) \rightleftharpoons 4NO(g) + 6H_2O(g) \qquad \Delta H = -909 \text{ kJ mol}^{-1}$$

- absorption in water in the presence of oxygen to make nitric acid:

$$2NO(g) + O_2(g) \rightleftharpoons 2NO_2(g)$$
$$4NO_2(g) + O_2(g) + 2H_2O(l) \rightarrow 4HNO_3(aq)$$

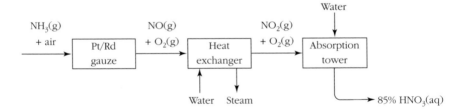

Flow diagram for the manufacture of nitric acid

Most of the nitric acid produced each year is converted to ammonium nitrate. Ammonium nitrate is largely used as a nitrogen **fertiliser**. Ammonium nitrate is also widely used as part of most **explosives** for mining and quarrying. Nitric acid is used to make other explosives such as nitrocellulose and nitroglycerine.

Nitric acid is an important reagent in the production of intermediates for making **plastics**, especially **polyamides** (such as nylon) and **polyurethanes**.

nitriles: are organic compounds with the functional group —CN. Their general formula is R—CN where R is an *alkyl* or an *aryl group.*

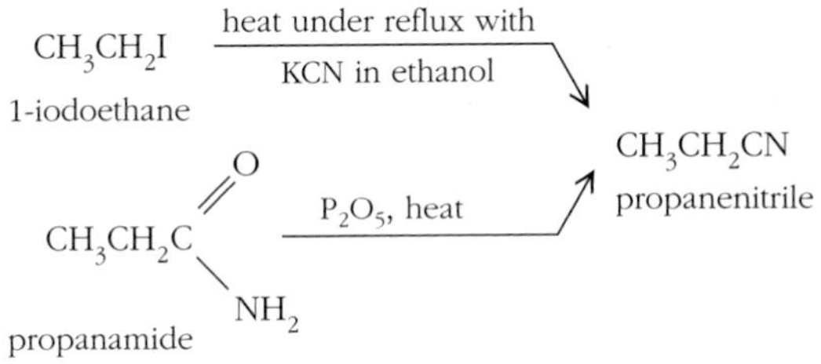

CH_3CH_2I
1-iodoethane

heat under reflux with KCN in ethanol

CH_3CH_2CN
propanenitrile

P_2O_5, heat

$CH_3CH_2C \begin{smallmatrix} O \\ \\ NH_2 \end{smallmatrix}$
propanamide

Two ways of making nitriles

Hydrolysis converts a nitrile to a *carboxylic acid*. Acids or alkalis can catalyse this reaction. Reduction with *lithium tetrahydridoaluminate(III)*, or with hydrogen and a nickel catalyst, converts nitriles to amines.

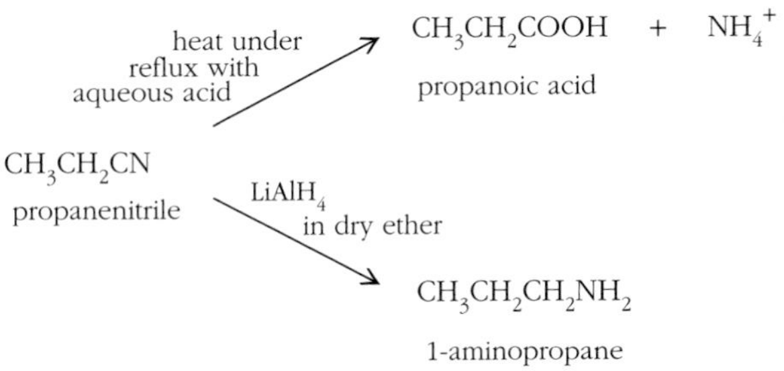

heat under reflux with aqueous acid

CH_3CH_2COOH + NH_4^+
propanoic acid

CH_3CH_2CN
propanenitrile

$LiAlH_4$
in dry ether

$CH_3CH_2CH_2NH_2$
1-aminopropane

Reactions of nitriles

Nitriles are useful intermediates in organic synthesis, especially when there is a need to add to the carbon skeleton of the molecule.

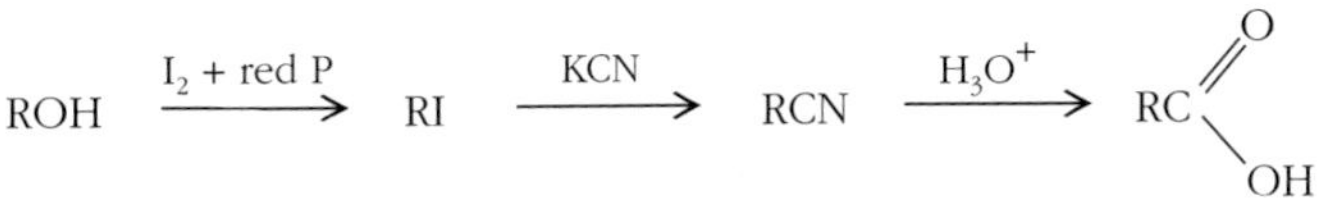

$ROH \xrightarrow{I_2 + \text{red P}} RI \xrightarrow{KCN} RCN \xrightarrow{H_3O^+} RC \begin{smallmatrix} O \\ \\ OH \end{smallmatrix}$

Lengthening the chain of a carbon skeleton

nitrites are salts of *nitrous acid*, also called nitric(III) acid. The two common examples are sodium nitrite ($NaNO_2$) and potassium nitrite (KNO_2). Sodium nitrite and potassium nitrite are preservatives and curing agents used in cooked meats and sausages. They are permitted food additives E250 and E249, respectively.

nitrogen (N_2) is an element which is chemically quite *inert* but which plays a vital part in the environment and forms many compounds of natural and economic importance. Nitrogen is the first element in group 5 and has the electron configuration $[He]2s^22p^3$. Nitrogen gas (N_2) is unreactive because the atoms are joined by a very strong triple bond and the molecule is non-polar.

The inertness of nitrogen makes it useful wherever oxygen must be excluded, as in food packaging, some chemical processing and the production of *semiconductor* chips.

Nitrogen makes up 79% of the air and can be separated by *fractional distillation* of liquid *air*. Alternatively, when only nitrogen is needed, the gas can be separated using:

- *selectively permeable membranes* which let though all the gases except nitrogen which is retained
- molecular sieves which absorb oxygen but let nitrogen pass through.

nitrogen cycle: the ways by which the element nitrogen circulates through the environment. The nitrogen cycle is chemically complex because nitrogen crops up in the environment in a range of oxidation states. Some of the nitrogen is inorganic as NH_3 and NH_4^+, N_2, N_2O, NO_2^- and NO_3^-. Some of it is organic as *amino acids*, *proteins* and *nucleic acids* in living things or their remains.

There are three main reservoirs of nitrogen:

- the air (about 3×10^{20} mol N)
- the crust of the Earth (about 2×10^{16} mol N)
- the oceans (about 6×10^{16} mol N)

The main types of chemical change in the nitrogen cycle are:

- *nitrogen fixation* – converting nitrogen from the air into inorganic and then organic nitrogen
- denitrification – converting inorganic nitrogen in the soil or oceans to nitrogen gas in the air
- assimilation – converting inorganic nitrates to organic nitrogen compounds
- mineralisation – converting organic nitrogen in living things into inorganic nitrites and nitrates.

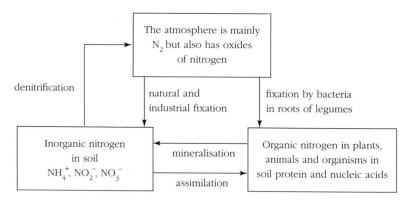

Diagram of the nitrogen cycle. Note that this version of the cycle leaves out the oceans which are major reservoirs of nitrogen compounds.

nitrogen fixation: any process that converts *nitrogen* gas in the *air* to nitrogen compounds which can be taken up by plants and converted to amino acids, proteins and nucleic acids. Nitrogen fixation happens naturally:

- during thunderstorms the energy in lightning allows nitrogen and oxygen to combine to form oxides of nitrogen which are washed into the soil by rain and taken up by plants as nitrates
- some soil bacteria have an enzyme called nitrogenase which can harness the energy of *ATP* to convert nitrogen from the air into ammonium compounds
- bacteria in the nodules of leguminous plants such as peas, beans and clover can also make ammonium salts from nitrogen.

As agriculture became more intensive, the demand for nitrogen by crops outstripped the rate it could be produced by natural processes. Manures could not meet the demand either, so the challenge for chemists was to find a practicable and economic way to fix nitrogen industrially. This is the problem which was solved by Fritz Haber, leading to *ammonia manufacture* by the Haber process. Some of the ammonia is used for *nitric acid manufacture* and the two chemicals combine to make nitrogen *fertilisers*.

nitrogen oxides: there are six compounds of nitrogen with oxygen. All the oxides have positive enthalpies of formation. They are unstable relative to the elements and are easily decomposed by heating. Two of the oxides, N_2O_3 and N_2O_5, are particularly unstable and of little practical importance. Some oxides of nitrogen are *atmospheric pollutants*. NO_x is a general formula used in accounts of air pollution to stand for any of the various nitrogen oxides when fuels burn at high temperatures in engines and furnaces. NO_x contributes to *acid deposition* and the formation of *photochemical smog*.

Oxidation state	Formula	Name	Type	$\Delta H_f^{\ominus} /$ kJ mol^{-1}	Notes
+4	$2NO_2 \rightleftharpoons N_2O_4$	nitrogen dioxide and dinitrogen tetroxide	acidic	+33.2 (for NO_2)	NO_2 is the brown gas formed on heating nitrates. It is used as an oxidant in rocket fuels.
					Lowering the temperature makes the gas paler as the equilibrium favours the colourless N_2O_4.
+2	NO	nitrogen monoxide	neutral	+90.2	Colourless but reacts with oxygen in the air to make NO_2; an intermediate in the manufacture of nitric acid from ammonia.
+1	N_2O	dinitrogen monoxide	neutral	+82.0	Colourless; also called laughing gas.

nitrous acid (HNO$_2$)is an unstable, *weak acid* which is an important reagent for producing the *diazonium salts* needed to make *azo dyes*. Nitrous acid is too unstable to be stored so it is made in solution when needed by adding a strong acid, such as hydrochloric acid, to a solution of sodium nitrite:

$$NO_2^-(aq) + H_3O^+(aq) \rightarrow HNO_2(aq) + H_2O(l)$$

The solution of the acid is blue. It starts to decompose at room temperature, giving off nitrogen monoxide (NO) which turns brown as it meets the air and turns into nitrogen dioxide (NO_2).

The systematic name of the acid is dioxonitric(III) acid which is sometimes abbreviated to nitric(III) acid. This shows that nitrogen is in the +3 oxidation state. When it decomposes it *disproportionates* to the +5 and the +2 states.

$$3HNO_2(aq) \rightarrow HNO_3(aq) + 2NO(g) + H_2O(l)$$
$$+3 \qquad\qquad +5 \qquad +2 \qquad\qquad\qquad \textit{Oxidation states of nitrogen}$$

NMR spectroscopy is a powerful analytical technique for finding the structures of carbon compounds. Two versions of the technique are *carbon-13 NMR spectroscopy* and *proton NMR spectroscopy*. They are used to identify unknown compounds, to check for impurities and to study the structure of molecules.

The name of the technique summarises its key features:

- nuclear – the technique detects nuclei of atoms such as hydrogen-1 (protons) or carbon-13
- magnetic – the nuclei detected by the technique are the ones which act like tiny magnets that can line up either in the same direction or in the opposite direction to an external magnetic field
- resonance – the absorption of energy is from radio waves with a frequency corresponding to the size of the energy jump as the nuclei flip from one alignment in a magnetic field to the other.

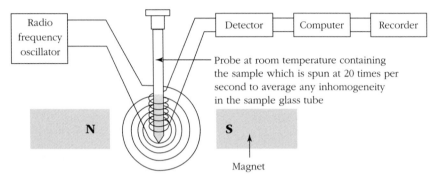

Schematic diagram of an NMR spectrometer

A tube with the sample is supported in a strong magnetic field. The operator turns on a source of radiation at radio frequencies. The radio-frequency detector records the intensity of the signal from the sample as the oscillator emits pulses of radiation across a range of wavelengths. Peaks are detected at frequencies corresponding to the energy jumps when the magnetic nuclei flip from one alignment to the other.

The sample is dissolved in a solvent. Also in the solution is some tetramethylsilane (TMS) which is a standard reference compound that produces a single, sharp absorption peak well away from the peaks produced by samples for analysis.

Each peak corresponds to one or more magnetic atoms in a particular chemical environment. Nuclei in different parts of a molecule experience slightly different magnetic fields in an NMR machine. This is because they are shielded to a greater or lesser extent from the field applied by the spectrometer by the tiny magnetic fields associated with the electrons of neighbouring bonds and atoms.

The recorder prints out a spectrum which has been analysed by computer to show peaks wherever the sample absorbs radiation strongly. The zero on the scale is fixed by the absorption of magnetic atoms in the TMS reference chemical. The distances of the sample peaks from this zero are called their 'chemical shifts' (δ).

noble gases: the unreactive gases in *group 8* (IUPAC group 18) of the periodic table. They were called the 'inert gases' until the discovery that *krypton* and *xenon* can form compounds. Even so the gases keep apart from most chemical changes and react only with highly reactive chemicals such as fluorine and oxygen.

The term 'noble' has been used for a long time to describe metals such as gold and silver.

These metals were *inert* to the reagents used by alchemists and early chemists. Like the nobility in society, these elements were seen to be special by standing apart from everyday changes and ordinary events.

nomenclature: see *names of carbon compounds*, *names of complex ions* and *names of inorganic compounds*.

non-aqueous solvent: any *solvent* other than water. A non-aqueous solvent has to be used if the *solute* will not dissolve in water. For this reason white spirit (a *hydrocarbon* solvent) is used to make oil paint while other organic solvents are used as nail-varnish removers, stain removers and dry-cleaning fluids. A non-aqueous dry-cleaning fluid removes dirt without causing shrinkage or other damage to fabrics harmed by water.

A non-aqueous solvent has to be used for a reaction in solution if one of the reactants or products is rapidly hydrolysed by water. For this reason *lithium tetrahydridoaluminate(III)* reductions take place in ether solution.

The choice of solvent can affect the outcome of a reaction. A solution of potassium hydroxide or sodium hydroxide in water hydrolyses *halogenoalkanes* to alcohols. If the solvent is ethanol it is much more likely that an *elimination reaction* will happen and the product will be an alkene.

non-metals are the elements towards the right-hand side of the *periodic table* which do not show the characteristic properties of metals. In general non-metals:
- consist of small molecules with the atoms linked by *covalent bonds* (H_2, S_8, P_4 and Cl_2); the exceptions include carbon, silicon and red phosphorus in which covalent bonding is continuous throughout the *giant structure* of atoms (see *crystal structures of non-metals*)
- are mostly gases at room temperature because of the weak *intermolecular forces* (for example hydrogen, oxygen, nitrogen, chlorine and all the noble gases)
- if solid, are not shiny like metals; they are brittle rather than bendable and do not normally conduct electricity or thermal energy
- tend to form negative ions by gaining electrons when they react with metals
- form *acidic oxides* which react with water to produce *oxoacids*
- form covalent molecular chlorides which are liquids and usually rapidly hydrolysed by water, such as PCl_3 and $SiCl_4$
- form covalent molecular hydrides which are gases at room temperature, such as CH_4, SiH_4, NH_3, HCl, HBr and HI – the notable exception is water which is a liquid because of *hydrogen bonding*.

non-polar solvent: a solvent in which the molecules are not *polar molecules*. Examples of non-polar solvents are the many liquids *hydrocarbons* and the mixtures of hydrocarbons obtained by refining crude oil. These are also *non-aqueous solvents*.

Following the general rule that 'like dissolves like', non-polar solvents dissolve many compounds which consist of small molecules. Chlorine, bromine and iodine, for example, dissolve freely in hexane. The stain removers and dry-cleaning fluids intended to remove oily grease from clothes consist of non-polar solvents.

Non-polar solvents do not dissolve ionic crystals. The interaction between non-polar molecules and ions is much too weak to surround ions and pull them away from the crystal against the *electrostatic forces* between ions with the opposite charge.

non-rechargeable cells: see *alkaline cell*.

non-stoichiometric compounds are compounds with formulae which do not have simple whole number ratios of atoms. The formulae of non-stoichiometric compounds vary. Many oxides and sulfides of *d-block elements* have variable formulae, for instance iron(II) sulfide ($Fe_{1.1}S$ to $FeS_{1.1}$) and iron(II) oxide ($Fe_{1.06-1.19}O$). This variation is a result of variable oxidation states and irregularities in the crystal lattice. Other examples are the *interstitial hydrides* of d-block elements such as vanadium hydride, $VH_{0.6}$.

NO_x: see *nitrogen oxides*.

nuclear fusion is the process which produces the energy of the Sun and other stars. It is also the process which accounts for the origin of all the elements.

In the Sun, at a temperature of ten million degrees or so, hydrogen atoms fuse to make helium atoms releasing about 10^9 kJ per mole of helium formed.

$$4\,{}^{1}_{1}H \rightarrow {}^{4}_{2}He + 2\ \text{positrons}$$

The pressure and temperature at the centre of very large stars is high enough for helium nuclei to fuse to make heavier elements such as carbon, silicon and iron. Since helium has an even number of *nucleons*, it turns out that elements with even *atomic numbers* are more abundant than other elements.

Iron is the final product of the series of *exothermic* fusion reactions. Energy is needed to make heavier elements. Heavier elements form during the highly energetic explosion when massive stars run out of nuclear fuel and start to collapse and then burst apart. The massive explosion is a supernova which scatters dust and gas through the universe where it mixes with hydrogen and helium.

In time the remains of 'dead' stars start to coalesce into new stars and the process starts all over again. The Sun is an example of a second generation star; it and the planets of the solar system formed from the remains of supernovae. This explains the variety of elements on Earth in a universe where over 90% of all atoms are hydrogen.

nuclear magnetic resonance spectroscopy: see *NMR spectroscopy*.

nuclear reactions: changes affecting the nuclei of atoms. Nuclear reactions are brought about by bombarding materials with high-energy particles in particle accelerators or with beams of neutrons from nuclear reactors.

Nuclear reactions can be used to make elements which do not occur naturally, with higher proton numbers than uranium. Bombarding uranium-238 with neutrons produces uranium-239 which decays by *beta decay* to neptunium. Neptunium in turn decays to plutonium.

$$\underset{\text{neutron}}{{}^{238}_{92}U + {}^{1}_{0}n} \longrightarrow {}^{239}_{92}U \xrightarrow{\text{decay}} {}^{239}_{93}Np + \underset{\beta\ \text{particle}}{{}^{0}_{-1}e}$$

$$\xrightarrow{\text{decay}} {}^{239}_{94}Pu + \underset{\beta\ \text{particle}}{{}^{0}_{-1}e}$$

Nuclear reactions producing the transuranium elements neptunium and plutonium. Note that the mass numbers and atomic numbers balance.

Nuclear reactions are used to make radioactive isotopes for use as tracers.

nucleic acids are the molecules which carry genetic information and control protein synthesis in cells. There are two main types of nucleic acid, *DNA* and *RNA*. Nucleic acids are polynucleotides formed as *nucleotides* link together in long chains.

nucleon: a particle in the nucleus of an atom – either a *proton* or a *neutron*.

nucleon number: the nucleon number for an atom is an alternative term to *mass number*.

nucleophiles are molecules or ions with a lone pair of electrons which can form a new covalent bond. Nucleophiles are reagents which attack molecules where there is a partial positive charge, δ+. So they seek out positive charges – they are 'nucleus-loving'.

Some examples of nucleophiles are:

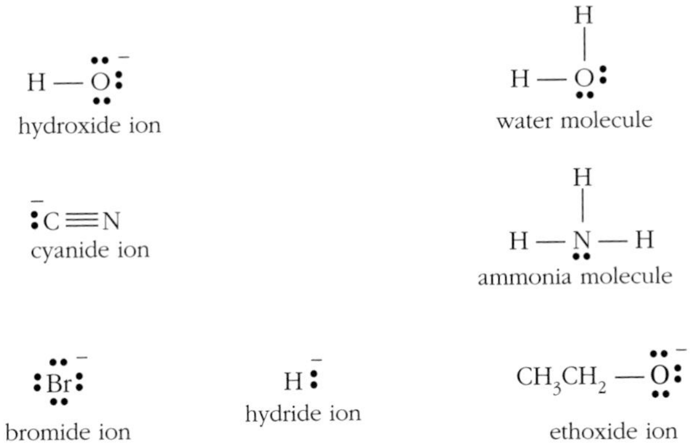

H — O:⁻
hydroxide ion

H — O: with H above
water molecule

:C≡N⁻
cyanide ion

H — N — H with H above
ammonia molecule

:Br:⁻
bromide ion

H:⁻
hydride ion

CH_3CH_2 — O:⁻
ethoxide ion

Examples of nucleophiles

Nucleophiles take part in nucleophilic addition and nucleophilic substitution reactions.

nucleophilic addition: the attack of a nucleophile on a *carbonyl compound* leading to an *addition reaction*. The *electronegative* oxygen draws electrons away from carbon so that the C=O bond in the aldehyde or ketone is *polar*. The incoming nucleophile uses its lone pair to form a new bond with the carbon atom. This displaces one pair of electrons in the double bond onto oxygen. Oxygen has thus gained one electron from carbon and now has a negative charge.

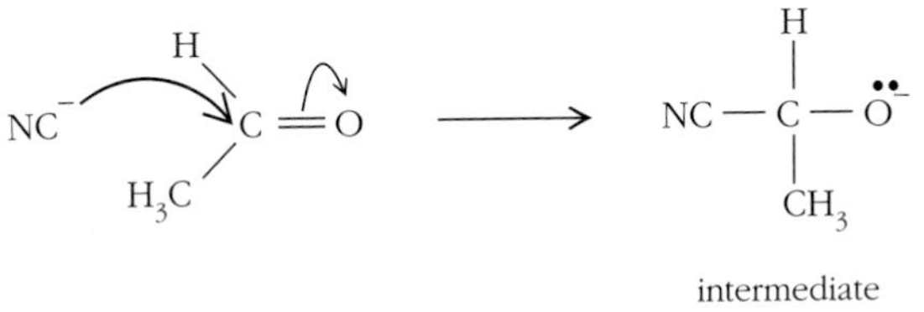

NC⁻ ... C=O (with H and H_3C) ⟶ NC — C — O⁻ (with H above and CH_3 below)
intermediate

First step of nucleophilic addition of hydrogen cyanide to ethanal

To complete the reaction, the negatively charged oxygen acts as a *base* and gains a proton.

intermediate

Second step of nucleophilic addition of hydrogen cyanide to ethanal

nucleophilic substitution in derivatives of carboxylic acids: nucleophilic attack on the carbon atom of the C=O bond in the functional group which leads to a *substitution reaction*. The first step is similar to *nucleophilic addition*.

The *electronegative* oxygen draws electrons away from carbon so that the C=O bond is *polar*. The incoming nucleophile uses its lone pair to form a new bond with the carbon atom. This displaces one pair of electrons in the double bond onto oxygen. Oxygen has thus gained one electron from carbon and now has a negative charge.

The first step in the attack of a nucleophile on a derivative of a carboxylic acid. If X is OC_2H_5 the compound is an ethyl ester, if X is Cl it is an acyl chloride, if X is NH_2 it is an amide.

At this point the negative oxygen does not gain a proton; instead it uses a pair of electrons to reform the double bond and displace X.

The second step which eliminates X as an X⁻ ion

Nucleophiles which react with esters, acyl chlorides and amides in this way are hydroxide ions and hydride ions from lithium tetrahydridoaluminate(III).

nucleophilic substitution in halogenoalkanes: *nucleophilic attack* on the carbon atom of the C—X bond in a *halogenoalkane*, RX, which leads to a *substitution reaction*, where X = Cl, Br or I. Study of the *rate equations* suggests that there are two different mechanisms.

Hydrolysis of primary halogenoalkanes such as bromobutane, is overall a *second order reaction*. The rate equation has the form: rate = $k[C_4H_9Br][OH^-]$.

The suggested *mechanism* shows the C—Br bond breaking as the nucleophile, OH⁻, forms a new bond with carbon.

The S$_N$2 mechanism. Substitution–Nucleophilic–2 (for bimolecular – that is two molecules or ions involved in the rate-determining step). Note the inversion of the molecule.

Hydrolysis of tertiary halogenoalkanes such as 2-bromo-2-methylpropane, is a *first order reaction*. The rate equation has the form: rate = $k[(CH_3)_3CBr]$.

The suggested mechanism shows the C—Br bond breaking first to form a *carbocation intermediate*. Then the nucleophile OH$^-$ forms a new bond with carbon.

The S$_N$1 mechanism. Substitution–Nucleophilic–1 (for unimolecular – that is one molecule or ion involved in the rate-determining step. Note that the carbocation intermediate is planar and may be attacked by the nucleophile on either side

nucleotide: the monomer which polymerises to make *nucleic acids, DNA* and *RNA*. A nucleotide consists of three parts:
- a five-carbon sugar – ribose in RNA and deoxyribose in DNA
- a phosphate group
- a nucleotide base – there are five different bases; adenine, cytosine, guanine, thymine and uracil, often abbreviated as A, C, G, T and U.

In the nucleotides which make up DNA the base is one of A, C, G or T. In RNA, thymine, T, is replaced by uracil, U.

nuclide: an *atom* defined by the numbers of protons and neutrons in its nucleus. The *isotopes* of a particular element are made up of nuclides with the same proton number (atomic number) but different neutron numbers. Some nuclides are radioactive. Each radionuclide has a specific *half-life* for radioactive decay.

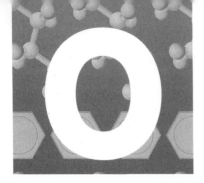

octane number is a measure of the performance of petrol. For smooth running, petrol has to burn smoothly without *knocking*. The higher the compression of fuel and air in the engine cylinders, the higher the octane number has to be to stop knocking.

The octane number scale was devised by Thomas Midgley (1889–1944) and he discovered anti-knock additives based on lead, which were used for many years. Leaded fuel has been phased out. There are two main objections to adding lead compounds to petrol. The first is that lead compounds are *toxic chemicals* and harmful, especially to young children, when released into the air from car exhausts. The second is that lead compounds 'poison' the catalysts in *catalytic converters*.

The oil companies now produce high-octane fuel by increasing the proportions of both branched alkanes and arenes plus blending in some oxygen compounds. The four main approaches are:

- *cracking* which not only makes more small molecules but also forms hydrocarbons with branched chains
- *isomerisation* which turns straight-chain alkanes into branched chain compounds by passing them over a platinum catalyst
- *reforming* which turns cyclic alkanes into *arenes* such as benzene and methylbenzene
- adding oxygenates such as *alcohols* or *ethers*.

octet rule: this was first suggested as a guide by the US chemist Gilbert Lewis in 1916. The rule says that atoms tend to gain, lose or share electrons when they combine with other atoms to acquire a stable octet of electrons. The 'stable octet' is the eight s^2p^6 electrons, corresponding to the outer *electron configuration* of the nearest noble gas in the periodic table.

Dot-and-cross diagrams for molecules and ions showing outer shells with eight electrons as in the nearest noble gas

There are many exceptions to the octet rule, so it is not a safe guide. The rule works pretty well for the elements Li to F in period 2 because there are only four orbitals in the second shell (one s and three p orbitals). The octet rule also works for ionic compounds of *s-block elements* with the halogens, oxygen and sulfur. Exceptions arise in period 3 and beyond, because *d orbitals* in the third shell can become involved in bonding.

Born–Haber cycles can help to account for the octet rule in ionic compounds. When magnesium, for example, forms ionic compounds with the Mg^{2+} ion, the extra *ionisation energy* needed to remove two rather than one electron from the outer shell is more than compensated for by the big increase in *lattice energy*. The lattice energy is much more negative for two reasons which increase the attraction between the positive and negative ions. An Mg^{2+} ion has a larger charge than an Mg^+ ion. Also an Mg^{2+} ion is much smaller than an Mg^+ ion because it has no electrons in the outer shell, so the ions can pack closer together.

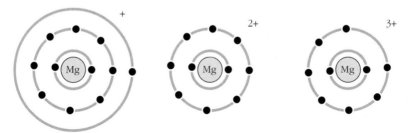

Electron configurations of Mg^+, Mg^{2+} and Mg^{3+} ions

Removing a third electron from a magnesium atom needs much more energy because it involves taking an electron from the next shell where the electrons are more strongly held by the nucleus. If it could form, an Mg^{3+} ion would be little smaller than an Mg^{2+} ion so the increased lattice energy is not enough to compensate for the very large third ionisation energy.

odd-electron compounds: stable compounds with an odd number of electrons in an outer shell. Examples are two *nitrogen oxides*, nitrogen oxide and nitrogen dioxide. Both are exceptions to the *octet rule*.

Dot-and-cross diagrams showing the odd electrons in molecules

oil: see *crude oil* and *vegetable oils*.

optical isomers: *isomers* which have opposite effects on *plane-polarised light*. One isomer rotates the plane of polarised light clockwise (the + isomer). The other isomer rotates the plane of polarised light in the opposite direction (the − isomer). Rotations are measured with monochromatic light in an instrument called a polarimeter (see *monochromatic radiation*). Optical activity is characteristic of *chiral compounds*.

Optical isomerism is shown by molecules or complex ions with the same structure but different three-dimensional shapes, so it is a type of *stereoisomerism*.

orbitals: see *atomic orbitals* and *molecular orbitals*.

order of reaction: see *rate equation*.

ore: a *mineral* mass which can be profitably mined and processed to produce a metal. Prospectors seek ore bodies which are local concentrations of ores. After mining or quarrying, the first stages of processing separate the valuable minerals from waste rock.

The work of the mineral industry would be impossible if minerals were evenly distributed throughout the *lithosphere*. Fortunately, natural processes have produced concentrations of valuable minerals with much higher percentages of rare metals than in the Earth's crust.

organic acids: see *carboxylic acids*.

organic analysis consists of chemical methods for *qualitative* and *quantitative analysis* to identify organic compounds and work out their composition and structure.

In a modern laboratory, organic analysis is based on a range of automated and instrumental techniques including *combustion analysis, mass spectrometry, infrared spectroscopy, ultraviolet spectroscopy* and *NMR spectroscopy*.

Melting points and *boiling points* also provide a check on the identity and purity of compounds.

Traditionally, chemical tests helped to identify *functional groups* in organic molecules. Preliminary tests typically involve observing:

- the state and appearance of the compound
- its solubility in water and the pH of any solution
- the type of flame when a small sample burns.

Organic analysis: tests for functional groups

Functional group	Test	Observations
$\diagdown\diagup$ $C = C$ $\diagup\diagdown$ in an *alkene*	• Shake with a dilute solution of bromine	• Orange colour of the solution decolorised
	• Shake with a very dilute, acidic solution of potassium manganate(VII)	• Purple colour fades and the solution turns colourless
$\vert$ $- C - X$ $\vert$ where X = Cl, Br or I	• Warm with a solution of sodium hydroxide, cool, acidify with nitric acid, then add silver nitrate solution	• Precipitate is white from a chloro compound, creamy yellow from a bromo compound and yellow from an iodo compound (*hydrolysis* produces ions from the covalent molecules)
$\vert$ $- C - OH$ $\vert$ in a primary *alcohol*	• Add a solution of sodium carbonate	• No reaction; unlike acids, alcohols do not react with carbonates
	• Add PCl$_5$ to the *anhydrous* compound	• Colourless, fuming gas forms (hydrogen chloride)
	• Warm with an acidic solution of potassium dichromate(VI)	• Orange solution turns green and gives a fruity smelling vapour. Alcohols are oxidised by dichromate(VI) but not by *Fehling's solution*

$\diagup^{\diagup}_{\diagdown}C={O}$ in *aldehydes* and *ketones*	• Add 2,4-dinitrophenyl-hydrazine solution	• Aldehydes and ketones both give yellow or red precipitates
	• Warm with fresh *Fehling's solution*.	• Solution turns greenish and then an orange-red precipitate forms with aldehydes, but not ketones
	• Warm with fresh *Tollens' reagent*	• Silver mirror forms with aldehydes but not with ketones
—COOH in *carboxylic acids*	• Measure the pH of a solution	• Acidic solution with pH about 3–4
	• Add a solution of sodium carbonate	• Mixture fizzes, giving carbon dioxide
	• Add a neutral solution of iron(III) chloride	• Methanoic and ethanoic acids give a red colour

organic chemistry is the study of carbon compounds (except the very simplest such as the oxides and carbonates). There are more carbon compounds than the compounds of any other element because carbon atoms have a remarkable ability to join up in stable chains, branched chains and rings. Organic chemists have devised ways to organise the mass of information about millions of compounds and reactions so that it makes sense.

One approach is to group compounds into series each with a distinctive *functional group*. Examples are the *alkanes*, *alkenes*, *arenes*, *carbonyl compounds* and *carboxylic acids*.

Another way of making sense of carbon chemistry is to classify reactions. One classification describes what happens to the molecules: *addition reactions*, *substitution reactions*, *elimination reactions* and *rearrangement reactions*. An alternative classification of reactions is based on the *mechanism of a reaction* showing the types of reagents, the method of bond breaking and the nature of any *intermediates*. This includes:
- *homolytic bond breaking* with *free radical* intermediates
- *heterolytic bond breaking* with *electrophiles* and *nucleophiles* as the attacking reagents leading to ionic intermediates.

Biochemistry and polymer chemistry are specialist aspects of organic chemistry.

Organic chemists have proved their understanding of organic structures and reactions by devising ways to synthesise increasingly complex molecules.

organic preparations take place in four main stages (see table overleaf) to make pure organic compounds.

Stage	Activities	Commonly used techniques
Planning	● Deciding on the reaction ● Working out reacting quantities from the equation ● Deciding on the conditions for reaction	● Consulting reference books ● *Risk analysis*
Carrying out the reaction	● Measuring out the reactants ● Mixing them in a suitable apparatus ● Controlling the temperature	● Heating in a flask with a *reflux condenser* ● Shaking *immiscible* reactants in a stoppered container
Separating the product from the reaction mixture	● 'Working up' the reaction mixture to obtain a crude product	● *Distillation* ● *Steam distillation* ● *Vacuum filtration*
Purifying the product and testing for purity	● Purifying liquids by shaking with reagents to extract impurities in a separating funnel, followed by drying and distillation ● Purifying solids by recrystallisation and filtration ● Using tests to check on the identity and purity of the product ● Calculating the percentage *yield*	● *Fractional distillation* and measuring boiling points ● *Solvent extraction* ● Recrystallisation ● Measuring *melting* points ● *Thin-layer chromatography* ● *Infrared spectroscopy* ● *Organic analysis*

organic routes: pathways from reactants to the required product in one or several steps. Organic chemists often start by examining the 'target molecule' and then work back through a series of steps to find suitable starting chemicals which are cheap enough and available. In recent years chemists have developed computer programs to help with the process of working back from the target molecule to a range of possible starting points in a systematic way.

All the reactions in organic chemistry convert one compound to another, but there are some reactions which are particularly useful for developing a synthetic route, such as:
- addition of hydrogen halides to *alkenes*
- substitution reactions to replace halogen atoms with other functional groups such as —OH or —NH$_2$
- substitution of a chlorine atom for the—OH group in an *alcohol* or *carboxylic acid*
- elimination of a hydrogen halide from a *halogenoalkane* to introduce a double bond
- oxidation of primary *alcohols* to *aldehydes* and then carboxylic acids
- reduction of *carbonyl compounds* to alcohols
- lengthening of a carbon chain by forming a *nitrile* or using a *Grignard reagent*.

organometallic compounds: organic compounds which contain a metal atom. *Grignard reagents* are organometallic compounds formed with magnesium.

osmosis is the movement of a solvent by diffusion through a *selectively permeable membrane*. The membrane allows small solvent molecules through but not dissolved molecules or ions.

The concentration of solvent molecules is higher in the pure solvent than in the solution. Solvent molecules diffuse in both directions, but overall they move from the solvent into the solution.

oxidant: an alternative term to *oxidising agent* which is convenient when describing half-equations for *redox reactions* and *standard electrode potentials*. By convention, when electrode potential values are being assigned the half-equation takes the form:

$$\text{oxidant} + ne^- \rightleftharpoons \text{reductant}$$

Every ionic half-equation involves an oxidant and a reductant.

oxidation: originally this meant combination with oxygen but the term now covers all reactions in which atoms, molecules or ions lose electrons. The definition is further extended to cover molecules as well as ions by defining oxidation as a change which makes the *oxidation number* of an element more positive, or less negative.

Oxidation and reduction always go together in *redox reactions*.

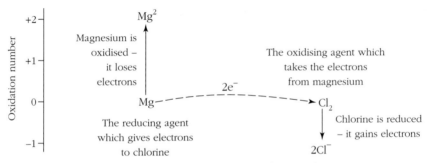

Chlorine oxidises magnesium by taking two electrons from each magnesium atom

Oxidation number rules apply in principle in organic chemistry, but it is often easier to use the older definitions. Oxidation is either addition of oxygen to a molecule or removal of hydrogen. (See also *functional group level*.)

Two-stage oxidation of an organic compound

oxidation numbers are used by chemists to define *oxidation* and *reduction* and to identify *redox reactions*. The advantage of oxidation numbers is that they apply to molecules and complex ions as well as to simple atoms and ions. The sum of the oxidation numbers in an uncharged compound is zero.

273

Chemists have agreed on a set of rules for deciding on the oxidation numbers of elements:
- the oxidation numbers of uncombined elements are zero
- in a simple ion the oxidation number is the charge on the ion
- the sum of the oxidation numbers in an uncharged compound is zero
- the sum of the oxidation numbers in an ion made of several atoms is equal to the charge on the ion
- some elements have fixed oxidation numbers.

Metals		Non-metals	
group 1 (Li, Na, K)	+1	hydrogen	+1
group 2 (Mg, Ca, Ba)	+2	(except in metal hydrides)	
group 3 (Al)	+3	fluorine	−1
		oxygen	−2
		(except in peroxides and compounds with fluorine)	
		chlorine	−1
		(except in compounds with fluorine and oxygen)	

Oxidation numbers appear in the *names of inorganic compounds* and the *names of complex ions*. They also help with writing balanced equations for *redox reactions*.

Oxidation number rules are a formal way of keeping track of electrons in redox reactions. For the s-block elements the oxidation numbers are the same as the charges on the ions. Magnesium has an oxidation state of +2 and the metal atoms turn into Mg^{2+} ions in its compounds. The higher oxidation states of *d-block elements*, however, are certainly not ionic. Manganese in the +7 state cannot form an Mn^{7+} ion. The bonding between manganese and oxygen in the MnO_4^- ion is covalent.

MgO	NH_4^+	CCl_4	MnO_4^-
+2 −2	−3 +1	+4 −1	+7 −2

Al_2O_3	SO_4^{2-}	H_2SO_4	$Cr_2O_7^{2-}$
+3 −2	+6 −2	+1 +6 −2	+6 −2

Examples of the application of oxidation number rules. Note that each element in a compound is assigned an oxidation number

oxidation states: the states of oxidation or reduction shown by an element in its chemistry. The states are labelled with the oxidation numbers of the element in each state.

Describing the chemistry of an element according to oxidation state is a useful way of making sense of a large number of compounds and reactions. See for example the entries for *chlorine, manganese, hydrogen peroxide* and *nitrogen oxides*.

oxides are compounds of elements with oxygen. The compounds of *metals* with oxygen are *basic oxides* or *amphoteric oxides*. The compounds of *non-metals* with oxygen are either *acidic oxides* or *neutral oxides*.

oxidising agents are chemical reagents which can oxidise other atoms, molecules or ions by taking electrons away from them. Common oxidising agents are *oxygen, chlorine, bromine, hydrogen peroxide*, the manganate(VII) ion in *potassium manganate(VII)*, and the dichromate(VI) ion.

Some reagents change colour when they are oxidised, which makes them useful for detecting oxidising agents. In particular, a colourless solution of iodide ions is oxidised to iodine which turns the solution to a yellow–brown colour.

$$2I^-(aq) \rightarrow I_2(aq) + 2e^-$$

The electrons are taken by the oxidising agent.

This can be a very sensitive test if **starch** is also present because starch forms an intense blue-black colour with iodine. Moistened starch–iodide paper is a version of this test which can detect oxidising gases such as chlorine and bromine.

oxidising substances: substances which are hazardous if they come into contact with materials that burn – especially if they are flammable liquids. The danger is that oxidising substances will start a fire in contact with materials that can burn. (See *hazards*.)

oxoacids are acids that form when acidic non-metal *oxides* react with water.

Oxoacid	Structure	Oxoanion	Formula
sulfuric acid [sulfuric(VI) acid] H_2SO_4		sulfate ion hydrogensulfate ion	SO_4^{2-} HSO_4^-
sulfurous acid [sulfuric(IV) acid] H_2SO_3		sulfite ion hydrogensulfite ion	SO_3^{2-} HSO_3^-
nitric acid [nitric(V) acid] HNO_3		nitrate ion	NO_3^-
nitrous acid [nitric(III) acid] HNO_2		nitrite ion	NO_2^-

The full systematic names for oxoacids are not widely used but they make it possible to work out the formulae from the names. The systematic name for sulfuric acid is tetraoxosulfuric(VI) acid.

Oxoacids ionise in solution by giving hydrogen ions to water molecules.

Ionisation of nitric acid in water showing delocalisation of the charge on the nitrate ion

Delocalised electrons spread the charge on an oxoanion, thus stabilising the ion and favouring ionisation. The more oxygen atoms doubly-bonded to the central atom, the stronger the acid. As a result, sulfuric acid is a stronger acid than sulfurous acid and nitric acid is a stronger acid than nitrous acid. So where an element forms two oxoacids, the one with the element in the higher oxidation state is the stronger.

oxonium ions are aqueous hydrogen ions or hydrated *protons* formed when an *acid* dissolves in water. A lone pair of electrons on a water molecule forms a *dative covalent* bond with a hydrogen ion from an acid.

The formula of the oxonium ion is H_3O^+. It is often convenient to write $H^+(aq)$ instead, but remember that the hydrogen ion is *hydrated*.

oxygen (O) is a colourless, odourless and highly reactive gas which makes up about a fifth of the *air* by volume and is essential to life. It is the first element in group 6 with the electron configuration $[He]2s^22p^4$. Most oxygen occurs in the form of O_2 molecules but *ozone*, O_3, is a second *allotrope*.

All living organisms need oxygen for respiration. Green plants produce oxygen by *photosynthesis* in sunlight. Many biochemical molecules include oxygen atoms, for example *carbohydrates*, *fats*, *proteins* and *nucleic acids*.

Oxygen is the most abundant element in the *lithosphere*, mainly in the silicon–oxygen giant structures of *silicates*.

Oxygen combines with most other elements to form *oxides*. Particularly important is *water*, the oxide of hydrogen.

Air separation plants use *fractional distillation* to separate oxygen and other gases. Alternatively, where only oxygen is required, the nitrogen from the air can be absorbed in a bed of a molecular sieve such as a *zeolite*.

Large volumes of oxygen are used to make *steel*. The manufacture of chemicals is another major use. Smaller quantities are used with a fuel for high-temperature metal cutting and welding. Oxygen is required medically by people with lung disease and is also used for *water treatment* to improve the quality of water contaminated with organic pollutants.

ozone (O_3) is a colourless, reactive and unstable gas which is an *allotrope* of oxygen.

Ozone is an *oxidising agent* and, like oxidising bleaches, it will destroy micro-organisms. Ozone is increasingly used to treat drinking water and swimming pools as an alternative to chlorine.

ozone in the stratosphere consists of a layer of the gas in the upper atmosphere at about 10–50 km above sea level. At this altitude ultraviolet (UV) light from the Sun splits oxygen molecules into oxygen atoms.

$$O_2 + UV \text{ light } \rightarrow O\bullet + O\bullet$$

These are *free radicals* that then convert oxygen molecules to ozone.

$$O\bullet + O_2 \rightarrow O_3$$

The ozone formed also absorbs UV radiation. This splits ozone molecules back into oxygen molecules and oxygen atoms destroying ozone. This is the reaction that protects living things from the damaging effects of UV

$$O_3 + UV \text{ light } \rightarrow O_2 + O\bullet$$

In the absence of pollutants, a *steady state* is reached with ozone being formed at the same rate as it is destroyed. Normally the steady-state concentration of ozone is sufficient to absorb most of the dangerous UV light from the Sun. Pollutants such as *CFCs* and *nitrogen oxides* destroy the ozone layer.

ozone in the troposphere forms when sunlight shines on the *atmospheric pollutants* from motor vehicles.

NO_2 + UV light $\rightarrow$ NO + O •

O • + O_2 $\rightarrow$ O_3

Examples of reactions which form ozone in polluted air

Ozone itself is damaging to people's lungs and to crop plants. Ozone can react with unburnt hydrocarbons to form *photochemical smog*.

A–Z Online

Log on to A–Z Online to search the database of terms, print revision lists and much more. Go to **www.philipallan.co.uk/a-zonline** to get started.

paint is a product formulated to protect and decorate surfaces. A paint has three ingredients:

- pigments, which scatter and absorb light so that the paint covers up the surface underneath and decorates it with colour
- polymers, which hold the pigment to the surface by forming a smooth plastic film as the paint dries and sets – in gloss paints the film-forming polymers are *polyesters* resins; in emulsion paints they are a vinyl or *acrylic* addition polymer
- a solvent, which is a liquid in which the other ingredients are dissolved or dispersed – in gloss paint the vehicle is traditionally a *hydrocarbon* solvent such as white spirit; in emulsion paints it is water.

paper chromatography: a type of *chromatography* in which the stationary phase is water in the fibres of paper and the moving phase is another solvent such as ethanol or propanone. The technique is used to analyse mixtures such as inks, food colours, dyes and amino acids.

During paper chromatography the chemicals in a mixture *partition* themselves between two solvents: water and the moving solvent. Each component has a different equilibrium constant (partition constant) which decides whether it has a greater tendency to dissolve in the stationary phase or in the mobile phase. As a result the mixture separates as the chemicals move at different speeds.

If the conditions are kept the same, each chemical in a mixture will move a fixed fraction of the distance moved by the solvent. The R_f value for the substance is a measure of this fraction.

If the chemicals in a mixture are colourless they are invisible on the paper. If so, the analyst has to 'develop' the plates with a suitable *locating agent*.

paracetamol is a mild painkiller (*analgesic*) which can also help to lower the body temperature during fever. It is readily available as an over-the-counter drug. Unlike aspirin it does not reduce inflammation.

Paracetamol is made by *acylating* 4-aminophenol with ethanoic anhydride.

The advantage of paracetamol over aspirin is that it does not irritate the lining of the stomach and it has few side effects.

The great danger is that an overdose with *medicines* containing paracetamol causes severe liver damage which can be fatal. The symptoms do not appear until after about two days. Unfortunately with paracetamol, the dose that causes poisoning is not many times the dose required to relieve pain or lower fever.

partial pressures are a useful alternative to concentrations when studying mixtures of gases and gas reactions. In a mixture of gases A, B and C the sum of the three partial pressures equals the total pressure.

$$p_{total} = p_A + p_B + p_B$$

The *ideal gas* equation is $pV = nRT$ and it is obeyed pretty well by real gases. In a mixture of gases, the gas molecules move around independently. It is the number of molecules that matters and not their chemical nature. So the partial pressure of each gas is the pressure it would exert if it was the only gas in the container under the same conditions; thus:

$$p_A = \frac{n_A RT}{V}, \quad p_B = \frac{n_B RT}{V} \quad \text{and} \quad p_C = \frac{n_C RT}{V}$$

Since R is a constant, these equations show that at constant temperature:

$$p_A \propto \frac{n_A}{V}, \quad p_B \propto \frac{n_B}{V} \quad \text{and} \quad p_C \propto \frac{n_C}{V}$$

So the partial pressures are proportional to the concentrations of the gases, since $n \div V$ is the concentration in moles per unit volume. This is the justification for working in partial pressures when applying the equilibrium law to gas reactions (see K_p).

The total pressure in a mixture can be 'shared out' between the gases according their *mole fractions*, X. So:

$$p_A = X_A p_{total}, \quad p_B = X_B p_{total} \quad \text{and} \quad p_C = X_C p_{total}.$$

Worked example

An experimental study of an equilibrium mixture of $N_2O_4(g)$ and $NO_2(g)$ found that it contained 9.01×10^{-3} mol of $N_2O_4(g)$ and 7.16×10^{-3} mol of $NO_2(g)$ at 298 K and 3.99×10^4 Pa pressure. Calculate the partial pressures of the two gases.

Notes on the method

First work out the mole fractions of the two gases. Multiply the total pressure by the mole fractions to get the partial pressures. Check that the sum of the partial pressures equals the total pressure.

Answer

Total number of moles $= 9.01 \times 10^{-3}$ mol $+ 7.16 \times 10^{-3}$ mol

$= 16.17 \times 10^{-3}$ mol

Mole fraction of $N_2O_4(g)$ $= \dfrac{9.01 \times 10^{-3} \text{ mol}}{16.17 \times 10^{-3} \text{ mol}} = 0.557$

Mole fraction of $NO_2(g)$ $= \dfrac{7.16 \times 10^{-3} \text{ mol}}{16.17 \times 10^{-3} \text{ mol}} = 0.443$

Partial pressure of $N_2O_4(g)$ $= 0.557 \times 3.99 \times 10^4$ Pa $= 2.22 \times 10^4$ Pa

Partial pressure of $NO_2(g)$ $= 0.443 \times 3.99 \times 10^4$ Pa $= 1.77 \times 10^4$ Pa

particles are extremely small pieces of matter ranging from electrons to grains of sand. Chemists use the term 'particle' in several ways. At the smallest end of the scale they refer to protons, neutrons and electrons as *fundamental particles*. But they also use the term as a general word to cover atoms, molecules and ions when discussing the movements of particles in solids, liquids and gases. In addition, chemists use the term particle when describing much larger solid specks of matter finely dispersed in *colloids* such as paint, muddy water and smoke.

particulates are solid *particles* suspended in the air. In rural areas the main sources of particulates are natural. They include soil and sand swept up by winds, as well as volcanic dust and salt from sea spray.

In urban areas pollution comes mainly from the incomplete combustion of *fuels* in engines and power stations. The solid particles consist of carbon and unburnt hydrocarbons. *Diesel fuel* engines are prone to producing particulates in their exhausts.

PM_{10} particles are of particular concern. These are very small particles which are less than 10 μm across. PM_{10} particles are small enough to penetrate deep into the lungs. Fine particles in the lungs can cause inflammation and be a threat to the people with heart and lung disease. They may also carry *carcinogens* into the lungs.

partition: the distribution of a dissolved substance between two *solvents* which do not mix (they are *immiscible liquids*). Partition is the basis for *solvent extraction* and *paper chromatography*.

Shaking a solid with two immiscible solvents produces an equilibrium system to which the *equilibrium law* applies. The ratio of the concentrations in the two layers is a constant, at a given temperature:

$$K = \frac{[X]_{\text{solvent 1}}}{[X]_{\text{solvent 2}}}$$

This special case of an equilibrium constant is called the partition constant or partition coefficient. (The alternative term is 'distribution constant'.)

Note that partition obeys the equilibrium law only so long as the dissolved chemical has the same molecular structure in both solvents.

parts per million (ppm) is a unit of *concentration* often used for low concentrations of pollutants in water. The density of water is 1 g cm^{-3} and dilute solutions have almost the same density. So the mass of 1 dm^3 (= 1000 cm^3) of water is 1000 g. For dilute solutions:

1 ppm = 1 mg substance dissolved in 1 dm^3 water

The unit ppmv means parts per million by volume and is used to measure the concentrations of *greenhouse gases* in the atmosphere (see *carbon cycle*).

Pauli exclusion principle: the fundamental principle which states that no two electrons in an atom can have the same value for all four *quantum numbers*. This rule accounts for the fact that no orbital can hold more than two electrons. The two electrons have opposite *spins*.

p-block elements are the elements in groups 3, 4, 5, 6, 7 and 8 in the *periodic table*. For these elements, the last electron added to the atomic structure goes into one of the three *p orbitals* in the outer shell.

PCBs (polychlorinated biphenyls) are a group of compounds formed from two linked benzene rings. They are liquids with valuable properties that lead to them being used in transformers, in hydraulic equipment, as *plasticisers* and *lubricants*. PCBs are suited to these applications because they are very stable, non-flammable, involatile and they are particularly suited for use in some electrical equipment. Some PCBs are also toxic pesticides.

Structure of a PCB

The problem with PCBs is that they escape into the environment where they are very persistent. They have been classified as are *persistent organic pollutants* and banned by international agreement. Countries that have signed the agreement were obliged to eliminate equipment and oils containing PCBs from use by 2025 and to destroy waste PCBs in an environmentally sound way by 2028

peptides are compounds made up of chains of *amino acids*. The simplest example is a dipeptide with just two linked amino acids. For chemists the link is an example of an *amide* bond but the tradition in biochemistry is to call it a 'peptide bond'. The formation of a peptide bond is an example of a *condensation reaction*.

Formation of a peptide bond between two amino acids

Polypeptides are long-chain peptides. There is no precise dividing line between a peptide and a polypeptide. A *protein* molecule consists of one or more polypeptide chains. Some experts, however, make a distinction between polypeptides and the longer amino acid chains in proteins. They restrict the definition of polypeptides to chains with 10 to 50 or so amino acids.

Heating a peptide with dilute sulfuric acid leads to hydrolysis of the peptide bonds. In time this process splits the peptide into separate amino acids. Alkaline hydrolysis produces salts of amino acids.

percentage composition: the percentage by mass of each of the elements in a compound. Percentage composition is one way of expressing the results of chemical analysis, such as *combustion analysis*. The *empirical formula* of the compound can be calculated from these results.

Worked example

What is the empirical formula of copper pyrites which has the analysis 34.6% copper, 30.5% iron and 34.9% sulfur?

Notes on the method

Follow the procedure in the worked example for finding an *empirical formula*.

The percentages show the combining masses in a 100 g sample.

Answer

	copper	iron	sulfur
Combining masses	34.6 g	30.5 g	34.9 g
Molar masses of elements	64 g mol^{-1}	56 g mol^{-1}	32 g mol^{-1}
Amounts combined	$\dfrac{34.6 \text{ g}}{64 \text{ g mol}^{-1}}$	$\dfrac{30.5 \text{ g}}{56 \text{ g mol}^{-1}}$	$\dfrac{34.9 \text{ g}}{32 \text{ g mol}^{-1}}$
	= 0.54 mol	= 0.54 mol	= 1.09 mol
Simplest ratio of amounts	1	1	2

The formula is $CuFeS_2$.

The percentage composition of a compound can be worked out from its chemical formula. This is a guide for people who formulate products such as fertilisers, medicines and cleaning agents.

Worked example

Two common nitrogen fertilisers are urea $(H_2N)_2CO$, and ammonium nitrate, NH_4NO_3. Compare the percentage of nitrogen in the two compounds.

Notes on the masses

Look up the relative atomic masses of the elements and use them to determine the two *relative formula masses*. (Note that when part of a formula is in brackets the number outside refers to all the atoms in the bracket.)

Answer

Relative formula mass of urea = $2 \times (2 + 14) + 12 + 16 = 60$
of which $(2 \times 14) = 28$ is nitrogen

Percentage of nitrogen in urea = $\dfrac{28}{60} \times 100\% = 46.7\%$

Relative formula mass of ammonium nitrate = $14 + 4 + 14 + (3 \times 16) = 80$ of which $(2 \times 14) = 28$ is nitrogen

Percentage of nitrogen in ammonium nitrate = $\dfrac{28}{80} \times 100\% = 35\%$

Urea contains the higher percentage of nitrogen.

percentage yield: see *yield*.

perfumes: fragrant chemicals used not only for expensive cosmetics but also to give a smell to everyday household products. Many natural perfumes are essential oils separated from plants by **steam distillation** or **solvent extraction**. Increasingly, however, perfumes now contain synthetic chemicals, some of which are identical to those extracted from natural sources.

Experts use an analogy with music to describe the odours from complex perfumes. 'Top' notes are the most volatile chemicals, the main effect depends on the 'middle notes' and the most long lasting are the 'end notes'. So a perfume chemist has to understand the volatility of perfume chemicals.

undecanal
top note; boiling point 117°C

geraniol
middle note;
boiling point 146°C

indane musk bottom note;
boiling point 393°C, melting point 53°C

Skeletal structures of some perfume chemicals showing top, middle and end notes

The connection between odour and chemical structure is little understood so there is no easy way of predicting smells by examining molecular models.

period 3: the third horizontal row of elements in the **periodic table** which starts with sodium and ends with argon. In each main group of the periodic table the element in period 3 is often regarded as the typical element in the group. Chemists then generalise about trends down the group by comparing the elements in periods 4, 5 and 6 with the period 3 element. The first member of each group (usually an element in period 2) is often exceptional in some way.

The changes in properties from Na to Ar (across period 3) show periodic patterns clearly (see **periodicity of physical properties** and **periodicity of chemical properties**).

periodic table: a table of the elements arranged in order of **atomic number** (proton number). Periods are the horizontal rows in the table. Each period ends with a **noble gas**. The vertical columns are **groups** arranged in blocks – the **s block**, **p block**, **d block** and **f block** based on the electron configurations of the elements.

The periodic table was a triumph for nineteenth-century chemistry. In 1869 Dmitri Mendeleev published a version of the table based on his periodic law that when elements are arranged in order of atomic mass, similar properties recur at intervals. He left gaps in his table for undiscovered elements and predicted their properties. His success with these predictions helped to persuade other scientists of the merits of his ideas when the missing elements were discovered.

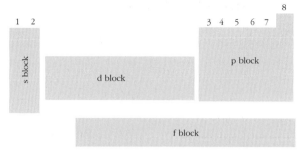

Outline periodic table showing the main blocks. Note that IUPAC now rules that the groups should be numbered from 1 to 18 across the table. In this system the old group 4 becomes group 14 and the old group 7 becomes group 17.

periodicity is a pattern which repeats itself. The most obvious repetition in the periodic table is from metals on the left of each period to *non-metals* on the right. Another repeating pattern from one period to the next is shown by the main *oxidation states* of the elements in their common compounds.

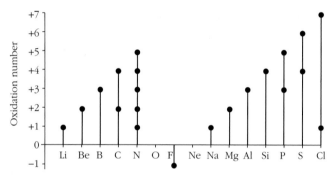

Periodicity of oxidation states of the period 2 and 3 elements in their oxides. Note that the pattern repeats but with variations

periodicity of chemical properties: the repeating patterns from one period to the next in the chemical properties of the elements and their compounds.

The *oxides* of reactive *metals* are ionic and basic. The oxides of non-metals are covalently bonded and acidic with *amphoteric oxides* in between.

		covalent giant structures		covalent molecular gases and solids		
ionic crystals						
$Li_2O(s)$	$BeO(s)$	$B_2O_3(s)$	$CO_2(g)$	$N_2O_5(s)$	$O_2(g)$	$F_2O(g)$
$Na_2O(s)$	$MgO(s)$	$Al_2O_3(s)$	$SiO_2(s)$	$P_4O_{10}(s)$ $P_4O_6(s)$	$SO_3(s)$ $SO_2(g)$	$Cl_2O(g)$
basic		amphoteric		acidic		

Periodicity: formulae, structures, bonding and acid–base character of oxides in periods 2 and 3

The chlorides of reactive metals are ionic. They dissolve in water to give neutral solutions. The chlorides of *non-metals* consist of covalent molecules and they are normally hydrolysed by water (see *hydrolysis of non-metal chlorides*).

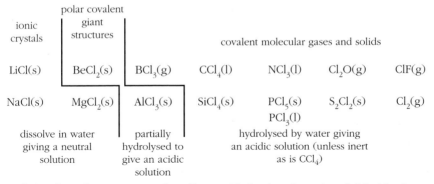

Periodicity: formulae, structures, bonding and behaviour in water of chlorides in periods 2 and 3

periodicity of physical properties: the repeating patterns in the *physical properties* of the elements and their compounds from one period to the next. Metals on the left of a period conduct electricity well. *Non-metals* to the right of a period are non-conductors. The in-between elements are the *metalloids*, many of which are *semiconductors*.

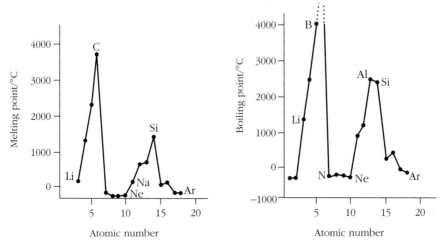

Plots of the melting points and boiling points against atomic number for periods 2 and 3. Note the repeating pattern. This is an example of periodicity. A plot of first ionisation energy against proton number is another example of a periodic pattern.

The periodicity in the physical properties of the elements in periods 2 and 3 reflects the underlying patterns of structure and bonding.

285

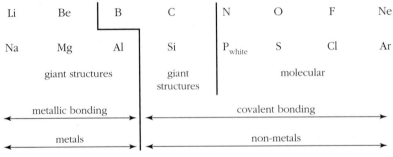

Li	Be		B	C		N	O	F	Ne
Na	Mg	Al	Si	P_{white}		S	Cl	Ar	

giant structures   giant structures   molecular

metallic bonding ⟷ covalent bonding ⟶

metals ⟶ non-metals ⟶

Periodicity: structure and properties of the elements in periods 2 and 3

permanent dipole: see *polar molecules* and *intermolecular forces*.

peroxides are compounds related to *hydrogen peroxide*, H_2O_2. The inorganic peroxides of *s-block elements* are ionic and contain the O_2^{2-} ion. When sodium burns in air it forms sodium peroxide Na_2O_2. Barium similarly produces barium peroxide, BaO_2.

Inorganic peroxides are powerful *oxidising agents*. The peroxide ion is a *strong base* so it also reacts with water and dilute acids to make hydrogen peroxide.

Organic peroxides, such as benzoyl peroxide, can act as a source of *free radicals* to initiate *addition polymerisation* of unsaturated compounds.

persistent organic pollutants (POPs) are toxic organic compounds that resist degradation and accumulate in food chains. They are transported by air, water and migrating species across international boundaries so that they are deposited far from their place of release, where they accumulate in terrestrial and aquatic ecosystems. Under a convention agreed in Stockholm in 2001, many countries have signed up to eliminate the twelve most serious POPs, the so-called 'dirty dozen'. These include a range of substances:
- some have been produced deliberately as *pesticides* for agriculture and to fight disease, while *PCBs* have a variety of industrial applications
- others are formed unintentionally, such as *dioxins*, which result from some industrial processes and from *incineration* of waste.

pesticides are *agrochemicals* which kill the living organisms that cause damage to crops, food, wood, fabrics and other materials. Some chemicals used as pesticides are entirely synthetic, others are natural (such as pyrethrums, obtained from chrysanthemums), while others are altered versions of natural chemicals.

Pesticides are used by farmers to:
- protect crops from insect pests, weeds and fungal diseases while they are growing
- prevent rats, mice, flies and other insects from contaminating foods while they are being stored
- safeguard human health, by stopping food crops being contaminated by fungi

Research chemists make and test many new compounds in search of pesticides which are:
- fast-acting
- specific so that they kill only the pests without harming other organisms
- effective at low doses
- biodegradable so that they quickly break down in the environment.

DDT is an example of a pesticide use to fight disease. DDT is one of the *persistent organic pollutants* but it is not toxic to humans while being very effective in controlling the mosquitoes that carry malaria. DDT has saved millions of lives but it can also have severe environmental consequences. Its use is still permitted for indoor spraying of homes as part of a systematic programme to control malaria infections.

petrochemical industry: the part of the *chemical industry* based on crude oil and natural gas. The industry has developed and expanded massively since the late 1930s.

petrol consists mainly of a mixture of *hydrocarbons* and is the fuel for motor engines. Petrol has to be carefully blended if modern engines are to start reliably and run smoothly. The proportion of *volatile* hydrocarbons added to petrol is higher in winter to help cold starting but lower in summer to prevent vapour forming before the fuel gets to the carburettor.

The starting point for making petrol is the gasoline fraction from the *fractional distillation of oil*. Hydrocarbons in this fraction are mainly *alkanes* with 5 to 10 carbon atoms. Fractional distillation does not provide enough gasoline and produces more than enough of the heavier fractions, so oil refineries run *catalytic crackers* to make more hydrocarbons of the right size. Other refining processes, and petrol additives, help to raise the *octane number* of petrol so that it burns smoothly in modern engines.

petroleum: a name for crude oil which is also used for some products from the *fractional distillation* and refining of oil. The *naphtha* fraction, which is an important feedstock for the chemical industry, is sometimes called the 'light petroleum fraction'.

Petroleum ether is a laboratory solvent which does not contain *ether*. Petroleum ether distils in the range 40–60°C and consists mainly of C_5 *hydrocarbons*. The solvent boiling in the 60–80°C range contains mainly C_6 hydrocarbons.

pharmacophore: the structural features of a *drug* molecule that are responsible for its biological activity. A pharmacophore is defined by the combination of functional groups and their three-dimensional arrangement that allows the molecule to bind to the target *active site* in a biological pathway. Since many biochemicals are *chiral compounds*, this means that pharmacophores are often chiral and only one *enantiomer* is active as a drug. Databases for *computational chemistry* in drug research classify molecules by their pharmacophores.

pH changes during acid–base titrations: the variation in *pH* as a solution of a base flows from a burette into a flask containing a measured volume of acid. Plotting a graph of pH against volume of alkali added gives a shape determined by the nature of the acid and the base.

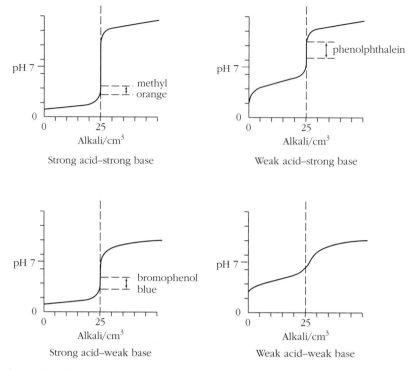

Graphs to show the pH change on adding 0.1 mol dm⁻³ alkali to 25 cm³ of 0.1 mol dm⁻³ acid. Note the sharp change of pH around the equivalence point except when both the acid and the base are weak. The graphs show the pH range for the colour change of an acid–base indicator for detecting the end-point of the titration.

The indicator chosen to detect the end-point must change colour completely in the pH range of the near vertical part of the curve. Note that at the equivalence point, when exactly equal amounts of acid and base have been added, the pH is not always neutral. (See also *neutralisation reactions*.)

It is possible to read off the value of K_a for a weak acid from the corresponding titration curve (see *Henderson–Hasselbalch equation*).

pH scale: a *logarithmic* scale for measuring the concentration of aqueous hydrogen ions in solutions.

$$pH = -\lg[H^+(aq)]$$

pH	0	1	2	3	4	5	6	7	8	9	10	11	12	13	14
[H⁺(aq)]/mol dm⁻³	10^0	10^{-1}	10^{-2}	10^{-3}	10^{-4}	10^{-5}	10^{-6}	10^{-7}	10^{-8}	10^{-9}	10^{-10}	10^{-11}	10^{-12}	10^{-13}	10^{-14}

increasingly acidic ◄————— neutral —————► increasingly alkaline

Worked example

What is the pH of 0.02 mol dm^{-3} hydrochloric acid?

Notes on the method

Hydrochloric acid is a strong acid so it is fully ionised. Note that 1 mol HCl gives 1 mol H$^+$(aq).

Use the log button on your calculator. Do not forget the minus sign in the definition of pH.

Answer

[H$^+$(aq)] = 0.02 mol dm^{-3}

pH = −lg (0.02) = 1.7

Worked example

What is the aqueous hydrogen ion concentration in a cola drink with pH 2.3?

Notes on the method

pH = −lg [H$^+$(aq)]

From the definition of logarithms this rearranges to [H$^+$(aq)] = 10^{-pH}

Use the inverse log button (10^x) on your calculator. Do not forget the minus sign in the definition of pH.

Answer

pH = 2.3

[H$^+$(aq)] = 10$^{-2.3}$ = 5 × 10^{-3} mol dm^{-3}

Strictly speaking, the full definition of pH is given by:

$$pH = -\lg_{10} \frac{[H^+(aq)]}{1 \text{ mol dm}^{-3}}$$

Mathematically it is only possible to take logarithms of numbers and not of quantities with units. Dividing by a standard concentration of 1 mol dm^{-3} does not change the value but cancels the units.

pharmaceuticals are chemicals used in *medicines*. In the UK the production of pharmaceuticals accounts for about a third of the wealth created by the *chemical industry*. (See also *chiral synthesis*, *drug development* and *pharmacophore*.)

phase: the three states of matter – solid, liquid or gas. Chemical systems often have more than one phase. Each phase is distinct but need not be pure:

- a solid in *dynamic equilibrium* with its saturated solution is a two-phase system
- in the reactor for *ammonia manufacture* the mixture of nitrogen, hydrogen and ammonia gases is one phase, with the iron catalyst being a separate solid phase
- *chromatography* separates mixtures by letting the components in a mixture come to equilibrium between two phases – the stationary phase and the mobile phase.

The same material broken up into little pieces still counts as one phase. In a *colloid* one phase is finely dispersed as specks, droplets or bubbles in a continuous phase.

phenols: a series of compounds with one or more —OH groups directly attached to a benzene ring. The simplest example is phenol, C_6H_5—OH.

Phenol is manufactured on a large scale. The main uses are to make:
- *thermosetting polymers*
- *epoxy resins* for adhesives
- *intermediates* for the manufacture of nylon.

phenol   benzene-1,2-diol   benzene-1,4-diol

Structure of phenol and other phenols

The hydroxyl group and the benzene ring in phenol interact. The benzene ring makes the hydroxyl group more acidic in phenol than it is in *alcohols*. Phenol is acidic enough to form salts with alkalis. So phenol dissolves easily in a solution of sodium hydroxide.

Phenol reacting with alkali. Note that the phenoxide ion is stabilised because the negative charge can be spread by delocalisation over the whole molecule. This cannot happen in alcohols.

Phenol is not as acidic as *carboxylic acids*. It does not ionise significantly in water and it does not react with carbonates to produce carbon dioxide. (See also *acid strength of organic hydroxy compounds*.)

Unlike alcohols, phenols do not react directly with carboxylic acids to form esters. It is, however, possible to convert a phenol to an ester with an *acyl chloride*.

The hydroxyl group makes the benzene ring more reactive. *Electrophilic substitution* takes place under much milder conditions with phenol than with benzene. Phenol can be nitrated with dilute nitric acid. Adding aqueous bromine to a solution of phenol produces an immediate white precipitate of 2,4,6-tribromophenol as the bromine colour fades.

Many phenols form coloured complexes with $Fe^{3+}(aq)$ ions in neutral iron(III) chloride solution. Phenol itself forms a violet complex. Halogenated phenols are used as *antiseptics* and *disinfectants*.

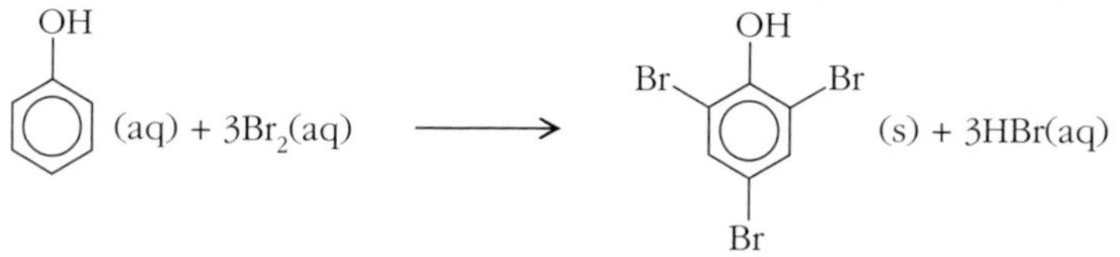

Reaction of phenol with bromine. The reaction is rapid at room temperature and there is no need for a catalyst.

phenyl group: the group C_6H_5— which is present in many compounds where one of the hydrogen atoms in benzene has been replaced by another atom or group.

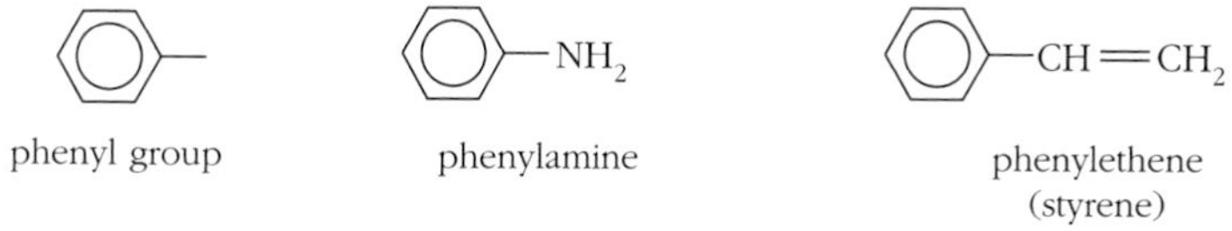

| phenyl group | phenylamine | phenylethene (styrene) |

Structure of the phenyl group and compounds derived from benzene

The use of phenyl in this way dates back to the first studies of benzene when 'phene' was suggested as an alternative name for the compound, based on a Greek word for 'giving light'. Benzene was found in the tar formed on heating coal to produce gas for lighting.

Note that 'benzyl' refers to aryl compounds with a carbon atom attached directly to the benzene ring.

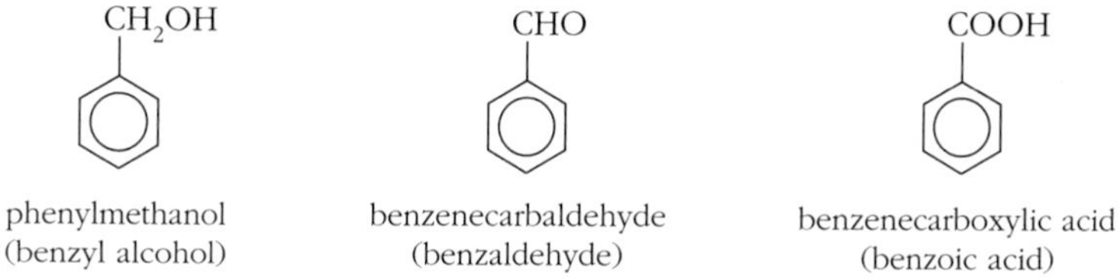

| phenylmethanol (benzyl alcohol) | benzenecarbaldehyde (benzaldehyde) | benzenecarboxylic acid (benzoic acid) |

Benzyl alcohol and other compounds related to $C_6H_5CH_2$—

phenylamine is a primary *amine* with an —NH_2 group attached to a benzene ring. There is a two-step laboratory route from benzene to aniline – first *nitration of benzene* then reduction:

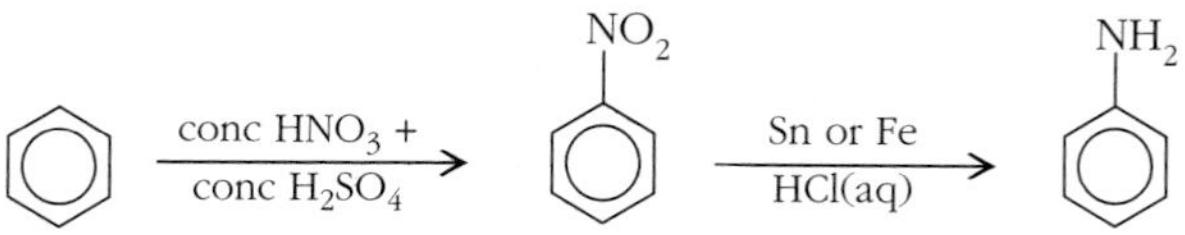

Formation of phenylamine from benzene

The chemical industry makes phenylamine from phenol and ammonia with an alumina catalyst. Phenylamine and other arylamines are important intermediates in the manufacture of *azo dyes*. This is because they react with nitrous acid below 10°C to produce *diazonium salts*.

Phenylamine is a weaker base than ammonia (see *base strength of amines*).

291

phosphorescence: see *luminescence*.

phosphoric(v) acid (H_3PO_4) is manufactured on a large scale from phosphate rock to make *fertilisers*, *detergent* phosphates and phosphates for food and drink. The pH of a typical cola drink is 2.3 because it contains 0.05% phosphoric acid. Another use of phosphoric acid is to make iron more resistant to corrosion.

The salts of phosphoric acid have many uses. These are a few examples:
- NaH_2PO_4 – used as a source of phosphate to make drugs and for pH control in toothpaste
- Na_2HPO_4 – used to make processed cheese (by a method patented by J. L. Kraft in 1916), also a *corrosion* inhibitor and an ingredient of enamels and glazes
- Na_3PO_4 – the most basic of the sodium salts used in heavy duty cleaners, paint strippers and as a water softener
- $Ca(H_2PO_4)_2$ – the acid in baking powders
- $CaHPO_4$ – the abrasive in toothpaste.

Organic phosphates play a vital part in biochemistry (see *ATP, nucleotides* and *nucleic acids*).

phosphorus (P) is a highly reactive element and important *non-metal*. It is the second element in group 5 of the periodic table, $[Ne]3s^22p^3$.

There are two common *allotropes*:
- white phosphorus, which is molecular (P_4), has to be stored under water otherwise it catches fire in air
- red phosphorus, which is made up of long chains of P_4 units, is more stable and does not catch fire in air at room temperature.

There are also scarlet and black allotropes.

White phosphorus condenses when phosphorus vapour cools. This happens in the manufacture of phosphorus from phosphate rock (fluorapatite). The red form is more stable and, at 540 K, white phosphorus changes rapidly to the red form without a catalyst.

Phosphorus forms two oxides. They are both molecular and both *acidic oxides*. The more important oxide is P_4O_{10} which forms as a white solid when white phosphorus burns in excess air (see *phosphoric(v) acid*).

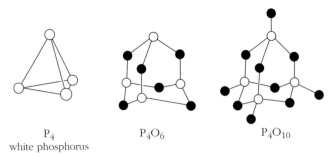

P_4
white phosphorus

P_4O_6

P_4O_{10}

Structures of white phosphorus, phosphorus(III) oxide, P_4O_6, and phosphorus(v) oxide, P_4O_{10}

As a laboratory reagent P_4O_{10} is used as a drying and dehydrating agent. It dehydrates *amides* to *nitriles*.

Phosphorus also forms two chlorides. Phosphorus trichloride (PCl_3) is made commercially from white phosphorus and chlorine. It is an important intermediate in the manufacture of medical drugs, insecticides and flame retardants. PCl_3 is a molecular compound. At room temperature it is a colourless fuming liquid which is rapidly hydrolysed by water (see *hydrolysis of non-metal chlorides*).

Phosphorus pentachloride (PCl_5) forms when PCl_3 reacts with excess chlorine. It is a volatile solid which in the solid state consists of $PCl_4^+PCl_6^-$. It too is rapidly hydrolysed by water and is a laboratory reagent for replacing hydroxyl groups with chlorine atoms in *alcohols* and *carboxylic acids*.

photodegradable: a term which describes any material, such as a plastic, that breaks down when exposed to light, especially ultraviolet light (UV). Some *polymeric materials* are transparent to UV light and so are not affected by photodegradation, Polyethylene, however, deteriorates relatively quickly when exposed to UV. Additives for plastics can include stabilisers to prevent the polymer becoming weak or brittle as a result of photodegradation.

photochemical smog is produced by sunlight and the *atmospheric pollutants* from the exhaust gases of motor vehicles. This type of *smog* forms on still, sunny days when there is no wind to blow away the gases. It is severe in cities such as Los Angeles where weather conditions and the local geography tend to trap pollutants in the city.

The primary pollutants are *nitrogen oxides* and unburnt *hydrocarbons* which are emitted in large amounts during the morning rush hour in a city. Bright sunlight during the middle of the day sets off photochemical reactions involving oxygen in the air. The products are the secondary pollutants which create smog.

The level of *ozone in the troposphere* rises and oxidising *free radicals* form. The unburnt hydrocarbons are then oxidised to *aldehydes*, *ketones* and other chemicals such as organic nitrates which irritate the eyes and lungs.

Catalytic converters are designed to reduce this kind of pollution.

photochemistry: the study of reactions caused by light or other forms of *electromagnetic radiation*. Photons of ultraviolet light have enough energy to break covalent bonds and form *free radicals*. These reactive intermediates can then start *free-radical chain reactions*. Photochemistry is important in the environment as well as in the laboratory (see *ozone in the stratosphere, ozone in the troposphere* and *photochemical smog*).

photon: a quantum of light energy (see *quantum theory* and *electromagnetic radiation*).

photosynthesis is the process by which plants use energy from the Sun to convert carbon dioxide and water to carbohydrates and oxygen (see *carbon cycle*). It is the starting point for the synthesis of all the organic molecules in plants, including *proteins*, *lipids* and *nucleic acids*. The nitrogen, phosphorus and other elements which plants need are taken in from the soil as soluble salts.

Chlorophyll, the green pigment in plants, absorbs light and transfers the energy to electrons. With the help of other molecules the energy is used to split water into hydrogen and oxygen and to make *ATP*. All the oxygen released into the air during photosynthesis comes from water.

Through a series of biochemical steps, the hydrogen atoms from water produce a reducing agent which, with ATP, converts carbon dioxide to the basic building block for carbohydrates and other chemicals. So all the carbon and oxygen in carbohydrates comes from carbon dioxide.

Overall, photosynthesis is an *endothermic* process.

physical chemistry: the branch of chemistry which provides the theories that help to make sense of inorganic and organic chemistry. Physical chemistry explores the states of matter and uses the *kinetic theory of gases* to explain the properties of solids, liquids and gases. Physical chemistry also investigates the interactions between *electromagnetic radiation* and matter to explain structures and *bonding*. Physical chemists use the full range of the electromagnetic spectrum in the various types of *spectroscopy*.

The methods and theories of physical chemistry help to answer four important questions about all sorts of chemical reactions:

- How much? – measuring *amounts* of chemicals makes it possible to find the *formulae* of compounds and the *balanced equations* for reactions, which in turn makes it possible to calculate the quantities of reactants and products involved in reactions.
- How far and in which direction? – the *equilibrium law, enthalpy changes, free energy changes* and *entropy* changes as well as *electrode potentials* help chemists to predict the direction and likely extent of chemical change.
- How fast? – the study of *rates of reaction* explores the factors which determine the speed of chemical change.
- How? – *rate equations*, the use of *isotopes* as *tracers* and the various types of spectroscopy help chemists to work out the *mechanism of a reactions*.

physical properties: properties which describe how a substance behaves when it is chemically unchanged. Examples include:

- appearance (colour, transparency)
- mechanical properties (*strength, hardness, ductility*)
- electrical properties (*electrical conductors*)
- thermal properties (*melting points* and *boiling points*, thermal conductivity, *specific heat capacity* and latent heats of melting and vaporisation).

pi bonds (π-bonds) are the type of bond found in molecules with double and triple bonds. The bonding electrons are in a π orbital formed by sideways overlap of two atomic p orbitals. In a π bond the electron density is concentrated on either side of the line between the nuclei of the two atoms joined by the bond.

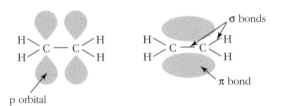

Pi bond in ethene. Note that the formation of a pi bonds holds the six atoms of the molecule in the same plane.

Pi bonds prevent rotation about the double bond and give rise to *cis–trans isomerism (E–Z isomerism)*. Electrons in π bonds are *delocalised electrons* in molecules such as benzene with alternating double and single bonds (see *conjugated systems*).

A π bond is not as strong as a *sigma bond* and so the double bond in ethene (or any other alkene) is not twice as strong as a single bond.

pigments are insoluble solids used in *paints*, cosmetics and printing inks which spread a thin layer of the pigment over a surface. Pigments may also colour the bulk of a material such as a plastic. *Titanium(IV) oxide* is a bright white pigment.

Natural pigments include chlorophyll (see *photosynthesis*), haem (see *haemoglobin*) and carotene (see *coloured compounds*).

pipettes are used to measure small volumes of liquid accurately. A volumetric pipette usually has a bulb for the main volume of the solution and then a single calibration line in the narrow tube above the bulb. This type of pipette can be used to deliver the same fixed volume of a solution again and again during titrations. (See *measurement uncertainty*.)

pK: the logarithmic form of an equilibrium constant which is particularly useful for pH calculations. Taking *logarithms* produces a conveniently small scale of values.

Examples:
- $pK_a = -\lg K_a$
- $pK_b = -\lg K_b$
- $pK_w = -\lg K_w$

For an acid and its conjugate base: $pK_a + pK_b = pK_w$

The *Henderson–Hasselbalch equation* is one illustration of the advantage of working with pK_a instead of K_a. Another is the use of pK_w (the logarithmic version of the *ionic product of water*).

$$K_w = [H^+][OH^-] = 1 \times 10^{-14} \text{ at 298 K}$$

Taking logarithms gives: $\lg K_w = \lg[H^+] + \lg[OH^-] = \lg 10^{-14} = -14$

Multiplying through by –1, reverses the signs:

$$-\lg K_w = -\lg[H^+] - \lg[OH^-] = 14$$

Hence: $pK_w = pH + pOH = 14$, where pOH is defined as $-\lg[OH^-]$ by analogy with pH.

So: pH = 14 − pOH which makes it easy to calculate the pH of alkaline solutions

---

Worked example

What is the pH of a 0.02 mol dm⁻³ solution of sodium hydroxide?

Notes on the method

Sodium hydroxide (NaOH) is a *strong base* so it is fully ionised. Find the values of logarithms with the lg button on a calculator.

Answer

$[OH^-] = 0.02$ mol dm⁻³

$pOH = -\lg 0.02 = 1.7$

$pH = 14 - pOH = 14 - 1.7 = 12.3$

planar molecules are flat molecules with all the atoms in the same plane (see *shapes of molecules* and *pi bonds*).

Planck's constant: see *quantum theory.*

plane-polarised light: a light beam in which all the waves are vibrating in the same plane.

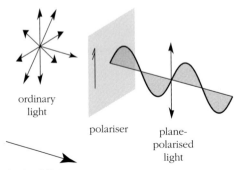

ordinary light

polariser

plane-polarised light

Ordinary light and polarised light

Light is plane-polarised after passing through a sheet of Polaroid. *Optical isomers* rotate the plane of polarised light.

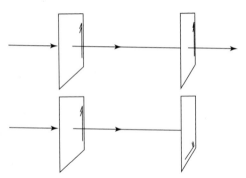

Effect of passing a light beam through two sheets of Polaroid. After the first sheet the light is polarised. The beam passes through the second sheet if it is aligned the same way as the first. No light gets through if the second sheet is rotated through 90°.

plastic materials are materials which can be moulded by gentle pressure. Examples are potter's clay and Plasticine. Once moulded, a plastic material keeps its new shape, unlike an *elastomer* which tends to spring back to its original shape.

plasticisers are additives mixed with *polymer* materials to make them more flexible. A plasticiser is usually a liquid with a high boiling point, often composed of large *ester* molecules. The molecules of the plasticiser get between the polymer chains and reduce the intermolecular forces between them so that they can slide past each other. Rigid uPVC (unplasticised) is suitable for window frames and guttering. Adding a plasticiser to PVC changes the polymer to a material suitable for squeeze bottles, hosepipes and the insulation on electric cables.

plastics: materials made of long-chain molecules which at some stage can be easily moulded into shape. *Thermoplastic polymers* become *plastic materials* when they are hot and harden on cooling. Some thermoplastics are rigid and brittle at room temperature such as polystyrene and uPVC; others, such as polythene, are flexible.

platinum (Pt) is a valuable metal which is the most abundant of the group of so-called 'platinum metals' which occur together in sulfide ores. The platinum metals are the *d-block elements* ruthenium and osmium, rhodium and iridium plus palladium and platinum.

Platinum is rarer than gold in the Earth's crust. Most of the annual production of the metal is from mines in South Africa where the yield of metal is about 30 g from 10 tonnes of rock.

Platinum is an attractive metal and it is very unreactive so it does not tarnish. This makes platinum attractive for jewellery. As well as being expensive, it is a *malleable* and *ductile* metal so it can be worked into shape.

Chemists take advantage of the inertness of the metal in electrochemistry. Platinum electrodes are required to measure standard electrode potentials. The conductor in a *standard hydrogen electrode* is platinum covered with a thin layer of finely divided metal (platinum black) deposited by electrolysis.

Platinum is an effective *catalyst*. Its catalytic activity is enhanced by alloying with rhodium. The second main use of platinum is to make *catalytic converters* for cars. Platinum–rhodium gauzes catalyse the oxidation of ammonia in *nitric acid manufacture*. Another use for a platinum–rhodium catalyst is 'platforming', which is platinum-catalysed *reforming* in oil refining.

Platinum forms a range of *complex ions* some of which, like *cisplatin*, are used to treat cancer.

polar covalent bonds are covalent bonds formed between atoms of different elements so that the shared electrons are drawn towards the more *electronegative* atom. One end of the bond has a slight excess of negative charge ($\delta-$). The other end of the bond has a slight deficit of electrons so that the charge cloud of electrons does not cancel the positive charge on the nucleus ($\delta+$).

Examples of molecules with polar bonds

Molecules with polar bonds are generally more reactive than non-polar molecules, especially with ionic reagents such as acids, alkalis and oxidising agents. The presence of polar bonds makes molecules open to attack by *electrophiles* and *nucleophiles*.

polar molecules have polar bonds which do not cancel each other out, so that the whole molecule is polar.

Molecules with polar bonds. Note that in the examples on the left the net effect of all the bonds is a polar molecule. In the examples on the right the overall effect is a non-polar molecule.

Polar molecules are little electrical dipoles (they have a positive pole and a negative pole). Dipoles tend to line up in an electric field. The bigger the dipole the bigger the twisting effect (dipole moment). By making measurements with a polar substance between two electrodes it is possible to calculate dipole moments. The units are Debye units named after the physical chemist Peter Debye (1884–1966).

The *intermolecular forces* are stronger between molecules with permanent dipoles.

Molecule	Dipole moment (in debye units)
HCl	1.08
H_2O	1.94
CH_3Cl	1.86
CCl_4	0
CO_2	0

Measuring dipole moments can distinguish between pairs of *cis-trans isomers* or *E–Z isomers*.

Polarity of cis–trans (E–Z) isomers

polar solvents are *solvents* made of *polar molecules*. Highly polar solvents such as water are needed to dissolve ionic compounds (see *hydration*). In organic chemistry the use of polar solvents favours *heterolytic bond breaking* and reactions with ionic intermediates.

polarisable: an indication of the extent to which the electron cloud in a molecule or ion is distorted by a nearby electric charge (see *polarisation of ions*).

Larger molecules with more electrons are generally more polarisable. The *intermolecular forces* which account for the weak intermolecular forces between non-polar molecules are stronger between molecules which are more polarisable.

polarisation of ions: this refers to the distortion of the electron cloud of a negative ion by a neighbouring positive ion. The extent of polarisation is greater if:

● the charges on the ions are large
● the radius of the positive ion is small and the radius of the negative ion large.

These points can help to decide the extent to which the bonding in a compound is ionic, covalent or intermediate between ionic and covalent (see *intermediate bonding*).

The approach is to start by picturing ionic bonding between two atoms, and then consider the extent to which the positive metal ion will polarise the neighbouring negative ions giving rise to some degree of electron sharing (that is a degree of covalent bonding).

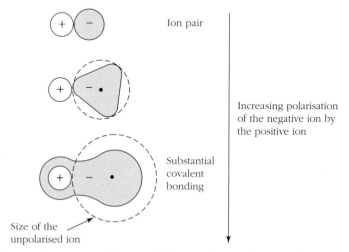

Ionic bonding with increasing degrees of electron sharing because the positive ion has polarised the neighbouring negative ion. Dotted circles show the unpolarised ions.

The smaller a positive ion is and the larger its charge, the greater the extent to which it tends to polarise a negative ion. So *polarising power* increases along the series Na^+, Mg^{2+}, Al^{3+}, Si^{4+}. Sodium chloride is an ionic, crystalline solid. The bonding in anhydrous *aluminium chloride* is largely covalent. Silicon chloride is a covalent, molecular liquid.

The larger the negative ion is and the larger its charge, the more *polarisable* it becomes. So iodide ions are more polarisable than fluoride ions. Fluorine, which forms the small, singly charged fluoride ion, forms more ionic compounds than any other non-metal.

polarising power: an indication of the extent to which a positive ion is able to distort the electron cloud around a neighbouring negative ion. The larger the charge is on a positive ion and the smaller its size, the greater its polarising power (see *polarisation of ions*).

Ionic radii for metal ions in the third period. Polarising power increases from left to right across this series.

pollution: contamination of the *atmosphere, hydrosphere* and *lithosphere* as a result of natural processes or events (such as excretion, the decay of dead organisms and volcanic eruptions) and human activities (such as agriculture, sewage discharges, combustion of fuels, extraction of resources, industrial processes and the disposal of *wastes*).

polyamides are *polymers* in which the monomer molecules are linked by amide bonds. The first important synthetic polyamides were formed by *condensation polymerisation* between diamines and dicarboxylic acids. (See also *Kevlar*.)

Condensation polymerisation to make nylon-6,6. The numbers in the name indicate the numbers of carbon atoms in the two monomers.

Peptides and proteins are natural polyamides.

polydentate ligands (multidentate ligands) form several *dative covalent bonds* with metal ions in *complexes*. An important example is *EDTA*. (See also *chelates*.)

polyesters are *polymers* formed by *condensation reactions* between acids with two carboxylic acid groups and alcohols with two or more hydroxyl groups. Units in the polymer chains are linked by a series of *ester* bonds.

Condensation polymerisation to produce a polyester

One pair of monomers for making polyester fibres is benzene-1,4-dicarboxylic acid and ethane-1,2-diol. The traditional names for these reactants are *ter*ephthalic acid and eth*ylene* glycol, hence the commercial name Terylene for one brand of polyester. An alternative name for the polymer is polyethylene terephthalate, which gives rise to the name PET when the same polymer is used as a plastic to make bottles for carbonated drinks.

Some polyesters, such as polylactic acid, are *biodegradable* because bacteria can break them down by hydrolysing the ester links.

polymers consist of long-chain molecules. Natural polymers include *proteins*, polysaccharides (see *carbohydrates*) and *nucleic acids*. Synthetic polymers include *polyesters*, *polyamides* and the many polymers formed by the *addition polymerisation* of compounds with C=C bonds.

polymeric materials are very varied and include *plastics*, *elastomers* and *fibres*. As polymer science has developed, chemists and materials scientists have learnt how to develop new materials with particular properties. Some of the ways of modifying polymeric materials include:

- altering the average length of the polymer chains
- changing the side groups along the chains, which alters the *intermolecular forces* (see also *polymer solubility*)
- varying the degree of *cross-linking* between chains
- *copolymerisation*
- controlling the three-dimensional shape of the polymer (see *isotactic polymers* and *atactic polymers*)
- changing the alignment of the polymer chains, for example by spinning the polymer into fibres and then cold drawing the fibres to align the molecules and make the polymer more crystalline
- adding *fillers*, *pigments*, *blowing agents* and *plasticisers*
- 'doping' an unsaturated polymer with iodine to give a *conducting polymer*
- making *composites*.

Molten polymers harden as they cool below their melting temperature (T_m). Many solid polymers are partly *amorphous* and partly crystalline. Even when solid, the molecules of a polymer retain some of their mobility. Rotation about covalent bonds is possible and the chains may even be able to slide past each other. This accounts for the flexibility and elasticity of these solid polymers. As polymer solids are cooled further below T_m there comes a point when all mobility of the molecules is lost. The solid then becomes glassy and brittle. The temperature at which this happens is the glass transition temperature (T_g).

polymerisation: a process in which many small molecules (*monomers*) join up in long chains (see *addition polymerisation* and *condensation polymerisation*).

polymer solubility: the extent to which *polymeric materials* can dissolve in solvents. Most polymers are insoluble in water but an exception is poly(ethenol). This polymer is used to make soluble laundry bags that dissolve in the wash. The polymer, cannot be made directly because the monomer that would be needed is unstable; however, it can be made either by alkaline *hydrolysis*, or by *transesterification* of another polymer called polyethenyl ethanoate, better known as PVA or polyvinyl acetate:

$$-[CH_2-CHOCOCH_3]_n- + nH_2O \rightarrow -[CH_2-CHOH]_n- + nCH_3COOH$$

The extent of hydrolysis can be controlled to vary the relative proportions of —OH groups and ester groups. The more —OH groups there are the greater the possibilities for *hydrogen bonding* and the more soluble the polymer can be in water depending on method of processing, the polymer chain length, the degree of crystallinity and the presence of plasticisers.

polymorphism: the existence of two or more crystalline forms of a substance. Some ionic compounds are polymorphic. Zinc sulfide, for example, can crystallise either with the zinc blende structure or with the alternative wurtzite structure. Polymorphism in solid elements is an example of allotropy (see *allotropes*).

polyols are compounds with more than one hydroxyl group. They include diols, such as ethane-1,2-diol and triols such as propane-1,2,3-triol (better known as glycerine or glycerol). Polyols are used to make *polyurethanes*.

polysaccharides: see *carbohydrates*.

polyurethanes are a varied range of cross-linked polymers made from two liquids – a *polyol* and an isocyanate. Polymerisation is exothermic and happens at room temperature. Adding a chemical to make a gas means that the polymer forms as a plastic foam. By choosing different isocyanates and polyols, manufacturers can vary the properties of the polyurethane. The products range through flexible foam for bedding and upholstery, rigid foam for wall panels and refrigerators, hard wearing but bendable soles for shoes and ingredients for paints and adhesives.

p orbitals: see *atomic orbitals*.

position of equilibrium: see *Le Chatelier's principle*, the *equilibrium law* and the *temperature effect on equilibria*.

potassium (K) is a very soft, shiny metal which rapidly tarnishes in moist air. It is the third member of *group 1* with the electron configuration $[Ar]4s^1$.

Like other group 1 metals, potassium:
- is stored in oil
- floats on water, melts and reacts violently forming hydrogen, which catches fire, and KOH, which is soluble and strongly alkaline
- forms an ionic, crystalline chloride K^+Cl^-.

Unlike lithium and sodium, potassium produces a *superoxide* (KO_2) when it burns in oxygen. This is an ionic compound with the O_2^- ion. One of the main uses of potassium is to make this oxide for use in emergency breathing apparatus. The oxide removes carbon dioxide from moist breathed-out air and replaces it with oxygen.

$$4KO_2(s) + 4CO_2(g) + 2H_2O(l) \rightarrow 4KHCO_3(s) + 3O_2(g)$$

Potassium, as potassium ions, is an essential nutrient for plants and an ingredient of NPK *fertilisers*. Large, underground deposits of potassium chloride are mined as the mineral sylvinite just south of Teesside in the UK.

potassium manganate(VII) ($KMnO_4$) consists of greyish-black crystals which dissolve in water to give a deep purple solution. It is used as a powerful *oxidising agent*.

Potassium manganate(VII) is an important reagent in *redox titrations* because it will oxidise many reducing agents in acid conditions, such as iron(II) ions. The reactions go according to their equations which makes them suitable for quantitative work.

$$MnO_4^-(aq) + 8H^+(aq) + 5e^- \rightarrow Mn^{2+}(aq) + 4H_2O(l)$$

No indicator is required for a manganate(VII) titration. When the solution is added from a burette, the manganate(VII) rapidly changes from purple to colourless (because the colour of the Mn^{2+} ion is so pale). At the end-point it takes only the slightest excess of manganate(VII) to give a permanent red-purple colour.

In organic chemistry potassium manganate(VII) is a reagent used to oxidise the side chains of arenes such as *methylbenzene*.

precious metals are expensive metals such as gold and silver. Chemists also use the term to describe other valuable metals which are used as catalysts, including the *platinum* metals.

precipitation: chemically this is a reaction which produces an insoluble product from soluble chemicals in solution. In the context of weather and climate, precipitation refers to rain, hail, sleet or snow falling from the atmosphere.

A common example of chemical precipitation is the formation of an insoluble salt on mixing solutions of two soluble salts (see *ionic precipitation*). This type of precipitation is the basis of many *anion tests* and *cation tests*.

In *organic analysis* the *2,4-dinitrophenylhydrazine* reagent for aldehydes and ketones forms a precipitate by an *addition–elimination reaction*. The *Fehling's solution* or *Tollens' reagent* tests, used to distinguish aldehydes from ketones, both make precipitates by redox reactions.

precision of data: data is precise if repeat measurements have values which are close to each other. Precise measurements have a small random *error*.

Higher precision is possible if measurements are taken with an instrument that allows readings to be taken with more *significant figures*.

Precise measurements may or may not be accurate. As a result of a systematic error, a series of precise measurements may give values which are almost the same but are not the true value.

pressure is defined as force per unit area. The *SI unit* of pressure is the pascal (Pa) which is a pressure of one newton per square metre (1 N m^{-2}). The pascal is a very small unit so pressures are often quoted in kilopascals (kPa).

When gases are being studied the standard pressure is *atmospheric pressure* which is:

$$101.3 \times 10^3 \text{ N m}^{-2} = 101.3 \text{ kPa}.$$

In accounts of chemical processes, multiples of atmospheric pressure give an indication of the extent to which gases are compressed (see for example *ammonia manufacture*).

The standard pressure for definitions in thermodynamics is now 1 bar which is:

$$100\ 000 \text{ N m}^{-2} = 100 \text{ kPa}$$

The *partial pressure* of a gas is a measure of its concentration in a mixture of gases.

primary, secondary and tertiary organic compounds are labels which distinguish different chemical situations for functional groups in *alcohols*, *halogenoalkanes*, *amines* and *carbocations*.

The labels have the same meaning for alcohols, halogenoalkanes and carbocations but a different meaning for amines.

primary standard: a chemical which can be weighed out accurately to make up a *standard solution* for *volumetric analysis*. A primary standard must:
- be very pure
- not gain or lose mass when exposed to the air (so it must not be *hygroscopic* or *deliquescent*)
- be soluble in water
- have a relatively high molar mass so weighing errors are minimised
- react exactly as described by the chemical equation.

A convenient primary standard for *acid–base titrations* is anhydrous sodium carbonate.

Primary standards for *redox titrations* include potassium dichromate(VI) and potassium iodate(V).

promoters are chemicals which make catalysts more effective. The iron catalyst in the Haber process for ***ammonia manufacture*** contains potassium hydroxide as a promoter. The vanadium(V) oxide catalyst used in ***sulfuric acid manufacture*** has potassium sulfate on a silica support to act as a promoter.

propagation is a step in a *free-radical chain reaction*.

propanone (acetone) is the simplest ***ketone***, CH_3COCH_3, which is widely used as a solvent. Three to four million tonnes of propanone are manufactured worldwide each year. Propanone is also used to make monomers for the manufacture of ***polymers***, including acrylics, polycarbonates and ***epoxy resins***. Most of the propanone is made in the cumene process, which also produces ***phenol***.

proportionality: a description of any relationship which takes the form:

$y \propto x$ or $y = $ constant $\times x$

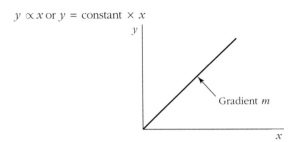

A graph showing that y is proportional to x. A straight line through the origin has the equation y = mx. The gradient of the line gives the value of the constant m.

The volume of a fixed amount of an ideal gas, for example, is proportional to the temperature on the ***kelvin*** scale at constant pressure. The rate of a ***first order reaction*** is proportional to the concentration of the reactant in the rate equation. (See also ***linear relationship***.)

proteins consist of long chains of ***amino acids*** joined by ***peptide*** bonds. Some fibrous proteins are vital to the structure of animals, including the proteins in muscles, ligaments, tendons, skin and hair. Other proteins coil up into a globular shape and dissolve in body fluids where they act as oxygen carriers, ***enzymes*** and ***hormones***.

Biochemists describe the structure at a series of levels:
- the primary structure is the sequence of amino acids in the polymer chains
- the secondary structure is the way in which the chains are arranged and held in place by ***hydrogen bonding*** within and between chains – this includes the coiling of chains into an α-helix in proteins such as keratin and the formation of layers of parallel chains as in the β-pleated sheets of silk
- the tertiary structure describes the three-dimensional folding of protein chains which gives some proteins, such as ***enzymes***, a definite three-dimensional shape held in place by hydrogen bonding, disulfide 'bridges' (see ***cross-linking***) and interactions between amino acid side chains with surrounding water molecules
- the quaternary structure describes the linking of two or more amino acid chains, as in the hormone ***insulin*** which consists of two chains linked by disulfide bonds, or in antibodies consisting of four chains.

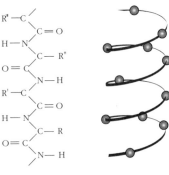  

Primary	Secondary	Tertiary	Quaternary
(sequence of amino acids)	(coiling of the amino acid chain)	(folding of coiled chain to create an active site)	(association of two or more coiled chains)

The four levels of protein structure

Proteins can be broken down to amino acids by acid or alkaline *hydrolysis*. The amino acids can be separated by *chromatography*.

proton: one of the two types of particle which make up the nucleus of an atom. The relative mass of a proton is 1 and its charge is +1. The number of protons in the nucleus of an atom is the *atomic number* (or proton number).

A hydrogen atom, H, normally consists of one proton in the nucleus and one electron in the first shell. So a hydrogen ion, H^+, is simply a proton. Chemists often use the terms 'hydrogen ion' and 'proton' interchangeably, especially when describing *acid–base reactions* as *proton transfer* reactions.

proton NMR spectroscopy is a type of *NMR spectroscopy* based on the magnetic properties of the hydrogen-1 isotope. The number of main peaks in the spectrum shows how many different chemical environments there are for hydrogen atoms. The values for the chemical shifts are a useful indication of the types of chemical environment for the hydrogen atoms corresponding to each peak.

In a proton NMR spectrum the area under a peak is proportional to the number of nuclei. This makes it possible to work out the number of hydrogen atoms in each environment. An NMR instrument is set up to integrate the curve and work out the ratios of the areas under the peaks. Sometimes the results of the calculation are shown by an integration trace with the spectrum.

Proton NMR spectra are usually recorded with the sample in solution. It is important that the solvent does not contain hydrogen atoms that would give peaks with chemical shifts similar to those in the sample. One possibility is to use a solvent that contains no hydrogen atoms, such a tetrachloromethane, CCl_4. The other is to use a solvent in which atoms of the hydrogen-1 isotope have been replaced by *deuterium* atoms. A compound that is often used is $CDCl_3$. Deuterium atoms produce peaks in regions of the spectrum well away from the chemical shifts for proton NMR.

The reference chemical for proton NMR is tetramethylsilane (TMS). There are twelve hydrogen atoms in a molecule of TMS and they all have the same chemical environment. TMS provides a single strong peak which marks the zero on the scale of chemical shifts.

At high resolutions, proton NMR spectra show that the spins of protons connected to neighbouring carbon atoms interact with each other. Chemists call this interaction 'spin-spin coupling' and they find that the effect is to split the main peaks into a number of lines.

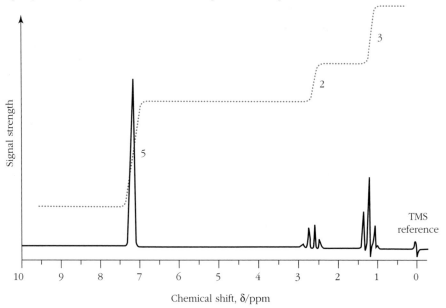

High-resolution proton-NMR spectrum of ethylbenzene, $C_6H_5CH_2CH_3$, with integration trace.

In simple examples the '$n + 1$' rule makes it possible to work out the numbers of coupled protons. A peak from protons bonded to an atom which is next to an atom with two protons splits into three lines. A peak from protons bonded to an atom which is next to an atom with three protons splits into four lines.

Labile protons do not couple with the protons linked to neighbouring atoms. This means that the NMR peak for a proton in an —OH group appears as a single peak in a high resolution spectrum.

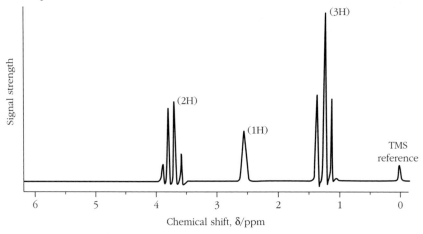

The high resolution NMR spectrum of ethanol. Note that there is no coupling between the proton in the —OH group and the protons in the neighbouring —CH₂ group

A useful technique for detecting labile protons is to measure the NMR spectrum in the presence of deuterium oxide (heavy water), D_2O. Deuterium nuclei can exchange rapidly with labile protons. Deuterium nuclei do not show up in the proton NMR region of the spectrum and so the peaks of any labile protons disappear.

proton number: the number of protons in the nucleus of an atom. An alternative to *atomic number*.

proton transfer describes the movement of a hydrogen ion (or *proton*) from an acid to a base during an *acid–base reaction*. According to the *Brønsted–Lowry theory* an acid is a proton donor and a base is a proton acceptor.

purification: a process used by chemists to remove impurities from the products of reaction. Common methods of purification include:
- *distillation* or *fractional distillation* for liquids
- *steam distillation* to separate a liquid with a high boiling point which does not mix with water
- *recrystallisation* to separate solids from solid impurities
- *solvent extraction* for liquids or solids.

Some types of *chromatography* are also used not only for analysis but also to separate pure products.

Chemists use a range of techniques to test for purity. A chemical is pure if it:
- melts exactly at its melting point
- boils at its boiling point
- gives only one spot or peak when analysed by chromatography
- gives a print out which matches the spectrum of a known pure sample when analysed by *mass spectrometry, infrared spectroscopy* or *NMR spectroscopy*.

Do you need revision help and advice?

Go to pages 388–407 for a range of revision appendices that include plenty of exam advice and tips.

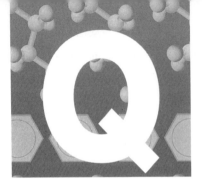

qualitative analysis is any method for identifying the nature of chemicals in a sample. Examples of qualitative analysis include:

- *gas tests* such as the use of limewater to detect carbon dioxide
- *anion tests* such as the use of silver nitrate solution to detect and distinguish chlorides, bromides and iodides
- the separation and identification of amino acids by *paper chromatography*
- the identifying of *functional groups* in an organic compound using *infrared spectroscopy*.

quantitative analysis: any method for determining the amount of a chemical in a sample. Examples of quantitative analysis include:

- an *acid–base titration* to determine the concentration of a solution of hydrochloric acid
- the use of a *colorimeter* to determine the concentration of a coloured *complex ion*
- testing the blood-alcohol concentration with a *breathalyser*
- finding the relative abundances of the *isotopes* of chlorine with a *mass spectrometer*.

quantum numbers identify the *energy level* or *orbital* occupied by an electron in an atom. Theory shows that four quantum numbers uniquely identify each electron in an atom:

- the principal quantum number indicates the distance of the electron from the nucleus
- the second quantum number identifies the type of *atomic orbital*, s, p, d or f
- the third quantum number shows the direction in space of the orbital, p_x, p_y or p_z
- the fourth quantum number states the alignment of the *spin*, spin up or spin down.

The principal quantum number identifies the main *shell* occupied by an electron. The first shell ($n = 1$) can hold 2 electrons, the second shell ($n = 2$) up to eight electrons and the third shell ($n = 3$) up to 18 electrons. Thus the maximum number of electrons in a shell is given by $2n^2$.

quantum theory states that radiation is emitted or absorbed in discrete amounts called energy quanta. Max Planck, the German physicist, put forward the theory in a paper published in 1900. Quanta have energy $E = h\nu$ where h is Planck's constant and ν is the frequency of the radiation.

The Danish physicist, Niels Bohr took up quantum theory in 1913 to explain the lines in hydrogen's *atomic emission spectra*. Bohr's theory could account very well for the frequencies of the lines in the spectrum of the hydrogen atom by making these assumptions:

- electrons in a hydrogen atom can only be at certain definite *energy levels*
- a photon of light is emitted or absorbed when an electron jumps from one energy level to another

- the energy of the photon, E, equals the difference between the two energy levels
- the frequency of the radiation emitted or absorbed is related to the energy of the photon by $E = h\nu$.

quartz is a crystalline form of **silicon(IV) oxide** (silica). Quartz has a high melting point because it consists of a covalently bonded **giant structure** in which each silicon atom is linked to four oxygen atoms and each oxygen atom to two silicon atoms.

quaternary ammonium cations are the cations formed when all the hydrogen atoms in an ammonium ion are replaced by **alkyl groups**.

$$
\begin{array}{c}
\text{R}' \\
| \\
\text{R} - \text{N}^+ - \text{R}'' \\
| \\
\text{R}'''
\end{array}
$$

General structure of a quaternary ammonium cation

Cationic **surfactants** used in fabric and hair conditioners are quaternary ammonium cations such as dodecyltrimethylammonium bromide, $CH_3(CH_2)_{11}N^+(CH_3)_3Br^-$.

quenching: the rapid cooling of a hot metal sample during heat treatment by plunging it into cold oil or water. Quenching a hot metal produces a different structure to that produced by slow cooling.

Similarly, chemists 'quench' reactions by sudden cooling. This effectively stops (or 'freezes') the reaction. In this way it is possible to find the composition of an equilibrium mixture or measure the concentration of a reactant or product during a study of rate of reaction.

Are you studying other subjects?

The *A–Z Handbooks (digital editions)* are available in 14 different subjects. Browse the range and order other handbooks at **www.philipallan.co.uk/a-zonline.**

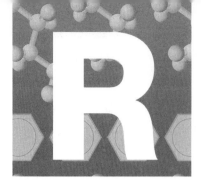

R_f values: ratios which measure how far a particular chemical moves during *paper chromatography* and *thin-layer chromatography* (TLC).

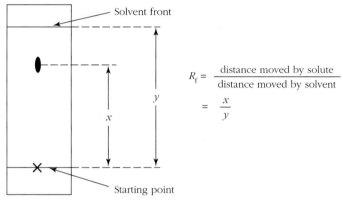

$$R_f = \frac{\text{distance moved by solute}}{\text{distance moved by solvent}}$$

$$= \frac{x}{y}$$

The R_f value is the ratio of the distance moved by the chemical in the mixture to the distance moved by the solvent front

R_f (retardation factor) values can help to identify components of mixtures so long as conditions are carefully controlled. The values vary with the type of paper or TLC plate and the nature of the solvent.

racemic mixture: a mixture of equal amounts of the two mirror image forms of a *chiral compound*. The mixture does not rotate *plane polarised light* because the two *optical isomers* have equal and opposite effects so they cancel each other out.

Lactic acid (2-hydroxypropanoic acid) from muscles is optically active because it consists of the (+) isomer. Lactic acid from sour milk is not optically active because it is a 50:50 racemic mixture of the two mirror image forms. Laboratory synthesis of lactic acid also produces the racemic mixture because there is an equal chance of the two isomers forming.

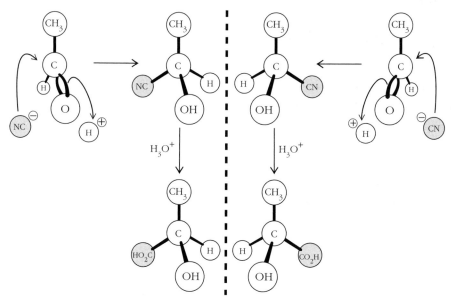

Addition of HCN to ethanal followed by hydrolysis produces a racemic mixture

radiation dose is a measure of the amount of energy transferred by *ionising radiation* to each kilogram of body tissue. The unit is the gray (Gy), 1 Gy=1 J kg⁻¹.

Some types of radiation are more damaging to living cells than others. Also some parts of the body are more at risk than others. This is allowed for by calculating an effective radiation dose which is a measure of the *risk* resulting from an exposure to ionising radiation. The effective dose is measured in *sieverts*.

radioactive decay: the decay of a radioactive nucleus to form the nucleus of another element by *alpha decay* or *beta decay*. Gamma rays may also be emitted.

radioactive nucleus → daughter nucleus + alpha or beta radiation

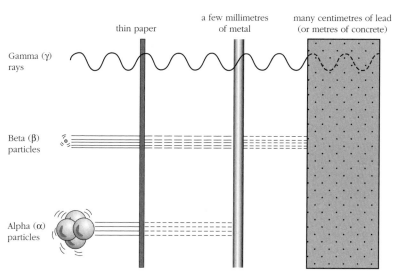

Radiations from radioactive materials differ in their ability to penetrate materials. The more penetrating the radiation, the more shielding is needed to protect people from dangerous exposure

Radioactive decay is a *first order reaction*. The rate of decay is proportional to the number of radioactive atoms. This means that each radioactive *nuclide* decays with a constant *half-life*.

Radionuclides with a short half-life decay more quickly and tend to give off more intense radiation. The unit of radioactivity is the *becquerel* (symbol Bq). 1 Bq is equal to the disintegration of one radioactive nucleus per second.

rare earths: the traditional name of the *lanthanide elements*.

rate constants: the constants in *rate equations*. The units of a rate constant depend on the overall order of reaction.

Overall order	Units of the rate constant
zero	$mol\ dm^{-3}\ s^{-1}$
first	s^{-1}
second	$mol^{-1}\ dm^3\ s^{-1}$
third	$mol^{-2}\ dm^6\ s^{-1}$

rate-determining step: the slowest step in the mechanism of a reaction, which therefore controls the overall rate of reaction. The rate-determining step in a multi-step reaction is the one with the highest activation energy. It is generally the molecules or ions involved (directly or indirectly) in the rate-determining step which appear in the rate equation for the reaction.

The *iodination* of propanone shows that it is not possible to deduce the rate equation from the balanced equation for the reaction:

$$I_2(aq) + CH_3COCH_3(aq) \rightarrow CH_2ICOCH_3(aq) + H^+(aq) + I^-(aq)$$

The rate equation is:

$$rate = k[CH_3COCH_3(aq)]^1[H^+(aq)]^1[I_2(aq)]^0 = k[CH_3COCH_3(aq)][H^+(aq)]$$

The iodine concentration does not affect the rate of reaction, showing that iodine is not involved in the rate-determining step. The reaction is catalysed by acid and the concentration of hydrogen ions appears in the rate equation.

The suggested mechanism for this reaction involves four steps. The rate-determining step involves propanone and hydrogen ions slowly producing an intermediate, which can rapidly rearrange and react with iodine.

rate equations are equations which show how changes in concentration, temperature and catalysts affect the rate of a reaction.

For the reaction of hydrogen peroxide with hydriodic acid:

$$H_2O_2(aq) + 2I^-(aq) + 2H^+(aq) \rightarrow 2H_2O(l) + I_2(aq)$$
$$rate = k[H_2O_2(aq)][I^-(aq)]$$

Note that the rate equation cannot be deduced from the balanced equation; it has to be found by experiment. The equation shows that this reaction is a *first order reaction* with respect to each reactant. Overall it is a *second order reaction*.

Methods for finding the order of reaction include:
- the *half-life* method
- the *initial-rate method*.

The rate constant, k, is constant only for a particular temperature. The value of k varies with temperature. A rise in temperature of 10 K doubles the value of k and so doubles the rate of reaction if the *activation energy* is about 50 kJ mol^{-1}. Adding a *catalyst* increases the value of k and hence the rate of reaction because the catalysed pathway has a lower activation energy.

The *Arrhenius equation* is the relationship between the rate constant, the activation energy and the temperature.

rates of reaction, in mol dm^{-3} s^{-1}, are found by measuring the rate of formation of a product or the rate of removal of a reactant. The usual procedure for finding rates is to measure some property of the reaction mixture, such as its volume, and to see how this property varies with time:

$$\text{rate} = \frac{\text{change in the measured property}}{\text{time}}$$

Rates are measured in practical units such as 'cm per second' and then converted to mol dm^{-3} s^{-1} if necessary.

Methods for studying rates include:
- collecting and measuring the volume of a gas formed
- removing measured samples of the mixture, stopping the reaction and then determining the concentration of one reactant or product by *titration*
- using a *colorimeter* to follow the formation of a coloured product or the removal of a coloured reactant
- using a conductivity cell and meter to measure the changes in electrical conductivity of the reaction mixture and the number or nature of the ions changes.

Factors which affect reaction rates are:
- the *concentration* of reactants in solution – in general the higher the concentration, the faster the reaction (see also *collision theory*)
- the *pressure* of gaseous reactants – high pressures compress gases and increase their concentration
- the surface area of solids – breaking a solid into smaller pieces increases the surface area in contact with a liquid or gas
- the temperature – raising the temperature increases the proportion of atoms, molecules or ions with enough energy to react when they collide
- catalysts – adding a *homogeneous catalyst* or a *heterogeneous catalyst* provides an alternative reaction mechanism with lower *activation energy*
- radiation – *electromagnetic radiation* can initiate *free-radical chain reactions* and the more intense the radiation the faster the reaction goes.

raw materials: see *chemical industry* and *renewable resources*.

reacting masses: the masses of elements and compounds which take part in a chemical reaction. Chemists work out *balanced equations* by measuring the masses of reactants and products. They use equations to work out how much of each chemical they need for a chemical process and the theoretical *yield* of the products.

Worked example

Carbon monoxide reduces iron(III) oxide to iron in a blast furnace. What mass of carbon monoxide is required to reduce 20 tonnes of the oxide?

Notes on the method

Start by writing the balanced equation for the reaction. Work out the *molar masses, M,* of the reactants in g mol⁻¹.

The reacting masses are in the same ratio whether measured in grams or in tonnes.

Answer

$$Fe_2O_3(s) + 3CO(g) \rightarrow 2Fe(s) + 3CO_2(g)$$

1 mol $Fe_2O_3(s)$ reacts with 3 mol $CO(g)$.

$$M(Fe_2O_3) = (2 \times 56 \text{ g mol}^{-1}) + (3 \times 16 \text{ g mol}^{-1}) = 160 \text{ g mol}^{-1}$$

$$M(CO) = (12 + 16) \text{ g mol}^{-1} = 28 \text{ g mol}^{-1}$$

So (1 mol × 160 g mol⁻¹) = 160 g iron reacts with

(3 mol × 28 g mol⁻¹) = 84 g carbon monoxide,

or 160 tonnes iron reacts with 84 tonnes carbon monoxide

Hence mass of CO needed to react with 20 tonnes iron $= \dfrac{20}{160} \times 84$ tonnes

$$= 10.5 \text{ tonnes}$$

reacting volumes of gases: see *gas volume calculations.*

reaction kinetics: the study of the factors that affect the *rate of reaction*. Rates are followed either by measuring the rate of formation of one of the products or the rate of removal of one of the reactants. The results of these studies are summed up in *rate equations* for reactions. *Collision theory* is one model which helps to explain the factors affecting rates. Studying the kinetics of a change can help to work out the *mechanism of a reaction*.

reaction mechanisms: see *mechanism of a reaction.*

reactors are reaction vessels used by the chemical industry for *synthesis*. There are two main types of reactor:

- batch reactors – for *batch processes* in which all the reactants are mixed in a vessel where they are heated or cooled until the process is complete
- continuous reactors – for *continuous processes* in which the reactants are fed continuously into a stirred tank or tube and the products drawn out at an equal flow rate.

reagent: any chemical, mixture or solution supplied for chemical analysis or synthesis.

'Reagent grade' chemicals match the minimum standards specified for normal laboratory use.

Analytical grade (AR) reagents are much purer than standard laboratory grade reagents. They are intended for accurate analytical work where impurities in the reagents would affect the results.

real gases are gases that deviate from *ideal gas* behaviour, as predicted by the ideal gas equation. In practice, many gases at room temperature follow the gas laws sufficiently closely for it to be reasonable to treat them as ideal gases.

The assumptions of the *kinetic theory of gases* explain why real gases tend to deviate from ideal behaviour at:

- low temperatures and moderate pressures – when gases are not far above their boiling points so that they are close to condensing because of *intermolecular forces*
- very high pressures – when the volume of the gas is so small that the volume of the molecules cannot be ignored and the gases are less compressible than the gas laws predict.

rearrangement reaction: a reaction which rearranges the atoms in a molecule to convert one *structural isomer* to another.

One example of a rearrangement reaction is the isomerisation of alkanes to make branched compounds to increase the *octane number* of *petrol*.

$$CH_3 - CH_2 - CH_2 - CH_2 - CH_3 \longrightarrow CH_3 - \overset{\overset{\displaystyle CH_3}{\displaystyle |}}{CH} - CH_2 - CH_3$$

pentane   2-methylbutane

Isomerisation rearranges the atoms in pentane to make 2-methylbutane

rechargeable cells: see *lead–acid cell*, *lithium polymer cells* and *nickel–cadmium cells*.

recrystallisation is a procedure often used to purify solid products of *organic preparations*. The procedure is based on a *solvent* which dissolves the product when hot but not when cold. The choice of solvent is usually made by trial and error. Use of a *Buchner flask and funnel* speeds filtering and makes it easier to recover the purified solid from the filter paper.

The procedure is as follows:

1 Warm the impure solid with the hot solvent.
2 If the solution is not clear, filter the hot solution though a heated funnel to remove insoluble impurities.
3 Cool the solution so that the product recrystallises leaving the smaller amounts of soluble impurities in solution.
4 Filter to recover the purified product.
5 Wash the solid with small amounts of pure solvent to wash away the solution of impurities still clinging to the solid.
6 Allow the solvent to evaporate in a stream of air and then in a *desiccator*.

recycling is a method of dealing with *wastes*. The economics of recycling often depend on the energy required to process wastes compared to the energy needed to produce materials directly from raw materials. The large amount of energy involved in *aluminium extraction* makes recycling particularly worthwhile. There are other benefits to recycling, such as reducing the environmental impact of extracting and processing raw materials.

Recycling metals – recycling *steel* can be done repeatedly because steel is as good as new after reprocessing. For every tonne of steel recycled there is a saving of 1.5 tonnes of iron ore and half a tonne of coal, as well as a reduction of 80% in the carbon dioxide emissions when compared to steel made from iron ore. There is also a big reduction in the total amount of water needed since large quantities of water are involved in mineral processing.

Worldwide, about half of all the steel produced is recycled metal. The basic oxygen process for making steel uses a minimum of 25% scrap steel. Electric arc furnaces make steel plate, beams and bars by remelting nearly 100% scrap steel. Much of the recycled steel is waste from various stages of manufacturing steel objects. Another large source of scrap is from old motor vehicles.

Some steel is recovered from the tin cans in household waste. The advantage of steel is that it is magnetic so magnets can easily separate cans from other waste materials.

Aluminium scrap from manufacturing processes is always recycled. Recycling aluminium requires only 5% of the energy and produces only 5% of the CO_2 emissions when compared with *aluminium extraction* from bauxite. It also reduces the waste going to landfill. Aluminium can be recycled indefinitely because reprocessing does not damage its structure.

Recycling plastics – plastics are much more difficult to recycle than iron or steel. One problem arises from the low density of plastic waste. It can take as many as 20 000 bottles to make up a tonne of waste. Also, there are many types of plastic which have to be sorted. The industry has introduced codes for labelling plastic products to help consumers to separate them for recycling.

Plastics vary greatly in their value. The *polyester* used to make sparkling-drink bottles, PET, is generally the most valuable. PET can be remelted and spun into fibres for carpets, bedding and clothing. Another plastic worth recycling is PVC, which can be recycled to make sewage pipes, flooring and even the soles of shoes.

An alternative approach to recycling plastics is to use cracking to break the polymer molecules into small molecules which can be added to the *feedstock* for oil refineries and chemical plants.

redox potential: see *standard electrode potential*.

redox reactions: reactions which involve **red**uction and **ox**idation. In every redox reaction an atom, molecule or ion is reduced while another atom, molecule or ion is oxidised. Reduction and oxidation always go together.

Redox reactions involve the transfer of electrons from a *reducing agent* to an *oxidising agent*. Writing *half-equations* for redox reactions helps to show electron transfer.

Oxidation numbers help to identify redox reactions. In any redox reaction the oxidation number of one element becomes more positive (or less negative) while the oxidation number of the other element becomes less positive (or more negative).

Equations for redox reactions, like other balanced equations, show the *amounts* (in moles) of reactants and products involved. Oxidation numbers help to balance redox equations because the total decrease in oxidation number for the element reduced must equal the total increase in oxidation number for the element oxidised.

Worked example

What is the balanced equation for the oxidation of iron(II) ions by manganate(VII) ions in acid solutions.

Answer

> **Step 1** – Write down the formulae for the atoms, molecules and ions involved in the reaction:
>
> $$MnO_4^- + H^+ + Fe^{2+} \rightarrow Mn^{2+} + H_2O + Fe^{3+}$$

Step 2 – Identify the elements which change in oxidation number and the extent of change:

$$\overset{\text{change of }-5}{\overbrace{MnO_4^- \; + \; H^+ \; + \; Fe^{2+} \; \longrightarrow \; Mn^{2+}}} + H_2O + Fe^{3+}$$

change of +1

Step 3 – Balance so that the decrease in oxidation number of one element equals the total increase of the other element.

In this example the decrease of –5 in the oxidation number of manganese is balanced by five iron(II) ions, which each increase their oxidation number by +1.

$$MnO_4^- + H^+ + 5Fe^{2+} \rightarrow Mn^{2+} + H_2O + 5Fe^{3+}$$

Step 4 – Balance for oxygen and hydrogen.

In this example the four oxygens from the manganate(VII) ion join with eight hydrogen ions to form four water molecules:

$$MnO_4^- + 8H^+ + 5Fe^{2+} \rightarrow Mn^{2+} + 4H_2O + 5Fe^{3+}$$

Step 5 – In an *ionic equation*, check that the positive and negative charges balance and add state symbols.

The net charge on the left is now 17+, which is the same as the net charge on the right.

$$MnO_4^-(aq) + 8H^+(aq) + 5Fe^{2+}(aq) \rightarrow Mn^{2+}(aq) + 4H_2O(l) + 5Fe^{3+}(aq)$$

redox titration: a practical technique used to determine the *concentration* of a solution of an *oxidising agent* or of a *reducing agent*. A *titration* measures the volume of a *standard solution* of oxidising agent or reducing agent needed to react exactly with a measured volume of the unknown solution. The procedure gives accurate results only if the reaction is rapid and is exactly described by the chemical equation. Titrations with *potassium manganate(VII)* can give very accurate results and do not need an indicator.

An acidic standard solution of the oxidising dichromate(VI) ions, for example, can be used to estimate the concentration of iron(II) ions. The end-point for a dichromate(VI) titration is detected using a redox indicator – a chemical which gives a clear colour change when oxidised. The procedure and method of calculating results is similar to those for other redox titrations such as *iodine–thiosulfate titrations*.

reducing agents: chemical reagents which can reduce other atoms, molecules or ions by giving electrons to them. Common inorganic reducing agents are metals such as zinc, iron and tin, often with acid; other common reducing agents are sulfur dioxide, iron(II) ions and iodide ions.

$$\overset{\text{reduction}}{\overbrace{2Fe^{3+}(aq) \; + \; 2e^- \; \longrightarrow \; 2Fe^{2+}(aq)}}$$

$$\underset{\text{oxidation}}{\underbrace{2I^-(aq) \; \longrightarrow \; I_2(aq) \; + \; 2e^-}}$$

Iodide ions acting as a reducing agent by giving electrons to iron(III) ions. A reducing agent is itself oxidised when it reacts. Reduction and oxidation always go together in redox reactions

Some reagents change colour when they are reduced, which makes them useful for detecting reducing agents.

Test	Observations	Explanation
Add a solution of potassium manganate(VII) acidified with dilute sulfuric acid	Purple solution turns colourless	Purple MnO_4^- ions are reduced to very pale pink Mn^{2+} ions
Add potassium dichromate(VI) solution acidified with dilute sulfuric acid	Orange solution turns green	Orange $Cr_2O_7^{2-}$ ions are reduced to green Cr^{3+} ions

Powerful reducing agents used in organic chemistry include *lithium tetrahydridoaluminate(III)* and *sodium tetrahydridoborate(III)*.

reducing sugars: *sugars* such as glucose and fructose which give a positive result with *Fehling's solution* or *Benedict's solution*. These sugars reduce the blue copper(II) complex in the test solution to an orange-red precipitate of copper(I) oxide.

All monosaccharides and many disaccharides are reducing sugars, but the disaccharide sucrose is not a reducing sugar.

reductant: an alternative term to *reducing agent* which is convenient when describing half-equations for *redox reactions* and *standard electrode potentials*. By convention, when electrode potential values are assigned, the half-equation takes the form:

$$\text{oxidant} + ne^- \rightleftharpoons \text{reductant}$$

Every half-equation involves an *oxidant* and a reductant.

reduction: originally this meant removal of oxygen or the addition of hydrogen but the term now covers all reactions in which atoms, molecules or ions gain electrons. The definition is further extended to cover molecules, as well as ions, by defining oxidation as a change which makes the *oxidation number* of an element more negative or less positive.

$$\overset{+6 \quad\quad\quad \text{sulfur reduced} \quad\quad +4}{2Br^- + 2H^+ + H_2SO_4 \longrightarrow Br_2 + SO_2 + 2H_2O}$$

$$\underset{-1 \quad\quad\quad \text{bromine oxidised} \quad\quad 0}{}$$

Bromide ions reducing concentrated sulfuric acid to sulfur dioxide

Oxidation and reduction always go together in *redox reactions*.

Oxidation number rules also apply in principle in organic chemistry but it is often easier to use the older definitions. Reduction is either removal of oxygen from a molecule or the addition of hydrogen. (See also *functional group level*.)

$$\begin{array}{c} H_3C \\ \diagdown \\ C{=}O + 2[H] \\ \diagup \\ H_3C \end{array} \xrightarrow{\text{NaBH}_4(\text{aq})} \begin{array}{c} H \\ | \\ H_3C{-}C{-}CH_3 \\ | \\ OH \end{array}$$

Reduction of a ketone to an alcohol by addition of hydrogen

reference electrodes: electrodes used to measure electrode potentials instead of the standard *hydrogen electrode*. A hydrogen electrode is difficult to use so it is much easier to use a secondary standard such as a silver/silver chloride electrode or a calomel electrode. These electrodes are available commercially and are reliable in use. They have been calibrated against a standard hydrogen electrode. (Calomel is an old-fashioned name for mercury(I) chloride.)

$$Hg_2Cl_2(s) + 2e^- \rightleftharpoons 2Hg(l) + 2Cl^-(aq) \quad E^{\ominus} = +0.27 \text{ V}$$

refining: the processes which separate, convert and purify chemicals in *crude oil* in an oil refinery, or the removal of impurities from metals (see *copper refining*).

reflux condenser: a condenser fitted to some chemical apparatus to prevent vapour escaping while heating a liquid. Vapour from a boiling reaction mixture condenses and flows back into the flask.

reforming: a process in oil refining which converts *alkanes* to *arenes* such as benzene and methylbenzene. This helps to raise the octane number of mixtures used in *petrol*.

$$CH_3CH_2CH_2CH_2CH_2CH_2CH_3 \xrightarrow[\text{heat, pressure}]{\text{Pt catalyst}}$$

heptane

methylbenzene

$+ \quad 4H_2(g)$

An example of reforming to make **methylbenzene**

refractories: materials with very high melting points used to line furnaces and make crucibles. *Ceramics* are refractory materials.

Fireclay is the raw material for making refractory bricks. Some industries need more specialised refractories. Molten *glass* and the *slags* formed during smelting are very corrosive when molten and would quickly destroy ordinary refractory bricks made of fireclay. These industries use refractories such as pure *silicon(IV) oxide* (an *acidic oxide*) or pure magnesium oxide (a *basic oxide*).

relative atomic mass (A_r) is the mean mass of the atoms of an element relative to the mass of atoms of the isotope carbon-12 for which A_r is defined as exactly 12. The values are relative so they do not have units.

Mass spectrometry is the technique for determining accurate relative atomic masses.

Amount of substance in chemistry is defined in such a way that the *molar mass* of an element is numerically equal to its relative atomic mass (but the latter is a ratio and has no units).

Relative atomic masses often do not have whole number values because elements have *isotopes*. Relative atomic masses are average values for the mixture of isotopes found naturally (see *isotopic abundance*).

relative formula mass (M_r) is a term used for the relative mass of ionic compounds or ions to avoid the suggestion that their formulae represent molecules. The scale is the same as for relative atomic and relative molecular masses, that is $^{12}C = 12$.

The relative formula mass of an ionic compound or ion is the sum of the *relative atomic masses* for the atoms in the formula. For anhydrous magnesium nitrate:

$$M_r[Mg(NO_3)_2] = 24 + (2 \times 14) + (6 \times 16) = 148$$

$$\uparrow \qquad \uparrow \qquad \uparrow$$

$$A_r(Mg) \quad A_r(N) \qquad A_r(O)$$

Amount of substance in chemistry is defined in such a way that the *molar mass* of an ionic compound or an ion is numerically equal to its relative formula mass (but the relative mass has no units).

relative isotopic mass: the mass of the atoms of an *isotope* of an element relative to the mass of atoms of the isotope carbon-12. Chemists determine relative isotopic masses with a *mass spectrometer*. The values are always very close to the mass numbers of the *nuclide*. It is usually accurate enough to assume that the value of the relative isotopic mass is the same as the *mass number*.

relative molecular mass (M_r) is the relative mass of the molecules of an element or compound on the scale $^{12}C = 12$.

The relative molecular mass of a molecular element or compound is the sum of the *relative atomic masses* for the atoms in the *molecular formula*. For ethanol:

$$M_r(CH_3CH_2OH) = (2 \times 12) + (6 \times 1) + 16 = 46$$

$$\uparrow \qquad \uparrow \quad \uparrow$$

$$A_r(C) \qquad A_r(H) \; A_r(O)$$

Amount of substance in chemistry is defined in such a way that the *molar mass* of a molecular element or compound is numerically equal to its relative molecular mass but as with other relative masses, the relative value has no units.

reliability: a measurement is reliable if repeating the measurement gives the same result. A burette is a highly reliable instrument for measuring volumes so long as it is kept clean. A pH meter is generally less reliable unless it is carefully recalibrated each time it is used.

A report of a set of results is more reliable if it is made by more than one person, perhaps using different methods.

renewable resources are energy and material resources that are replaced by natural processes at least as fast as they are being used in human activities. Renewable energy resources include solar, wind and hydro power. Renewable materials include the crops and biomass used to make *biofuels*. One of the aims of *green chemistry* is to develop renewable *feedstocks* for manufacturing processes. A biorefinery can extract sugars, starch, cellulose and other *carbohydrates*, as well as *vegetable oils*, from plant materials and then convert them to starting points for synthesis in the form of alkenes, alcohols, carboxylic acids and other compounds.

reserves and resources describes the quantities of *minerals* and *fossil fuels* available in the Earth's crust. Estimates of reserves and resources include some which are proven because they have been identified and evaluated, and others which seem likely to exist by a mixture of inference and guesswork.

The proven reserves of a mineral have been identified and evaluated so that it is known that it is economical to extract and process them at the current market price of the product. The resources of a mineral are all the possible sources, including those that it is not economical to extract at the time.

Judgements about changes in technology, in the demand for materials and in the economy can all shift the balance of reserves and resources. Figures for reserves have to be interpreted with care because commercial companies do not find it economic to establish reserves for more than about 25 years ahead.

residence time: a measure of the time a chemical stays in any part of a system where chemicals are continuously flowing in and out of containers or reservoirs. In environmental systems the reservoirs are parts of the Earth system such as the *atmosphere*, oceans or land. In a *steady-state system* the residence time equals the amount of the chemical in the reservoir divided by the rate of addition (or removal) of the chemical. The residence time of calcium ions in the oceans is a million years. The residence time of oxygen in the atmosphere is 7600 years. The residence time of a *greenhouse gas* in the atmosphere is one of the factors that determines its overall *global warming* potential.

residues: materials left over at the end of a chemical process. The term sometimes refers to a solid caught in a filter paper. Laboratories which use large quantities of valuable chemicals may have a 'residues bottle' to collect wastes for recovery of the expensive substances. Residues worth collecting include silver compounds, iodine compounds and solvents.

resins: originally these were natural materials from the sap of trees or from insects Examples are the rosin used for violin bows, shellac used in varnishes and myrrh which is added to perfumes. These gummy materials consist of long-chain molecules and on warming they soften before they melt.

Now the term resin refers to synthetic materials made by *polymerisation*. *Thermosetting polymers*, for example, are produced in two stages. The first stage produces a polymer resin. Compressing this resin in a hot mould shapes it and at the same time creates *cross-links* between the polymer chains so that the material sets to a hard plastic.

Similarly, one component of an *epoxy resin* adhesive is a polymer resin which sets on mixing with a hardener to start the chemical changes which cross-link the chains.

Synthetic *ion-exchange* resins consist of polymer beads which have been treated chemically so that they can hold cations or anions.

respiration: biochemical processes in cells which provide living organisms with the energy for growth, movement and warmth.

Aerobic respiration is respiration with oxygen. In this exothermic process, glucose is *oxidised* to carbon dioxide and water. Overall:

$$C_6H_{12}O_6(aq) + O_2(g) \rightarrow 6CO_2(g) + 6H_2O(l) \qquad \Delta H = -2816 \text{ kJ mol}^{-1}$$

This is not a one-step reaction; it takes place in a complex series of biochemical reactions which harness the energy from the oxidation process to produce *ATP*.

Anaerobic respiration is respiration without *oxygen*. Anaerobic respiration in yeast converts *sugars* to carbon dioxide and ethanol. This is the process of *fermentation* used in baking, brewing and winemaking. Anaerobic respiration in animal cells produces lactate ions (the anions of lactic acid). Anaerobic respiration produces much less ATP than aerobic respiration and so is a much less efficient source of energy for processes.

restricted rotation: see *pi bonds*, *cis–trans isomers* and *E–Z isomers*.

retention times: see *gas chromatography*.

reversible changes are processes which can be reversed by altering the conditions.

Melting is an example of a reversible change of state. Changing the temperature alters the direction of change. Ice melts above 0°C. Water freezes below 0°C. Water and ice are in equilibrium at 0°C. This is an example of *dynamic equilibrium*.

$$H_2O(s) \rightleftharpoons H_2O(l)$$

Many chemical reactions are reversible, including the reactions used in *ammonia manufacture* and *sulfuric acid manufacture*. *Acid–base reactions* and *redox reactions* are generally reversible too. There are many important reversible processes in living things. *Haemoglobin*, for example, picks up oxygen from the air in the lungs and then releases the oxygen for *respiration* in body tissues.

The direction and extent of change for a reversible process may vary with the temperature or the concentration of chemicals. *Pressure* is an important variable affecting reactions of gases.

Chemists predict the direction and extent of change for reversible processes when they consider *feasibility*.

rhodium is one of the *platinum* metals which is largely used as a catalyst in the platinum–rhodium alloys found in *catalytic converters* and in chemical plants for *nitric acid manufacture*.

risk in chemistry is an estimate of the chance that a hazardous substance or process will cause harm.

There are three questions to consider when assessing the risk of a chemical process:
- what are the hazards?
- what is the likelihood that someone will come to harm because of the hazards?
- what can be done to control and reduce the risks?

When a chemical investigation is being planned, the possibilities for controlling risks include:
- choosing a non-practical approach using secondary sources, models and simulations
- adopting an alternative safer procedure with less hazardous chemicals
- modifying the experimental design with different apparatus or a different method of heating
- creating a barrier between the apparatus and people, using safety screens or a fume cupboard
- wearing personal protection such as eye protection and gloves.

(See also *hazard*.)

RNA: a type of *nucleic acid* in which the five-carbon sugar is ribose and the four nitrogenous bases are adenine, cytosine, guanine and uracil. Two types of RNA are important in *protein* synthesis:
- messenger RNA (mRNA) which takes the genetic code from *DNA* molecules in the nucleus to the sites of protein synthesis in the cell

- transfer RNA (tRNA) which helps to assemble amino acids in the right order during the translation of the genetic code into protein molecules on the ribosomes.

In some viruses, such as HIV, the genetic material is RNA rather than DNA.

roasting converts sulfides to oxides during *metal extraction*. Oxides are much easier to reduce to metals than sulfides. Roasting is exothermic so once the reaction has begun it needs little fuel to keep the process going. Roasting produces hot metal oxides which may be fed directly to a furnace for reduction to the metal.

$$2PbS(s) + 3O_2(g) \rightarrow 2PbO(s) + 2SO_2(g)$$

In the past, roasting sulfide ores was a major cause of air pollution. The sulfur dioxide was simply released into the air where it caused acid rain. Now the sulfur dioxide is recovered and used for *sulfuric acid manufacture*. (See also *zinc extraction*.)

Most of the copper now extracted worldwide is produced by a flash process that combines roasting and smelting in a single high-temperature process to produce metal ready for *copper refining*.

rotation about covalent bonds is the rotation of one part of a molecule relative to another around a single covalent bond.

Double bonds stop free rotation and this accounts for the existence of *cis–trans isomers (E–Z isomers)*.

R–S names for *optical isomers* describe the precise three-dimensional arrangement of each of the atoms or groups around each chiral centre. To find the label for a particular chiral carbon atom:

1 Use the *Cahn–Ingold–Prelog priority rules* to determine the order of priority for the four groups around the carbon atom.

2 Look at the molecule so that the atom or group of lowest priority is on the far side. Trace a path from the group of highest priority through the group of second priority to the group of third priority. If this path is clockwise, the chiral centre has the *R* configuration. If the path is anticlockwise, the chiral center has the *S* configuration.

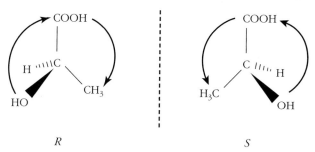

R   S

R–S names of the enantiomers of lactic acid (2-hydroxypropanoic acid). Priorities:
$O > C(OOH) > C(H_3) > H$

It is important to note that the configuration and the direction of rotation of the *plane of polarised light* are not related. The *R* isomer sometimes rotates the plane of polarised light clockwise (+) and sometimes anticlockwise (−).

Traditionally the enantiomers of the *amino acids* in proteins have been named according to the *D–L* system. All the natural acids are *L* isomers. They all have the *S* configuration with the exception of *L*-cysteine and *L*-cystine.

rubber: an elastic material which is easily stretched or bent but which tends to spring back to its original shape when the force is removed.

Rubbers are *elastic polymers* often called *elastomers*.

Elastomer molecules tend to coil up in such a way that they can uncoil as they stretch and coil up again when unstretched. The molecules are *cross-linked* by strong covalent bonds so that they cannot permanently slide past each other unless pulled much harder.

Natural rubber comes from the sap of the rubber tree. This white latex coagulates to a soft material. Vulcanising converts natural rubber into a useful elastomer by cross-linking the polymer chains with sulfur atoms. This was the process discovered by Charles Goodyear in 1839.

The science of polymer chemistry has helped to develop a range of alternatives to natural rubber suited for specific purposes. Examples are:
- neoprene – used for rubber dinghies
- styrene-butadiene rubber – used for motor vehicle tyres
- *polyurethanes* – used for furniture foams and stretch fabrics.

Tyres are black because the rubber is mixed with carbon *filler* which makes the material more resistant to wear. Carbon also absorbs *ultraviolet radiation* helping to slow down the rate at which rubber degrades in sunlight.

rusting is the *corrosion* of a ferrous metal.

Aiming for a grade A*?

Don't forget to log on to **www.philipallan.co.uk/a-zonline** for advice.

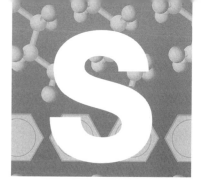

sacrificial protection: an electrochemical method of preventing *corrosion* used with pipelines, oil rigs and the hulls of ships. Steel corrodes where it is oxidised. This happens wherever the metal is an *anode*. In anodic regions the iron atoms give up electrons turning into ions. Attaching a more reactive metal, such as zinc or magnesium, to iron creates an electrochemical cell in which the iron is the *cathode*, so the more reactive metal corrodes while protecting the steel.

Steel can also be protected from corrosion by dipping it in molten *zinc*. This is galvanising. The layer of zinc on the steel corrodes in preference to steel to protect the steel. The corrosion products from the zinc are deposited on the steel resealing it from the atmosphere which helps to stop corrosion.

salt bridge: the ionic connection between the solutions of the two *half-cells* which make up an *electrochemical cell*. A salt bridge makes an electrical connection between the two halves of the cell by allowing ions to flow while preventing the two solutions from mixing. At its simplest a salt bridge consists of a strip of filter paper soaked in potassium nitrate solution. Potassium salts and nitrates are soluble so the salt bridge does not react with the ions in the half-cells.

In commercial *reference electrodes* the salt bridge is often a small plug of porous glass.

salt hydrates: see *hydration*.

salting-out effect: this is the effect of making a chemical much less soluble in water by adding a high concentration of a salt. Adding common salt (sodium chloride) helps to separate *soap* from water after *hydrolysis* of fats or oils.

salts are ionic compounds formed when an *acid* reacts with a *base*. Salts therefore have two 'parents'. Salts are related to a parent acid and to a parent base.

Acid	Salts
hydrochloric acid, HCl	sodium chloride, $NaCl$
	calcium chloride, $CaCl_2$
	ammonium chloride, NH_4Cl
sulfuric acid, H_2SO_4	sodium sulfate, Na_2SO_4
	calcium sulfate, $CaSO_4$
	ammonium sulfate, $(NH_4)_2SO_4$
ethanoic acid, CH_3COOH	sodium ethanoate, CH_3COONa
	calcium ethanoate, $(CH_3COO)_2Ca$
	ammonium ethanoate, CH_3COONH_4

Base	Salts
sodium hydroxide, NaOH	sodium chloride, NaCl
	sodium sulfate, Na_2SO_4
	sodium ethanoate, CH_3COONa
calcium oxide, $Ca(OH)_2$	calcium chloride, $CaCl_2$
	calcium sulfate, $CaSO_4$
	calcium ethanoate, $(CH_3COO)_2Ca$
ammonia, NH_3	ammonium chloride, NH_4Cl
	ammonium sulfate, $(NH_4)_2SO_4$
	ammonium ethanoate, CH_3COONH_4

A *neutralisation reaction* is not the only way to make a salt. Some metal chlorides, for example, are made by heating metals in a stream of chlorine. This is useful for making anhydrous chlorides, such as *aluminium chloride* or iron(III) chloride, both of which are hydrolysed by water and so cannot be dehydrated by heating.

Insoluble salts are conveniently prepared by *ionic precipitation*.

salts of strong and weak acids: see *hydrolysis of salts* and *neutralisation reaction*.

sampling for analysis is a vital first step in any analysis because it is important to make sure that the sample is representative of the whole specimen. This is easy when analysing solutions which are well mixed. It is more difficult when sampling an *ore* made up of lumps of rock each with variable but small amounts of a precious mineral. In these circumstances an analyst has the difficult task of preparing a sample with a mass of about 1 g from a batch of ore with a mass of several tonnes.

Analysts generally prepare multiple samples from the same specimen and work through the analysis with all of them to check on the *measurement uncertainty* of the results.

saturated compounds are compounds containing only single bonds between the atoms in their molecules. Examples of saturated *hydrocarbons* are the *alkanes* and cycloalkanes (see *cyclic hydrocarbons*).

The term 'saturated' is also used for compounds with saturated hydrocarbon chains such as the saturated *fats* and *fatty acids*, even though these compounds include C=O bonds. Saturated compounds do not undergo *addition reactions*.

saturated solutions are solutions which contain as much of the dissolved substance as possible at a particular temperature. A saturated solution is in *dynamic equilibrium* with undissolved excess of the substance in solution. The concentration of a saturated solution is the *solubility* of the substance at that particular temperature.

saturated vapour pressure: see *vapour pressure*.

s-block elements: the elements in groups 1 and 2 in the *periodic table*. For these elements the last electron added to the atomic structure goes into the *s orbital* in the outer shell. All the elements in the *s* block are reactive metals. (See also *atomic orbitals*.)

scaling-up processes: converting a small-scale synthesis in a laboratory or pilot plant to a process which can operate successfully and safely on a commercial scale.

second law of thermodynamics: see *thermodynamics (laws of)*.

second order reaction: a reaction is second order with respect to a reactant if the rate of reaction is proportional to the concentration of that reactant squared. The concentration term for this reactant is raised to the power two in the *rate equation*. At its simplest the rate equation for a second order reaction is:

$$\text{rate} = k[X]^2$$

It can be shown that for a second order reaction, with a rate equation in this form, a plot of $1/[X]_t$ against time is a straight line where $[X]_t$ is the concentration of X at time t. The gradient of the line is equal to k, the rate constant.

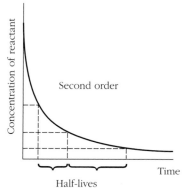

Variation of the concentration of a reactant plotted against time for a second order reaction

The half-life for a second order reaction is not a constant. The time for the concentration to fall from c to c/2 is half the time for the concentration to fall from c/2 to c/4. The half-life is inversely proportional to the starting concentration.

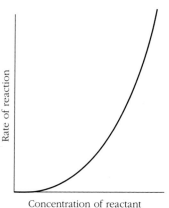

Variation of reaction rate with concentration for a second order reaction

The rate of reaction of 1-bromopropane with hydroxide ions is overall second order. It is a *first order reaction* with respect to the halogenoalkane and first order with respect to hydroxide ions. The overall order is the sum of the powers in the rate equation.

$$\text{rate} = k[CH_3CH_2CH_2Br][OH^-]$$

secondary organic compounds: see *primary, secondary and tertiary organic compounds*.

seed crystal: a crystal added to a *supersaturated solution* to encourage it to crystallise. *Recrystallisation* of an organic product sometimes produces a solution which is reluctant to crystallise. Scratching the sides of the container with a glass rod can encourage crystals to form. If this fails, adding a minute crystal of the product may be enough to start rapid crystallisation.

semiconductors are not *electrical insulators* but they do not conduct electricity as well as metals. Examples are elements such as *silicon* and *germanium* and compounds such as gallium arsenide. Semiconductors consist of covalent *giant structures* in which a few of the bonding electrons can break free and become *delocalised* in the structure. The conductivity of semiconductors is enhanced by 'doping' with traces of impurity atoms. Doping silicon (in group 4) with an element such as arsenic (group 5) increases the number of conducting electrons. An arsenic atom has one more electron in its outer shell than is needed for bonding in the giant structure.

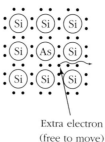

Silicon with an impurity atom from a group 5 element

Extra electron
(free to move)

selectively permeable membrane: a membrane which allows solvent molecules and some other small molecules or ions to pass through but is impermeable to larger molecules and other ions. *Osmosis* and *dialysis* both depend on selectively permeable membranes. So does the manufacture of chlorine by *electrolysis of brine* in a membrane cell.

separating funnel: a tap funnel used to separate liquids which do not mix (*immiscible liquids*). Separating funnels are used to:
- separate an organic product by *solvent extraction* with a solvent such as ether (ethoxyethane) – the product is more soluble in the organic solvent but the ionic reagents and ionic by-products remain in the water
- purify an impure organic product by 'washing' it with aqueous reagents (such as dilute acids or alkalis) which dissolve impurities.

Pressure can build up in a separating funnel when a mixture is being shaken if one liquid is very volatile or if a reaction produces a gas. Hence the technique of inverting the funnel while holding in the stopper and releasing excess gas or vapour through the tap.

sequestering agent: a ligand which forms such stable complexes with metal ions that it effectively makes them chemically inactive. Examples of sequestering agents are:
- *EDTA* used to treat people suffering from metal poisoning
- tripolyphosphates added to some *washing powders or liquids* to complex with calcium or magnesium ions in *hard water*.

shapes of complex ions depend on the number of *ligands* around the central metal ion. There is no simple rule for predicting the shapes of complexes from their formulae. Some octahedral complexes with *bidendate ligands* are *chiral compounds*.

Typically complexes with:

- six smaller ligands, such as H_2O and NH_3, are octahedral, as in $[Mn(H_2O)_6]^{2+}$, $[Cr(NH_3)_6]^{3+}$ and $[Fe(H_2O)_5(SCN)]^{2+}$
- four larger ligands, such as Cl^-, are usually tetrahedral, as in $[CuCl_4]^{2-}$ and $[CrCl_4]^-$ but may be planar, as in *cisplatin* $[Pt(NH_3)_2Cl_2]^{2+}$.
- two ligands are linear, as in $[CuCl_2]^-$, $[Ag(NH_3)_2]^+$, $[Ag(S_2O_3)_2]^{3-}$ and $[Ag(CN)_2]^-$.

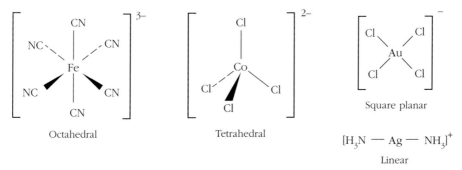

Octahedral   Tetrahedral

Square planar

$[H_3N - Ag - NH_3]^+$

Linear

Shapes of complex ions

shapes of molecules and ions can be predicted by examining the number of bonding pairs and non-bonding *lone pairs of electrons* in the outer shell of the central atom. The expected shape for a molecule is the one which minimises the repulsion between electron pairs by keeping them as far apart as possible in three dimensions. This is sometimes called the *VSEPR theory*.

Cl — Be — Cl
Linear

Trigonal planar

Shapes of molecules with two and three electron pairs around the central atom

Non-bonding lone pairs are held closer to the central atom. The result is that the order of repulsion between electron pairs is:

    lone pair–lone pair > lone pair–bond pair > bond pair–bond pair

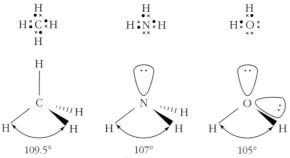

109.5°   107°   105°

Bond angles in molecules with four electron pairs around the central atom. The greater repulsion between lone pairs, and between lone pairs and bonding pairs, means that the angles between the covalent bonds decrease as the number of lone pairs increases.

A molecule with six electrons pairs around the central atom is octahedral. An octahedron has eight faces but six corners.

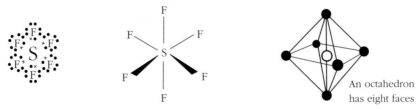

An octahedral molecule

An octahedron has eight faces

A PCl_5 molecule has five electron pairs around the central phosphorus atom. In the vapour phase this molecule is a trigonal bipyramid.

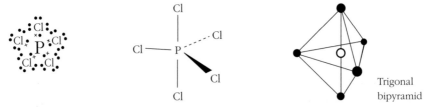

Trigonal bipyramid

A trigonal bipyramid – a shape based on two pyramids with triangular bases

The two electron pairs in a double bond count as 'one' when it comes to predicting molecular shapes.

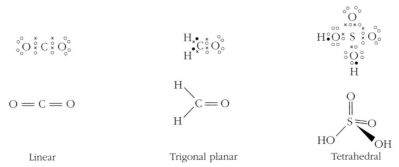

| Linear | Trigonal planar | Tetrahedral |

Shapes of molecules with double bonds

The same principles apply to the shapes of ions.

| Trigonal planar | Trigonal planar | Tetrahedral |

Shapes of ions consisting of more than one atom

shells are the main *energy levels* in atoms and are numbered 1, 2, 3 and so on. These numbers are the principal *quantum numbers*.

shielding is an effect of inner electrons which reduces the pull of the nucleus on the electrons in the outer shell of an atom. Thanks to shielding, the electrons in the outer shell are attracted by an 'effective nuclear charge' which is less than the full charge on the nucleus.

Shielding accounts for the fall in the first *ionisation energy* for the elements down a group in the periodic table.

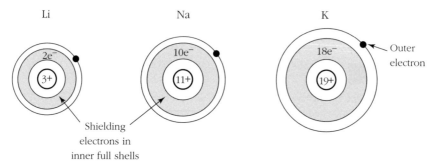

Shielding in the atoms of the elements in group 1 means that the 'effective nuclear charge' is 1+. Down the group the outer electron is held less strongly, being further from the same effective nuclear charge.

SI units are the internationally agreed units for measurement in science. There are seven base units in the system. All other units are derived from the base units.

Every physical quantity in the system has a symbol. A physical quantity has a value and a unit. In calculations it is good practice to substitute both the value and the unit in formulae as shown in the worked examples in this book. The six units in this table are the base units used in chemistry. Note that in print the symbols of physical quantities appear in italics but the units do not.

Physical quantity	Symbol	Unit	Unit symbol
length	l	metre	m
mass	m	kilogram	kg
time	t	second	s
electric current	I	ampere	A
temperature	T	kelvin	K
amount of substance A	n_A	mole	mol

side reactions are unwanted reactions which reduce the *yield* of the product being formed by the main reaction. Side reactions create *by-products* which are wasteful if they have no use.

sievert (symbol Sv) is the unit of effective *radiation dose*. One sievert is a large dose, so effective doses are often quoted in millisieverts (mSv) or microsieverts (μSv). The average dose from background radiation for people in the UK is about 2600 μSv per year. The dose for an individual, however, depends on where they live, how they travel and whether or not they have certain medical treatments. One dental X-ray may involve a dose of about 20 μSv. Radiotherapy can involve very large doses such as 40 Sv.

sigma bond: a single covalent bond formed by a pair of electrons with the electron density concentrated between the two nuclei. Free rotation is possible about single bonds unlike about *pi bonds*.

Sigma (σ) bonds can form by overlap of two *s orbitals*, an *s* orbital and a *p orbital* or two p orbitals.

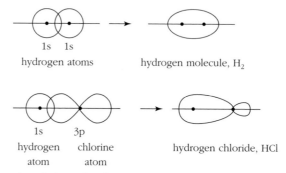

Examples of sigma bonds in molecules

significant figures: the number of significant figures quoted for a measurement shows the degree of uncertainty. Measuring instruments vary in their accuracy and there is uncertainty associated with any measurement. (See *accuracy of data* and *measurement uncertainty*.)

Putting values into *standard form* provides the clearest way of indicating the number of significant figures. This removes any doubt about any zeros, which sometimes just indicate the position of the decimal point, as in 0.0056 g which is a very small mass quoted to two significant figures. This is clearer when the value is in the form 5.6×10^{-3}.

In other values the zeros are included to show the number of significant figures. When writing the distance of 3500 m as 3.50×10^3 m it becomes clear that this is a quantity quoted only to three significant figures.

Generally, when multiplying or dividing values, the number of significant figures in the result is the same as the measurement with the least number of significant figures.

When values are being added or subtracted in a calculation, the number of significant figures in the result is determined by the *least* precise measurement included in the calculation. In other words, the measurement with the fewest figures after the decimal point when the measurement is written out in non-standard form.

silica gel: a non-crystalline, hydrated form of silica. Heating silica gel produces hard granules which absorb water strongly, so silica gel is used as a drying agent. Self-indicating silica gel contains enough of an anhydrous cobalt(ii) salt to colour the granules blue. After the gel has absorbed a certain quantity of moisture the cobalt ions become hydrated and turn pink as a warning that the gel is no longer as effective as a drying agent. Heating drives off the water so that the gel can be used again. Silica gel is also used as the stationary phase in *chromatography*.

silicates are the minerals that make up most of the Earth's crust. The basic building block for silicates is an SiO_4^{4-} tetrahedron. Zircon ($ZrSiO_4$) is an example of a simple silicate mineral with metal ions and silicate ions.

Silicate tetrahedra can join up in chains or strands, as in asbestos. They can also join up in sheets as in mica, talc or clay minerals. In these minerals the negative charges on the silicate part is balanced by positive charges from metal ions such as Na^+, Ca^{2+}, Mg^{2+} or Al^{3+}.

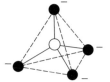

A single SiO_4^{4-} tetrahedron

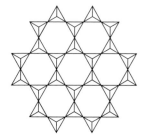

A fragment of a sheet of silicate tetrahedra

In *zeolites* the silicate tetrahedra are built up into three-dimensional networks.

A very wide range of silicate minerals is possible because some of the silicon atoms in the tetrahedra can be replaced by aluminium, boron or beryllium atoms.

silicon (Si) is the second most abundant element in the Earth's crust. Combined with oxygen it forms many minerals including *silicon(IV) oxide* (SiO_2) and *silicates*. Silicon is the second element in *group 4* of the periodic table. Its electron configuration is $[Ne]2s^22p^2$.

Solid silicon has a *diamond* structure. It is a shiny grey material made by reducing silicon dioxide with carbon in an electric furnace. The silicon formed in this way is not pure enough for use in electronics as a *semiconductor*.

Silicon compounds are typical of the compounds of a *non-metal*:
- the oxide SiO_2 is an *acidic oxide* though relatively unreactive because it has a *giant structure*
- the chloride ($SiCl_4$) is a molecular liquid which is rapidly hydrolysed by water to hydrated silica and hydrogen chloride (see *hydrolysis of non-metal chlorides*)
- the hydrides, such as SiH_4 and Si_2H_6, are molecular gases (silanes).

silicones are *polymers* based on chains of alternating silicon and oxygen atoms.

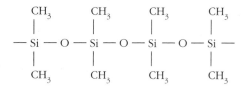

A silicone polymer

By controlling the chain length and degree of *cross-linking* it is possible to make a range of silicones for use as oils, greases and rubbery materials. Silicones are water repellent and can be more safely used at higher temperatures than polymers based on carbon atoms. They are electrical insulators and other materials do not stick to them. Silicones are colourless, have no smell and are inert.

silicon(IV) oxide (SiO_2), also known as silica, is the oxide of silicon which is abundant in rocks as quartz. Silica has a very high melting point at 1710°C, turning to a viscous liquid. On cooling the liquid becomes a *glass*.

Amethyst is crystalline quartz coloured purple due to the presence of iron(III) ions. Sandstone and sand consist mainly of silica. Flint is a non-crystalline form of silica.

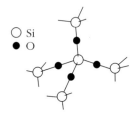

*Giant structure of silicon(IV) oxide. Each Si atom is at the centre of a tetrahedron of oxygen atoms. The arrangement of silicon atoms is the same as the arrangement of carbon atoms in **diamond** but there is an oxygen atom between each silicon atom.*

silver halides are used in qualitative analysis and are the light-sensitive ingredient used in traditional photography. The three insoluble silver salts form on mixing solutions of silver nitrate and a soluble chloride, bromide or iodide. Silver fluoride is soluble so there is no precipitate when silver nitrate is added to a solution of fluoride ions.

$$Ag^+(aq) + X^-(aq) \rightarrow AgX(s) \text{ where } X = Cl, Br \text{ or } I$$

This precipitation reaction is used as a test for halide ions (see *anion tests*). The three silver compounds can be distinguished by their colour and the ease with which they redissolve in ammonia solution. The values for the *solubility product constants* show the trend in solubility. Silver chloride is the most soluble.

Silver halide	Colour	Effect of adding ammonia solution to a precipitate of the compound	Solubility/ mol dm^{-3}	K_{sp}/mol^2 dm^{-6}
silver chloride, AgCl	white	redissolves readily in ammonia solution forming a complex ion	1.4×10^{-5}	2×10^{-10}
silver bromide, AgBr	cream	redissolves but only in concentrated ammonia	5.5×10^{-7}	3×10^{-13}
silver iodide, AgI	yellow	does not redissolve in ammonia solution	2.8×10^{-9}	8×10^{-18}

sizes of atoms and ions: see *atomic radius, covalent radius, ionic radius* and *van der Waals radius*.

skeletal formulae: outline formulae for carbon compounds which are a useful shorthand for complex molecules such as many natural products. The formulae need careful study because they represent only the hydrocarbon part of the molecule with lines for the bonds between carbon atoms, leaving out the symbols for the carbon and hydrogen atoms. *Functional groups* are included.

2-methylbutane

but-1-ene

Cl

2-chlorobutane

OH

butan-2-ol

O

H

butanal

O

butanone

O

OH

butanoic acid

O

O

ethylbutanoate

Skeletal formulae for a range of compounds with a chain of four carbon atoms

slag: the unwanted waste material from *metal extraction* at high temperature (pyrometallurgy). Slag is tapped from a furnace, such as a *blast furnace*. The slag solidifies as it cools and can be dumped or crushed for use in construction or road building.

smart materials are able to change their properties when there is a change in conditions, such as a change of temperature or pH. Shape-memory polymers, for example, are first moulded into their 'permanent' form by one of the usual moulding techniques for plastics. Then the polymer is deformed to a new 'temporary' shape which it retains until the polymer is heated above a certain temperature, at which point it reverts to its 'permanent' shape. The control of the molecular structure which allows the polymer to switch from its 'temporary' to is 'permanent' form is based on the transition between glassy and more flexible states of *polymeric materials* at their glass transition temperature.

smelting is a process of *metal extraction* at high temperature (pyrometallurgy). Smelting involves melting the concentrates from metal ores to remove impurities and to reduce metal compounds to metals. Examples of smelting include the extraction of iron in a *blast furnace* and *aluminium extraction* by electrolysis of a melt.

smog: a smoky fog caused by *atmospheric pollutants*. Burning coal in homes and industry created the dense city smogs of the nineteenth and first half of the twentieth centuries. The damaging components of these smogs were sulfur dioxide and soot. Many people affected by these acidic smogs died of lung disease. Smokeless zones and the shift of fuels from coal to oil and natural gas have largely eliminated this type of smog. Today cities are affected by another type of smog caused by motor traffic. This is *photochemical smog.*

smoke: a *colloid* with specks of solid dispersed in a gas. Smokes are examples of *aerosols.*

S_N1 and S_N2 reactions: see *nucleophilic substitution in halogenoalkanes.*

soaps are made by the *hydrolysis* of *fats* or *vegetable oils* with alkali. Soaps are the sodium or potassium salts of *fatty acids.*

S

$$H - \overset{\overset{\displaystyle H}{|}}{\underset{|}{C}} - O - \overset{\overset{\displaystyle O}{\|}}{C} - (CH_2)_{16}CH_3$$

(chemical structure of triglyceride)

$$H - C - O - C - (CH_2)_{16}CH_3 + 3Na^+OH^- \longrightarrow$$

$$H - \overset{\overset{\displaystyle H}{|}}{\underset{|}{C}} - O - H$$

$$H - C - O - H + 3Na^+ \ ^-O - \overset{\overset{\displaystyle O}{\|}}{C} - (CH_2)_{16}CH_3$$

propane-1,2,3-triol   sodium octadecanoate
(glycerol)   (sodium stearate, soap)

Saponification – the hydrolysis of a fat or vegetable oil (triglyceride) with alkali to make soap

Soaps are **surfactants** which help to remove greasy dirt because they have an ionic (water-loving) head and a long (water-hating) hydrocarbon tail.

Most toilet soaps are made from a mixture of animal fat and coconut palm oil. Soaps from animal fat are less soluble and longer lasting. Soaps from palm oils are more soluble so that they lather quickly but wash away more quickly. Bars of soap also contain a dye and perfume together with an antioxidant to stop the soap and air combining to make irritant chemicals.

sodium (Na) is a soft, shiny metal which rapidly tarnishes in moist air. It is the second member of **group 1** with the electron configuration [Ne]$4s^1$.

Like other group 1 metals, sodium:
- is stored in oil
- floats on water, melts and reacts violently forming hydrogen, which catches fire, and NaOH, which is soluble and strongly alkaline
- forms an ionic, crystalline chloride Na^+Cl^-.

Sodium produces a mixture of the oxide, Na_2O, and peroxide, Na_2O_2, when it burns in air.

Sodium is a powerful reducing agent used for the extraction of some metals such as zirconium and titanium.

sodium chloride structure: the cubic crystal structure of the ionic compound sodium chloride, NaCl. Each positive ion is surrounded by six nearest neighbours and each negative ion is surrounded by six positive ions, so the **coordination numbers** are 6 and 6.

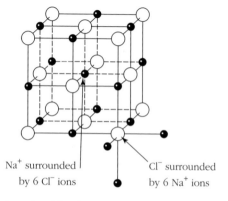

Structure of sodium chloride showing 6:6 coordination. The structure consists of a face-centred cubic array of negative ions with all the octahedral holes filled by positive ions.

Na^+ surrounded by 6 Cl^- ions

Cl^- surrounded by 6 Na^+ ions

Many other compounds have this structure, including the chlorides, bromides and iodides of Li, Na and K; the oxides and sulfides of Mg, Ca, Sr, Ba as well as the fluoride, chloride and bromide of Ag.

sodium hydroxide (NaOH) is a white, translucent solid supplied as flakes or pellets. It is a *deliquescent substance*. Sodium hydroxide is a *strong base*; it dissolves in water to form a highly alkaline solution. It is fully ionised both in the solid and in solution. Note that the solution is alkaline because of the hydroxide ions (not because of the sodium ions).

Sodium hydroxide is a useful test reagent (see *anion tests, cation tests* and *organic analysis*).

The traditional name for the alkali is caustic soda, which is a reminder that it is highly corrosive. Sodium hydroxide is more hazardous to the skin and eyes than many acids.

Sodium hydroxide is manufactured by the *electrolysis of brine*. It is widely used for manufacturing other chemicals including *soaps* and *detergents*, rayon *fibres* as well as *aluminium*, sodium cyanide and sodium peroxide.

sodium tetrahydridoborate(III) ($NaBH_4$) is a *reducing agent* used in organic chemistry. It is a milder reducing agent than *lithium tetrahydridoaluminate(III)* and has the advantage that it can be used in aqueous solution. $NaBH_4$ reduces *aldehydes* and *ketones* to alcohols.

The reduction of a carbonyl compound by $NaBH_4$ is an example of a *nucleophilic addition* reaction in which the hydride ion acts as the nucleophile.

Mechanism of nucleophilic addition of propanal with a hydride

sodium thiosulfate ($Na_2S_2O_3$) is used in *iodine–thiosulfate titrations*.

A milky precipitate of sulfur forms on adding dilute acid to sodium thiosulfate. This is an example of a *disproportionation reaction*.

$$S_2O_3^{2-}(aq) + 2H^+(aq) \rightarrow SO_2(aq) + S(s) + H_2O(l)$$

solid: one of the *states of matter*. (See also *crystal structures of ionic compounds, crystal structures of non-metals* and *crystal structures of metals*. See also *ceramics, glasses, metals* and *polymeric materials*.)

solubility: the mass (g) or amount (mol) of a substance that will dissolve in 100 g water.

As a general rule, 'like dissolves like'. *Polar solvents*, such as water, dissolve polar or ionic compounds. *Non-polar solvents*, such as hexane, dissolve other hydrocarbons and non-polar elements or compounds. Solids generally become more soluble in water as the temperature rises. Gases become less soluble as the temperature rises. Boiling water, for example, removes gases dissolved from the air. Bubbles of gas appear around the edge of a saucepan before the water boils. The bubbles contain air coming out of solution as the gases become less soluble with the rise in temperature. Gases become more soluble as their pressure rises.

No chemical is completely soluble and none is completely insoluble (see *solubility product constant*). Even so, chemists find it useful to use a rough classification of solubility based on what they see on shaking a little of the solid with water in a test tube:

● very soluble, like potassium nitrate; plenty of the solid quickly dissolves
● soluble, like copper(II) sulfate; crystals visibly dissolve to a significant extent
● sparingly or slightly soluble, like calcium hydroxide; little solid seems to dissolve but the solution becomes quite strongly alkaline
● insoluble, like iron(III) oxide; there is no sign that any of the material dissolves.

A similar rough classification applies to gases dissolving in water. Ammonia and hydrogen chloride are very soluble. Sulfur dioxide is soluble. Carbon dioxide is slightly soluble. Nitrogen is insoluble.

Some generalisations about solubilities help to interpret observations during qualitative analysis. The generalisations in the table apply to solutions in water at room temperature. Adding acid or alkali changes the patterns of solubility.

	Soluble in water	Insoluble in water
Acids	All common *acids* are soluble	
Bases	*Alkalis:* the hydroxides of sodium and potassium (calcium hydroxide which is slightly soluble), ammonia, plus the carbonates of sodium and potassium	All other metal oxides, hydroxides and carbonates
Salts	All nitrates	
	All chlorides... *except*	silver and lead chlorides
	All sulfates... *except*	barium sulfate, lead sulfate and calcium sulfate which is slightly soluble
	All sodium and potassium salts	All other carbonates, chromates, sulfides and phosphates

solubility product constants: equilibrium constants for almost insoluble salts in equilibrium with solutions of their own ions. Even salts which are insoluble for practical purposes do dissolve to a very slight extent. There is an equilibrium between the solid and its ions in solution.

$$AgCl(s) \rightleftharpoons Ag^+(aq) + Cl^-(aq)$$

The *equilibrium law* applies. As with other *heterogeneous equilibria*, the concentration of the solid silver chloride is constant and does not appear in the equilibrium law equation. In this context the equilibrium constant, K_{sp}, is the solubility product constant.

$$K_{sp} = [Ag^+(aq)][Cl^-(aq)] = 2 \times 10^{-10} \text{ mol}^2 \text{ dm}^{-6} \text{ for equilibrium concentrations}$$

K_{sp} values can be used to predict whether or not a precipitate will form on mixing two solutions.

solute: a substance which dissolves in a *solvent* to make a *solution*. In a sugar solution the solvent is water and the solute is sucrose.

solutions are formed when solids, liquids or gases dissolve in a *solvent*. Water is so abundant on Earth that solutions in water (*aqueous solutions*) are particularly important to the natural environment, to life and to chemistry in laboratories and in industry.

Most solutions are solids, liquids or gases dissolved in a liquid but there are also 'solid solutions'. Nickel–copper *alloys* are examples of solid solutions. In a solid solution, atoms of one metal replace atoms of the other metal in the crystal lattice.

solvation takes places when solvent molecules bond to ions or molecules as they dissolve. The bonding may be through *weak intermolecular forces*, attraction between ions and polar molecules or via covalent bonds. *Hydration* describes solvation when the solvent is water.

solvent: a liquid used to dissolve things. Chemists find it helpful to quote the rule 'like dissolves like'. What this means is that *non-polar solvents* dissolve non-polar substances while *polar solvents* dissolve ionic and polar compounds.

Water is the commonest solvent. It is a polar solvent and dissolves many ionic compounds. Water molecules also dissolve compounds with which they can form *hydrogen bonds* such as glucose molecules.

White spirit is a non-polar solvent consisting of a mixture of *hydrocarbons.* It dissolves oily and greasy materials, including oil paints.

solvent extraction: a technique for separating and purifying substances with a *solvent,* which dissolves the product required but leaves all other compounds dissolved in the original solvent. The solvents must not mix, so that the two liquids separate after being shaken up together.

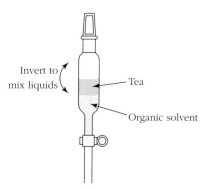

Use of a tap funnel for solvent extraction. Caffeine is more soluble in dichloromethane than in water. The other chemicals in tea are much more soluble in water. After solvent extraction the caffeine is recovered by distilling off the dichloromethane.

The substance being extracted *partitions* itself between the two solvents until the two solutions are in equilibrium. From the *equilibrium law* it is possible to show that it is more efficient to extract with two or three smaller volumes of solvent than to add all the solvent at once in a single extraction.

Solvent extraction is used in the *perfume* industry to extract fragrant oils from chopped-up plant blossom. Solvent extraction is also used to decaffeinate coffee. In both these processes it helps to use a liquefied gas as the solvent. One possibility is carbon dioxide under pressure (see also *supercritical carbon dioxide*). After the extraction the solvent is easily removed by lowering the pressure. The solvent turns back to a gas at a low temperature. This allows the use of a non-toxic solvent. It also means that there is no need for heating to distil off the solvent. Heating can easily destroy organic compounds.

s orbitals: see *atomic orbitals*.

space-filling models: atomic models which show the space taken up by atoms in molecules or crystals. They show the sizes of atoms, molecules or ions and how they pack together. They do not show the bond angles or the numbers of bonds between atoms in molecules as clearly as *ball-and-stick models*.

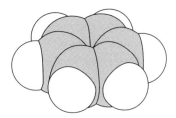

A space-filling model of benzene, C_6H_6

species: a useful collective noun used by chemists to refer generally to the atoms, molecules or ions taking part in a chemical process. A pure chemical species is a collection of identical chemical *entities*.

specific heat capacity: the energy transferred per unit mass when the temperature of a material changes by one degree kelvin. The SI unit is J kg^{-1} K^{-1}. When working on a small scale with a *calorimeter* chemists often work with values in J g^{-1} K^{-1}.

energy transfer = mass × specific heat capacity × temperature change
  (J)   (g)   (J g^{-1}K^{-1})   (K)

In symbols this becomes:

$q = mc\Delta T$

spectator ions: ions in a solution which do not take part in the chemical change during a reaction. For clarity, chemists omit spectator ions from *ionic equations*. Adding silver nitrate to a solution of potassium chloride produces a precipitate of insoluble silver chloride. Both silver nitrate and potassium chloride are ionised in solution.

$Ag^+(aq) + NO_3^-(aq) + K^+(aq) + Cl^-(aq) \rightarrow AgCl(s) + NO_3^-(aq) + K^+(aq)$

The potassium and nitrate ions remain in solution unchanged. They are the spectator ions left out of the ionic equation:

$Ag^+(aq) + Cl^-(aq) \rightarrow AgCl(s)$

spectroscopy: a range of practical techniques for studying the composition, structure and bonding of elements and compounds. These instrumental techniques have been developed in the last 80 years and continue to become more powerful. Spectroscopic techniques are now the essential 'eyes' of chemistry.

The instruments used are variously called spectroscopes (emphasising the uses of the techniques for making observations) or spectrometers (emphasising the importance of measurements).

Spectroscopy uses the full range of the *electromagnetic radiation* spectrum to study atoms, molecules, ions and the bonding between them:

- radio waves in *NMR* (*carbon-13 NMR spectroscopy* and *proton NMR spectroscopy*)
- microwaves to study the rotations of polar molecules
- infrared radiation in *infrared spectroscopy*
- visible and UV radiation in *atomic absorption spectroscopy*, *atomic emission spectroscopy* and *ultraviolet spectroscopy*
- X-ray spectroscopy to study electron jumps between the electron shells in atoms.

spin: the property of electrons which accounts for their behaviour in a magnetic field. Electrons behave like tiny magnets. In a magnetic field electrons either line up with the field or against the field.

An *atomic orbital* can hold only two electrons and they must have opposite spins. Arrows pointing up or down represent electrons in *energy level* diagrams.

Spin is also a property of any atomic nuclei with an odd numbers of *nucleons*. These nuclei are the basis of *NMR spectroscopy*.

spontaneous reaction: a reaction which tends to go without being driven by any external agency. Spontaneous reactions are the chemical equivalent of water flowing down-hill. Any reaction which naturally tends to happen is spontaneous in this sense even if it is very slow, just as water has a tendency to flow down a valley even when held up behind a dam. The chemical equivalent of a dam is a high activation energy for a reaction.

In thermochemistry, spontaneity has the same meaning as a *feasibility*. So strictly speaking any reaction which naturally tends to happen is spontaneous even if it is very slow.

The control of spontaneous reactions is of vital importance in metabolism. The hydrolysis of *ATP* is spontaneous. However, if all the ATP in cells were to react rapidly with water, the energy from respiration would be wasted and life would cease. Hydrolysis happens only with *enzymes* to speed up the reaction. With enzymes the energy from hydrolysis can be harnessed to growth and movement.

In practice chemists sometimes use the word spontaneous in its everyday sense to describe reactions which not only tend to go but also go fast when the reactants are mixed at room temperature. Here is a typical example:

> 'The hydrides of silicon catch fire spontaneously in air, unlike methane which has an ignition temperature of about 500°C.'

The reaction of methane with oxygen is also spontaneous in the thermodynamic sense, even at room temperature. However, the *activation energy* for the reaction is so high that nothing happens until the gas is heated with a flame. A mixture of methane and oxygen at room temperature is kinetically inert (see *inert chemicals*).

stability of benzene: the greater stability of *benzene* because of *delocalised electrons*. Benzene is more stable than expected of a compound which is often shown with three double bonds as in the *Kekulé structure*.

A measure of the greater stability of benzene comes from a comparison of the experimental enthalpy change on adding three moles of hydrogen to benzene (*hydrogenation*) with three times the enthalpy change on adding a mole hydrogen to cyclohexene.

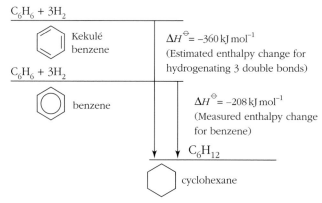

benzene   + 3H$_2$ ⟶ $\Delta H^{\ominus} = -208$ kJ mol^{-1}

cyclohexene   + H$_2$ ⟶ $\Delta H^{\ominus} = -120$ kJ mol^{-1}

Enthalpy changes for hydrogenating benzene and cyclohexene

Real benzene is more stable than might be expected by about 150 kJ mol^{-1}.

C$_6$H$_6$ + 3H$_2$

Kekulé benzene $\Delta H^{\ominus} = -360$ kJ mol^{-1}
(Estimated enthalpy change for hydrogenating 3 double bonds)

C$_6$H$_6$ + 3H$_2$

benzene $\Delta H^{\ominus} = -208$ kJ mol^{-1}
(Measured enthalpy change for benzene)

C$_6$H$_{12}$

cyclohexane

Comparing the enthalpy change for hydrogenating real benzene with the calculated enthalpy change for hydrogenating a ring compound with three normal double bonds

stable compounds: compounds are stable if they do not decompose into their elements or other compounds. A compound which is stable at room temperature and pressure may become more or less stable as conditions change (see *thermal stability of carbonates*).

Chemists often use standard *enthalpy changes* as an indicator of stability. Strictly they should use either the total *entropy* change or the standard *free energy change*, $\Delta G^{\ominus}_{f}$, values but in many cases $\Delta G^{\ominus}_{f} \approx \Delta H^{\ominus}_{f}$.

When discussing the stability of a compound it is important to specify the decomposition reaction. At a high enough temperature, for example, *calcium carbonate* becomes unstable relative to decomposition into calcium oxide and carbon dioxide. The *nitrogen oxides*, however, are unstable relative to decomposition into the elements.

A compound such as the gas N$_2$O is thermodynamically unstable; the compound tends to decompose into its elements. The values of $\Delta H^{\ominus}_{f}(= + 82$ kJ mol$^{-1})$ and $\Delta G^{\ominus}_{f}(= +104$ kJ mol$^{-1})$ are both positive. This shows that the change from elements to compound is endothermic so the decomposition reaction is exothermic. However, the rate is very slow under normal conditions. N$_2$O exists at room temperature because it is kinetically inert. (See also *inert chemicals*.)

ΔS_{total}	ΔG	Activation energy	Change observed	
negative	positive	high	no reaction	reactants stable relative to products
positive	negative	high	no reaction	reactants unstable relative to products but kinetically inert
negative	positive	low	no reaction	reactants stable relative to products
positive	negative	low	fast reaction	reactants unstable relative to products

stability constants (K_{stab}) are equilibrium constants which are a measure of the stability of *complex ions*. The greater the value of a stability constant the more stable the complex.

Stability constants show the position of equilibrium when a new ligand replaces water molecules in the hydrated ions.

$$[Co(H_2O)_6]^{2+}(aq) + 6NH_3(aq) \rightleftharpoons [Co(NH_3)_6]^{2+}(aq) + 6H_2O(aq)$$

For simplicity this is often written as:

$$Co^{2+}(aq) + 6NH_3(aq) \rightleftharpoons [Co(NH_3)_6]^{2+}(aq)$$

The usual rules for writing an equilibrium constant and its units apply.

$$K_{stab} = \frac{[Co(NH_3)_6^{2+}]}{[Co^{2+}][NH_3]^6} = 1 \times 10^5 \text{ mol}^{-6} \text{ dm}^{18}$$

Complex	$K_{stab}/(\text{mol dm}^{-3})^{-n}$ where n = the number of ligands
$[Ag(NH_3)_2]^+$	1.6×10^7
$[Cu(NH_3)_4]^{2+}$	1×10^{12}
$[Ni\ en_3]^{2+}$ where en is the bidentate ligand 1,2-diaminoethane	2×10^{18}
CuY^{2-} where Y is the hexadentate ligand EDTA	6×10^{18}

The values for stability constants show that *chelate* complexes formed by bidentate and hexadentate *ligands* are more stable than complexes formed by monodentate ligands.

standard conditions: the conditions used in *thermochemistry* to define standard enthalpy and free energy changes. These conditions are the temperature 298 K and a *pressure* of 100 000 N m^{-2} = 10^5 Pa (1 bar).

When *standard electrode potentials* are being measured, the standard conditions are the temperature 298 K, solutions at a concentration of 1.0 mol dm^{-3} and a pressure of 10^5 Pa if gases are involved.

standard electrode potentials are the basis of the *electrochemical series* for predicting the direction of *redox reactions*. They are also used to calculate the $E^{\ominus}_{cell}$ values for *electrochemical cells*. The standard electrode potential for a *half-cell* is measured relative a standard *hydrogen electrode* under *standard conditions*.

standard enthalpy changes: see *enthalpy change*.

standard form: the form which mathematicians and scientists use to write very large or very small numbers. Standard form is based on powers of 10.

S

343

$10^{-9} = 0.000\ 000\ 001$	$10^1 = 10$
$10^{-6} = 0.000\ 001$	$10^2 = 100$
$10^{-3} = 0.001$	$10^3 = 1\ 000$
$10^{-2} = 0.01$	$10^4 = 10\ 000$
$10^{-1} = 0.1$	$10^6 = 1\ 000\ 000$
$10^0 = 1$	$10^9 = 1\ 000\ 000\ 000$

Worked example

The dissociation constant for ethanoic acid, $K_a = 0.000\ 017$ mol dm^{-3}

$$= 1.7 \times 0.000\ 01 \text{ mol dm}^{-3}$$

$$= 1.7 \times 10^{-5} \text{ mol dm}^{-3}$$

which is standard form.

Worked example

The *Faraday constant* = 96 480 C mol^{-1} = 9.648 × 10 000 C mol^{-1}

$$= 9.648 \times 10^4 \text{ C mol}^{-1}$$

which is standard form.

standard free energy changes: see *free energy change.*

standard hydrogen electrode: see *hydrogen electrode.*

standard molar entropy, $S^{\ominus}$: the *entropy* per mole for a substance under *standard conditions.* Chemists use values for standard molar entropies to calculate entropy changes and so predict the direction and extent of chemical change.

In a perfectly ordered crystal at 0 K the entropy is zero. The entropy of a chemical rises as the temperature rises. Increasing the temperature raises the number of energy quanta to share between the atoms and molecules. There are also jumps in the entropy of the chemical wherever there is a change of state. This is because energy is added to change a solid to a liquid or a liquid to a gas. There is also an increase in the number of ways of arranging the atoms or molecules as a solid changes to a liquid and then to a gas.

The units for standard molar entropy are joules per mole per kelvin (J mol^{-1}K^{-1}).

Solids	$S^{\ominus}$/J mol^{-1} K^{-1}	Liquids	$S^{\ominus}$/J mol^{-1} K^{-1}	Gases	$S^{\ominus}$/J mol^{-1} K^{-1}
carbon (diamond)	2.4	mercury	76.0	hydrogen	130.6
magnesium oxide	26.9	water	69.9	carbon dioxide	213.6
copper	33.2	ethanol	160.7	propane	269.9

The table shows standard molar entropies at 298 K. Gases have higher standard molar entropies than solids. The number of ways of distributing the particles and energy in a system of gas molecules is much higher than in an ordered crystalline solid. The more rigid and regular a crystal, the lower its entropy.

Liquids have intermediate values of standard molar entropies. The more atoms in a molecule the greater the opportunities for vibration and rotation so the higher the entropy value.

standard solution: a solution with an accurately known concentration. The direct method for preparing a standard solution is to dissolve a weighed sample of a *primary standard* in water and to make the solution up to a definite volume in a graduated flask.

Standard solutions are used in *titrations* to determine the concentrations of other solutions. They are also needed for the *calibration* of instruments such as colorimeters.

standard state: the stable state of an element or compound under the *standard conditions* that apply in thermochemistry. Standard *enthalpy changes* and *free energy changes* are defined for substances in their standard states.

When an element such as carbon has two *allotropes*, diamond and graphite, the standard state of carbon is graphite because it is the more *stable* form.

Standard states normally refer to substances in their stable physical states under standard conditions. Carbon dioxide and methane are gases but water is a liquid.

starch is a *carbohydrate* which consists of long chains of glucose units. It is a polysaccharide. Starch is insoluble in cold water. In hot water, starch gelatinises forming a colloidal dispersion. Starch solution gives an intense blue-black colour with iodine. The solution is used as an indicator to detect the end-point in *iodine–thiosulfate titrations*.

state functions: measurable properties which depend only on the state of a system, not on how the system got to that state. Examples are *pressure*, *volume* and temperature, as are thermochemical quantities such as enthalpy, *entropy* and free energy.

Changes in state functions depend only on the initial and final states of the system and not on the pathway from one state to the other. This is the basis of *Hess's law*.

Equations of state show the mathematical connections between state functions. An important example is the *ideal gas* equation which relates the pressure, volume and temperature of an amount of a gas.

states of matter: the solid, liquid and gaseous states (see *changes of state*) as well as intermediate states such as *liquid crystals* and mixtures, including aqueous solutions and the colloidal state (see *colloids*).

state symbols in equations indicate the states of the chemicals: (s) solid, (l) liquid, (g) gas, and (aq) aqueous (dissolved in water).

stationary phase: see *chromatography*.

steady-state systems: systems in which the concentrations of chemicals stay constant because they are being supplied or formed as fast as they are removed or destroyed. The temperature is constant too because energy is transferred to the system as fast as it is lost to the surroundings.

The blue flame of the Bunsen burner is a steady-state system. There is a constant supply of gas and oxygen which burn to release energy. The products of burning and energy are constantly lost to the surroundings.

Steady-state systems are important in *environmental chemistry*. Problems arise when human activity on a large scale upsets one or more of the natural processes. *Ozone in the stratosphere*, for example, is naturally formed and destroyed:

rate of formation of ozone = rate of destruction of ozone

The release of chlorine compounds such as *CFCs* has increased the rate of destruction of ozone, upsetting the steady state and lowering the ozone concentration, especially over the poles, hence the 'hole' in the ozone layer.

steam cracking: see *thermal cracking*.

steam distillation is a useful technique for separating oils from plant materials such as rose petals, cloves, lavender, thyme and fennel. It is an important technique in the *perfume* industry. Steam distillation makes it possible to separate compounds that decompose if heated at their boiling points. The technique works only for compounds that do not mix with water.

Steam distillation is also sometimes used to separate products of organic preparations, leaving behind reagents and products which are soluble in water.

Steam distillation works because the *vapour pressure* of a mixture of *immiscible liquids* is the sum of the separate vapour pressures when they are shaken up together. The mixture boils when the total vapour pressure equals atmospheric pressure. So the mixture of steam and oily product distils at a little below the boiling point of water and well below the boiling point of the oil.

steam reforming is the reaction of steam with methane (or other hydrocarbon) in the presence of a nickel oxide *catalyst* at 800°C under pressure.

$$CH_4(g) + H_2O(g) \rightleftharpoons 3H_2(g) + CO(g)$$

It is followed by further reaction with excess steam. This 'shift reaction' converts the CO into CO_2. The reaction happens in the presence of an iron(III) oxide catalyst at 400°C.

$$H_2O(g) + CO(g) \rightleftharpoons H_2(g) + CO_2(g)$$

This mixture of hydrogen and carbon dioxide can be converted directly to *methanol*. Alternatively, absorbing the carbon dioxide in alkali provides hydrogen for *ammonia manufacture*.

steels are *alloys* of iron with carbon and often other metals.

Mild steel contains about 0.2% carbon as iron carbide. Crystals of iron carbide in the metal structure make the steel strong but is still *malleable*. Mild steel is used for car bodies. As the carbon content increases the steel becomes stronger and harder so that it is suitable for rail and tram lines.

Steel is made by the basic oxygen steelmaking (BOS) process which removes impurities from *iron extraction* in a blast furnace. During the process a blast of oxygen converts impurities in the liquid metal, such as carbon, silicon and phosphorus, into their oxides. Carbon dioxide escapes as a gas. The oxides of the non-metals silicon and phosphorus are *acidic oxides*. They are converted to a molten *slag* by adding the *basic oxides* of calcium and magnesium. The slag floats on the surface of the liquid steel and can be poured off separately.

$$SiO_2(s) + CaO(s) \rightarrow CaSiO_3(l)$$

Alloy steels consist of iron with small amounts of carbon together with up to 50% of one or more of these metals: aluminium, chromium, cobalt, manganese, molybdenum, nickel, titanium, tungsten and vanadium. The presence of other metals distinguishes alloy steels from carbon steels. Examples of alloys steels are:

- stainless steels, which contain chromium and nickel
- tool steels with tungsten or manganese which make the alloy harder, tougher and keep their properties at higher temperatures so that they are suitable for drill bits and cutting tools.

stereochemistry is the study of molecular shapes and the effect of shape on chemical properties. Stereochemistry is especially concerned with the study of the contrasting properties of *stereoisomers*.

Smell and taste seem to depend on molecular shape. One mirror-image form of limonene smells of oranges while the other smells of lemons.

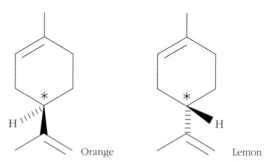

The enantiomers of limonene and their smells. Asterisks mark the chiral centres.

Orange   Lemon

Molecular shape can also subtly alter the physiological effects of drugs, as tragically illustrated by the two enantiomers of thalidomide (see *chiral compounds* and *chiral synthesis*).

Molecular shape affects the *physical properties* of materials such as polymers. This is illustrated by the differences between poly(propene) as an *isotactic polymer* and as an *atactic polymer*.

Steric factors can also affect the mechanisms of reactions and the rates of chemical change.

stereoisomers are distinct compounds with the same molecular formula and structural formula but with different three-dimensional shapes. There are two kinds of stereoisomerism: *cis–trans isomerism (E–Z isomerism)* and *optical isomerism*.

stereoregular polymer: a polymer with a regular three-dimensional shape. Poly(propene) can be an *isotactic polymer*. This is the useful stereoregular form. Or it can be an *atactic polymer* which is irregular and does not have useful properties.

stereospecific reactions are reactions which involve one of the *optical isomers* of a *chiral compound* but not the other. All the biochemical processes involving *amino acids* are stereospecific. The enzymes involved can act on the *L*-amino acids but not on the mirror-image *D*-forms (see also *chiral synthesis*).

steroids are *lipids*. An important steroid is *cholesterol*. Cholesterol can be converted to biologically active steroids such as the sex hormones oestrogen and progesterone.

stiffness is the opposite of flexibility. It is a very important property of materials measured by the Young modulus for the material. Materials scientists seek to develop materials which are both stiff and strong while not being too dense.

Steel is stiff and strong, a digestive biscuit is stiff but weak, nylon is flexible and strong, jelly is flexible and weak.

stoichiometry: a stoichiometric equation is the *balanced equation* for a reaction. It shows the amounts, in moles, of the substances involved in a reaction.

A stoichiometric compound has a composition which corresponds exactly to its formula. A stoichiometric reaction is one which uses up reactants and produces products in amounts exactly as predicted by the balanced equation.

The word stoichiometry sounds mysterious but is simply based on Greek words meaning 'element-measure'. Stoichiometry is the basis of *quantitative analysis* where amounts are measured in moles.

storage cells: see *lead–acid cell, lithium polymer cells* and *nickel–cadmium cells*.

stp (standard temperature and pressure) are the standard conditions for describing the properties of gases. The standard temperature is 273 K (0°C) and the standard *pressure* is 101.3 kPa (1 atmosphere).

The molar volume of an ideal gas at stp is 22 400 cm^3 (22.4 dm^3).

Note the distinction between these values and the values for defining *standard states* in thermochemistry.

strength is measured by the stress needed to deform and break a material in tension or compression. Stress is the force per unit area of the cross-section and is measured in N m^{-2}. Defining stress in this way allows for the fact that it is easier to stretch a thin sample than a thick one.

Metals such as steel have high tensile strength. Ceramics are weak in tension but have high strength in compression.

For many applications engineers look for materials which combine high strength with high *stiffness*.

strong acids are *acids* which are fully ionised when they dissolve in water. Examples are *hydrochloric acid, nitric acid* and *sulfuric acid*.

strong bases are *bases* which are fully ionised when they dissolve in water. Examples are the hydroxides of *sodium* and *potassium*.

structural formulae show the arrangements of atoms and *functional groups* in molecules.

Sometimes it is enough to show structure in a condensed form, such as $CH_3CH_2CH_2COOH$ for butanoic acid. Often it is clearer to write the full structural formula, showing all the atoms and bonds. This type of formulae are also called *displayed formulae*.

propene

butanoic acid

cyclohexane

Examples of displayed structural formulae. Drawn like this, the formulae do not show the true shape in three dimensions.

structural isomers: see *isomers.*

sublimation is the change of a solid directly to a gas on heating. Heating iodine crystals makes them sublime to a purple vapour which condenses to shiny crystals on a cold surface. This process is used to purify iodine.

Another substance which sublimes is solid carbon dioxide ('dry ice') because it turns to gas at −78°C without melting.

substitution reactions are chemical changes which replace an atom or group of atoms by another atom or group of atoms. An example is the reaction of butan-1-ol with hydrogen bromide (from sodium bromide and concentrated sulfuric acid).

$$CH_3CH_2CH_2CH_2OH + HBr \rightarrow CH_3CH_2CH_2CH_2Br + H_2O$$

In organic chemistry, substitution reactions are characteristic of *halogenoalkanes* (see *nucleophilic substitution in halogenoalkanes*) and *arenes* (see *electrophilic substitution*).

The *ligand substitution reactions* of complex ions are examples of inorganic substitution reactions (see *ligands*).

substrates in biochemistry are the molecules on which *enzymes* act as they catalyse change. An enzyme is specific because it acts only on the substrate which fits into its *active site*.

Literally, the word 'substrate' means a lower layer on which something else can form or grow. One way of creating a large surface area for an expensive *catalyst* is to spread it over the surface of an inert substrate. This helps to keep down the cost of *catalytic converters* in car exhausts.

sugars are water soluble *carbohydrates* such as glucose and sucrose. Sugars vary in their *sweetness*. Sugar molecules have many —OH groups and can *hydrogen bond* with each other and with water. This means that they are solid at room temperature and very soluble in water. (See also *reducing sugars.*) Sugars are a source of ethanol by *fermentation.*

Ribose and deoxyribose sugars are important in the formation of nucleic acids such as *DNA* and *RNA.*

sulfates are salts of *sulfuric acid.* Soluble sulfates such as blue copper(II) sulfate and green iron(II) sulfate are familiar laboratory reagents.

Some insoluble sulfates are natural minerals, including calcium sulfate which exists in a hydrated form as gypsum and in an anhydrous form called anhydrite. Heating gypsum produces plaster of Paris. Barytes is barium sulfate, called 'heavy spar' because of its density. (See also **solubility** and **anion tests**.)

sulfides of metals are salts of hydrogen sulfide, H_2S. Some valuable metals occur as sulfide ores, including copper, silver, mercury and iron (see also **roasting** and **zinc extraction**). Sodium sulfide is soluble in water but most metal sulfides are insoluble.

sulfites are salts of sulfurous acid, which is formed when sulfur dioxide dissolves in water. Sulfurous acid is unstable and exists only in solution:

$$SO_2(g) + H_2O(l) \rightarrow H_2SO_3(aq)$$

The most important sulfite is sodium sulfite (Na_2SO_3), made by dissolving sulfur dioxide in sodium hydroxide solution.

Sulfur dioxide and the sulfite ion are **reducing agents**. They reduce chlorine, iron(III) ions, dichromate(VI) ions and manganate(VII) ions.

Sulfur dioxide and sulfites are used as preservatives. They are **antioxidants** used in foods such as lemon juice. They inhibit the growth of bacteria. (See also **anion tests**.)

sulfonation of benzene is an **electrophilic substitution** reaction which takes place in the presence of fuming sulfuric acid, which is a solution of sulfur trioxide in concentrated sulfuric acid. The product is benzene sulfonic acid.

SO₃H

SO_3, H_2SO_4

warm

Sulfonation of benzene

benzene sulfonic acid

The **electrophile** is sulfur trioxide. A sulfur atom attached to three more **electronegative** oxygen atoms is electrophilic.

Sulfonation of arenes is important because it helps to produce a range of useful products including **surfactants**, **ion exchange** resins, **dyes** and sulfonamide drugs.

sulfur (S) is a yellow crystalline solid which normally consists of S_8 molecules. Sulfur is a **non-metal**, coming below oxygen in group 6 of the periodic table. Its electron configuration is $[Ne]3s^23p^4$.

Sulfur has two solid **allotropes**. At atmospheric pressure, rhombic sulfur is the stable form of the element at room temperature and up to 95.5°C. Above this temperature the stable form is monoclinic sulfur. At the transition temperature the two allotropes are in equilibrium. Both forms consist of S_8 molecules. Heating sulfur crystals produces a runny, pale yellow liquid which darkens as the temperature rises, producing a highly **viscous liquid**. The S_8 rings break and form long, tangled chains of sulfur atoms. Further heating below the boiling point (445°C) makes the liquid more fluid because the chains start to break into shorter lengths. Pouring this dark red liquid into cold water produces an elastic, non-crystalline mass of plastic sulfur. In time plastic sulfur hardens as the sulfur chains gradually reform S_8 rings and produce rhombic sulfur again.

Sulfur is a reactive element but it is a less powerful *oxidising agent* than oxygen. It combines with most other elements with the exception of nitrogen, iodine, the noble gases and some of the less reactive metals such as gold.

Hydrogen sulfide (H_2S) is a toxic, foul-smelling compound. Despite having a higher relative molecular mass than water, hydrogen sulfide is a gas, because sulfur is not sufficiently electronegative for *hydrogen bonding* to occur between H_2S molecules.

Sulfur forms two *acidic oxides* SO_2 and SO_3, which are both important in the process of *sulfuric acid manufacture*. Sulfur dioxide is a *reducing agent*, as are sulfurous acid and its salts, the *sulfites*.

Bonding in oxides of sulfur. Note that the rules for predicting the shapes of molecules apply.

The reactions of *halide ions* with concentrated sulfuric acid illustrate the range of oxidation states of sulfur.

The presence of sulfur compounds means that *desulfurisation* is necessary to prevent *fossil fuels* producing *atmospheric pollutants* when they burn, leading to *acid deposition*.

sulfuric acid (H_2SO_4) is a highly corrosive but important chemical reagent because it can act as a *strong acid*, a *dehydrating agent*, an *oxidising agent* and as a sulfonating agent. Pure sulfuric acid is a colourless *viscous liquid*. The molecules of this *oxoacid* are attracted to each other by *hydrogen bonding*.

Structure of sulfuric acid molecules showing hydrogen bonding

In its different roles sulfuric acid acts as:
- *a strong acid* – sulfuric acid ionises fully as it dissolves in water:

$$H_2SO_4(l) + H_2O(l) \rightarrow H_3O^+(aq) + HSO_4^-(aq)$$

  Dilute sulfuric acid neutralises bases. It forms two series of soluble salts called *sulfates* and hydrogensulfates. Hydrogensulfates are *acid salts*.
- *a powerful dehydrating agent* – concentrated sulfuric acid can:
  - remove the elements of water from *sugars* and other *carbohydrates* so that they char and turn to black charcoal
  - dehydrate *alcohols* turning them into *alkenes*
  - act as a drying agent to remove water from moist gases
- *an oxidising agent* – it reacts with *halide ions*, especially bromide and iodide ions
- *a sulfonating agent* – it is used for the *sulfonation of benzene* and other *arenes*.

sulfuric acid manufacture: the Contact process for making *sulfuric acid* from *sulfur* which produces over 150 million tonnes of the acid in the world each year.

The process operates in three main stages:

Stage 1: Burning sulfur to make sulfur dioxide

$$S(s) + O_2(g) \rightarrow SO_2(g)$$

This is a highly **exothermic** reaction. The hot gas is cooled in heat exchangers which produce steam used to generate electricity. Sulfuric acid plants do not have electricity bills. Electricity and steam exported from the process help to make the production of sulfuric acid economic.

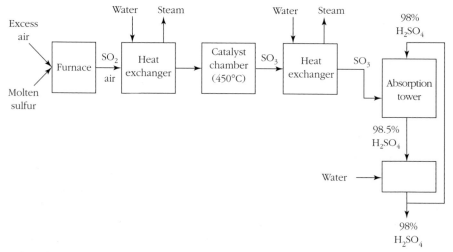

Outline flow diagram for the manufacture of sulfuric acid

Stage 2: Conversion of sulfur dioxide to sulfur trioxide

$$2SO_2(g) + O_2(g) \rightleftharpoons 2SO_3(g) \qquad \Delta H = -297 \text{ kJ mol}^{-1}$$

This is an exothermic, reversible reaction which takes place on the surface of vanadium(v) oxide, a **heterogeneous catalyst**. The temperature effect on equilibrium means that raising the temperature lowers the percentage conversion to SO_3, but the temperature must be high enough to make the reaction go fast enough. The catalyst is not active below 380°C and works best at a higher temperature.

Increasing the pressure would increase the conversion to sulfur dioxide but the extra cost is not usually justified.

Typically the gas mixture passes through four catalyst beds. Between each bed the gas mixture is cooled in a heat exchanger and cold air with more oxygen is added to the mixture. After the third bed the level of conversion is 98%. To ensure conversion of the remaining SO_2, sulfur trioxide is absorbed from the gas stream and more air is added before the gases flow through the fourth bed of catalyst. So three factors contribute to converting as much of sulfur dioxide as possible to sulfur trioxide – cooling, adding more of one of the reactants and removing the product.

Stage 3: Absorption

$$H_2O(l) + SO_3(g) \rightarrow H_2SO_4(l)$$

Sulfur trioxide cannot be dissolved in water directly because the violent reaction produces a hazardous mist of acid. Instead the gas is absorbed in 98% sulfuric acid. Sulfur trioxide passes up a tower packed with pieces of ceramic, down which the concentrated acid trickles. The circulating acid is kept at the same strength by drawing off product while adding water.

Environmental legislation requires negligible emissions of acid gases into the air to avoid *acid deposition*.

Sulfuric acid is needed on a large scale to make many other chemicals, including phosphate *fertilisers*, *paints* and *pigments*, *detergents*, *plastics*, *fibres* and *dyes*.

sunscreens protect the skin from harmful ultraviolet radiation from the Sun. Suntan lotions include chemicals such as 4-aminobenzoic acid which absorbs the UV from the Sun's *electromagnetic radiation*.

Sunblocks are heavily pigmented so that they form a barrier to sunlight by reflecting and scattering light. The white pigments are zinc oxide and *titanium(IV) oxide*.

superconductors are materials which have no electrical resistance. The resistance of a superconducting material falls suddenly to zero when the material is cooled below its critical temperature. Once this has happened, an electric current flows through a loop of the superconducting material for ever without any power source.

The first superconducting wire, a niobium–titanium alloy, was developed for commercial use in 1962. From that time, until 1986, scientific theory suggested that superconductivity would prove impossible above about 30 K. Then the discovery of ceramic superconducting solids based on lanthanum or yttrium with barium, copper and oxygen showed that there could be superconducting materials at temperatures as high as 92 K. The advantage of these materials is that they can be cooled below their critical temperature with liquid nitrogen, which is much cheaper than using liquid helium. Subsequent research has led to the development of materials with even higher critical temperatures for superconductivity.

supercritical carbon dioxide: see *carbon dioxide*.

superoxides: compounds with oxygen containing the O_2^- ion (see *potassium*).

supersaturated solution: a solution which contains more dissolved solute than a saturated solution. A supersaturated solution is unstable. Crystals form rapidly when the edge of the container is scratched or a tiny *seed crystal* is added.

The way to make a supersaturated solution is to cool a hot, saturated solution of a salt such as sodium thiosulfate or sodium ethanoate. Sometimes an unwanted supersaturated solution forms when a solid product is purified by *recrystallisation*.

Some 'instant hand warmers' consist of a supersaturated solution of sodium ethanoate in a plastic bag. Flexing a metal disc in the side of the bag starts crystallisation, which is exothermic and heats the bag to about 60°C. The hand warmer can be reused after heating in boiling water to redissolve the salt and then allowing it to cool to room temperature so that the solution is once more supersaturated.

surface tension: the tendency of surfaces to contract to the minimum surface area. Surface tension accounts for the spherical shape of soap bubbles.

Surface tension arises because of *intermolecular forces*. Molecules inside the liquid are pulled in all directions by surrounding molecules. Molecules near the surface experience a net pull inwards. This means that as many molecules as possible leave the surface, which tends to shrink.

The surface tension of many organic liquids is in the range 15 to 30 mN m⁻¹. The surface tension of water is higher because of strong *hydrogen bonding*. For water the value is about 73 mN m⁻¹. *Surfactants* lower the surface tension of water so that it can wet greasy surfaces (see *wetting*).

surfactants are surface-active agents. They are chemicals with molecules which seek out the boundary surface between two liquids or between a liquid and a gas. One of the important effects of surfactants is that they change the *surface tension* of liquids.

A surfactant molecule has an ionic or polar group attached to a hydrocarbon chain. The polar group is water-loving (hydrophilic). The hydrocarbon chain is water-hating (hydrophobic).

The hydrophobic chains of surfactant molecules tend to escape from water by moving to the surface or by forming *micelles*. Surfactant molecules at the surface lower the surface tension of the water so that it wets surfaces more effectively. Surfactant molecules adsorbed at the boundary between air and water stabilise bubbles of foam by stopping all the water draining away.

Non-polar;
hydrophobic 'tail'

Ionic;
hydrophilic 'head'

Structure of a surfactant molecule

The *detergents* in *washing powders or liquids* are surfactants that help to separate greasy dirt from surfaces. They keep dirt dispersed in water so that it rinses away. They also help to prevent dirt re-attaching itself to the surface of fabrics.

surroundings: see *system*.

suspension: particles of a solid suspended by shaking or stirring in a liquid or gas. In time the particles of a suspension settle out, unlike the smaller particles in *colloids*.

sustainable development: a development which help societies to live within the means of the environment in order that the Earth's natural resources are available for future generations. In other words, sustainability involves meeting the needs of the present without compromising the ability of future generations to meet their own needs. *Life-cycle assessment* is one method for analysing a product or process to see if it is sustainable.

sweetness: a taste sensation on the tip of the tongue produced by *sugars* and a number of synthetic chemicals such as saccharin, cyclamate and aspartame. Sugars vary in their sweetness. Fructose is sweeter than sucrose, which is sweeter than glucose.

Most synthetic sweeteners have been discovered when chemists have accidentally tasted chemicals made for other purposes. Synthetic sweeteners are taken in small amounts, especially by people trying to cut the quantity of energy foods in their diet. Some people find that cyclamates and saccharin have an unpleasant aftertaste.

saccharin   cyclamate

Chemical formulae of two sweeteners. Saccharin is 300 times sweeter than sucrose when equal quantities are compared.

Cyclamates are banned in the US because some people metabolise the sweetener to another chemical which causes bladder cancer if fed in large doses to rats.

Aspartame is an artificial sweetener which is about 200 times sweeter than sucrose. It consists of the methyl *ester* of a dipeptide consisting of two *amino acids* linked by a *peptide* bond.

Structure of aspartame   aspartame (Nutrasweet®)

Aspartame has no aftertaste but cannot be used in cooking because *hydrolysis* on heating destroys its sweetness. There is a warning on the labels of soft drinks and other foods sweetened with aspartame that they contain a source of phenylalanine. This can harm some people with a genetic disorder. Hydrolysis of aspartame produces phenylalanine.

synthesis: a process of making compounds from simpler starting materials. Synthesis puts things together. It is the opposite of analysis, which takes things apart to see what they are made of.

The Haber process is an example of synthesis: two elements (nitrogen and hydrogen) combine to make a compound (ammonia). This is the method of *ammonia manufacture* on a large scale in industry.

Synthesis of more complex molecules often takes several reaction steps.

synthetic routes: see *organic routes.*

system: a term used in *thermochemistry* to describe the material or mixture of chemicals being studied. Everything around 'the system' is 'the surroundings', which includes the apparatus with maybe a waterbath, the air in the laboratory – in theory, everything else in the universe.

An open system can exchange energy and matter with its surrounding. Most chemical reactions in laboratories take place in open systems.

A closed system cannot exchange matter with its surroundings. Energy can transfer in or out of a closed system but not the reactants or products.

An isolated system cannot exchange energy or matter with its surroundings. A mixture of chemicals in a vacuum flask with a stopper comes close to being an isolated system.

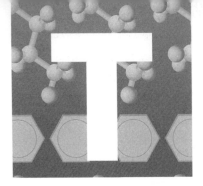

temperature effect on equilibria: the shift in the position of equilibrium which happens when the temperature changes. The effect depends on the enthalpy change for the reaction. *Le Chatelier's principle* predicts that raising the temperature makes the equilibrium shift in the direction that is endothermic. In *sulfuric acid manufacture*, for example, raising the temperature lowers the percentage of sulfur dioxide at equilibrium.

$$2SO_2(g) + O_2(g) \rightleftharpoons 2SO_3(g) \qquad \Delta H = -98 \text{ kJ mol}^{-1}$$

The equilibrium shifts to the left as the temperature rises because this is the direction in which the reaction is endothermic. This happens because the equilibrium constant, K_c, varies with temperature.

$$K_c = \frac{[SO_2(g)]^2}{[SO_2(g)]^2[O_2(g)]}$$

Temperature/K	K_c/mol^{-1} dm^3
293	9.9×10^{25}
500	1.0×10^{12}
700	1.7×10^{6}

temperature effects on reaction rates: see *activation energy* and the *Arrhenius equation*.

teratogens: substances which may cause harm to an unborn child if inhaled, swallowed or absorbed through the skin of the mother.

termination: a step in a *free-radical chain reaction* which tends to stop the process by removing free radicals. In a termination step two free radicals combine to form a molecule.

tertiary organic compounds: see *primary, secondary and tertiary organic compounds*.

theoretical yield: see *yield*.

thermal cracking breaks up bigger *hydrocarbon* molecules into smaller molecules by heating them at a high temperature. The process is especially useful for converting the *naphtha* fraction from crude oil distillation into starting materials for synthesis in the petrochemical industry.

Thermal cracking with steam converts ethane to *ethene* which is an important starting compound for chemical synthesis in industry. Ethane comes from the *natural gas* piped out of North Sea gas fields.

A mixture of the hydrocarbon vapour and steam passes through tubes in a furnace where they are heated to about 1000°C. Thermal cracking involves *free-radical chain reactions*, unlike *catalytic cracking* which has an ionic mechanism. The first step in the thermal cracking of ethane is for the molecules to split into methyl free radicals.

thermal decomposition is a reaction in which a compound decomposes on heating. An important example for industry and agriculture is the thermal decomposition of *calcium carbonate* (limestone):

$$CaCO_3(s) \rightarrow CaO(s) + CO_2(g)$$

Sometimes heating causes decomposition because a compound is stable at room temperature but becomes unstable at a higher temperature because the decomposition reaction becomes *feasible*. This is the case with the calcium carbonate and the other carbonates of *group 2* metals.

Sometimes heating causes decomposition of a compound which is unstable at room temperature but does not decompose because the rate of reaction is so slow. This is true of the *nitrogen oxides* which, at room temperature, are examples of unstable but *inert chemicals*. They all tend to decompose into nitrogen and oxygen but they do so only on heating.

thermal dissociation: a reversible process which splits a compound into fragments as the temperature rises, but re-forms the starting material on cooling.

The brown gas which forms on heating some metal *nitrates* contains an equilibrium mixture of N_2O_4 and NO_2.

$$\underset{\text{colourless}}{N_2O_4} \rightleftharpoons \underset{\text{brown}}{2NO_2}$$

The mixture darkens on heating as more colourless N_2O_4 dissociates into brown NO_2. The colour fades on cooling as N_2O_4 reforms.

thermal stability of carbonates: the ease with which the carbonates of metals decompose on heating.

Chemists relate differences in the properties of the compounds to two factors:
- the charges on the metal ions
- the sizes of the metal ions.

Group 2 carbonates and nitrates are generally less stable than the corresponding *group 1* compounds. This suggests that the larger the charge on the metal ion the less stable the compounds.

Down either group 1 or group 2 the carbonates become more stable. This suggests that the larger the metal ion the more stable the compounds.

Chemists explain the trend in thermal stability by analysing the energy changes. Two of the energy quantities they take into account are:
- the energy needed to break up the carbonate ion into an oxide ion and carbon dioxide
- the *lattice energy* given out as the positive and negative charges get closer together when the larger carbonate ions break up into smaller oxide ions and carbon dioxide.

It turns out that all the carbonates are thermally stable at room temperature but become less stable as the temperature rises. The key factor is the energy released as the ions get closer together. This is greater when the metal ion is small than when the metal ion is large.

357

Detailed analysis of the energy changes is complex, so chemists find it convenient to correlate the stability of compounds such as carbonates and nitrates with the *polarising power* of the metal ions. Generally, the greater the polarising power of the metal ion the less stable the carbonate and the more easily it decomposes to the oxide.

thermite reaction: a highly *exothermic* reaction used for welding rails and in some incendiary devices. A magnesium fuse heats and sets off the reaction in a mixture of iron(III) oxide and aluminium. The aluminium reduces the oxide to a glowing mass of molten iron

$$Fe_2O_3(s) + 2Al(s) \rightarrow 2Fe(s) + Al_2O_3(s)$$

thermochemistry is the study of energy changes during chemical reactions, including *enthalpy changes, free energy changes* and *entropy* changes. It is a major part of chemical thermodynamics. Thermochemistry is important for theory because it helps chemists to explain why some chemicals are *stable compounds* and to predict the likely direction of chemical change. With the help of thermochemistry, chemists can decide on the *feasibility* of reactions.

Practically, thermochemistry is important because of the significance of:
- calculating the energy from burning fuels
- keeping large-scale reactions under control in the *chemical industry*
- estimating the impact of energy changes in the environment.

thermodynamic control operates where there is a choice of possible products and the main product is the most *stable compound* (according to thermodynamics). This contrasts with *kinetic control* which operates when the main product is the one that forms fastest.

thermodynamic stability: see *stable compounds.*

thermodynamics: see *thermodynamics (laws of)* and *thermochemistry.*

thermodynamics (laws of): the laws which govern energy transfers and the direction of change. The laws originally grew out of the study of heating, mechanical working and steam engines. They have now been given a molecular interpretation which makes them applicable to chemical systems.

The **first law of thermodynamics** says that the total quantity of energy for a system and its surroundings is the same before and after any change. In other words, energy cannot be created or destroyed. This is the law of conservation of energy. Chemists are interested in energy transfers:
- by heating, as energy flows from a higher to a lower temperature
- by working as a system expands or contracts against atmospheric pressure
- by the flow of an electric current
- by the absorption or emission of radiation.

The value of the first law of thermodynamics is that it provides a check that all energy transfers have been accounted for.

The **second law of thermodynamics** shows which changes can go (are *feasible*). The law states that a change tends to happen if the total entropy of the system plus the surroundings increases. The *entropy* change, ΔS_{total}, is positive. Chemists often find it more convenient to work with *free energy changes*. The rule is that a process tends to happen if the free energy change, ΔG, is negative.

Thermodynamics indicates the possible direction of change but says nothing about the rate of change. A reaction which tends to go may in practice be so slow that it shows *kinetic inertness*.

The **third law of thermodynamics** states that as a system approaches absolute zero, the entropy of the system approaches a minimum value, which is zero for a perfect, pure crystal. One of the implications is that it is not possible to reach absolute zero.

thermoplastic polymers are solid when cold but they soften and become *plastic* on warming so that they can be moulded. Most *addition polymers* are thermoplastics. They consist of long-chain molecules with no cross-linking.

Thermoplastics such as polythene, polypropene and PVC are supplied by the chemical industry as small chips or as powders which can be melted and forced under pressure into closed moulds. Plastic buckets, crates, combs and many other articles are made by this process of injection moulding.

Thermoplastics can also be extruded by squeezing the heat-softened material through a die to create a continuous strip with a constant cross-section. This is the technique for making curtain tracks and pipes.

Extrusion through the fine holes of a spinneret converts a thermoplastic such as nylon into *fibres*. Cold drawing the fibres by stretching them when they are cold lines up the molecules and increases the tensile *strength*.

thermosetting polymers are *resins* which set to strong, stiff materials once they have been heated in a mould. Heating forms *cross-links* between the polymer chains with strong covalent bonds. Once formed, a thermosetting *plastic* cannot be melted again.

Bakelite® was an early example. It is a cross-linked thermoset made from phenol and *methanal* (formaldehyde). Modern thermosets include melamine–methanal and *urea*–methanal plastics.

Saucepan handles and electrical fittings are made from thermosets because these plastics do not melt on heating and are electrical insulators.

thin-layer chromatography (TLC) is a type of *chromatography* in which the stationary phase is a thin layer of a solid supported on a glass or plastic plate while the mobile phase is a liquid. The rate at which a sample moves up a TLC plate depends on the equilibrium between *adsorption* on the solid and solution in the solvent. The position of equilibrium varies from one compound to another so the components of a mixture separate.

Thin-layer chromatography is quick and cheap. Only a very small sample is needed. The technique is widely used both in research laboratories and in industry. It can be used quickly to check that a chemical reaction is going as expected and making the required product. After an attempt has been made to purify a chemical, TLC can show whether or not all the impurities have been removed.

Coloured compounds are easy to see on a TLC plate. However, the technique is often used with colourless compounds as well. A quick way of finding the position of colourless organic spots is to stand the plate in a covered beaker with iodine crystals. The iodine vapour is a *locating agent* that stains the spots.

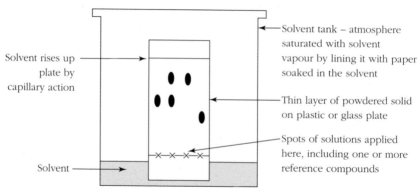

Solvent rises up plate by capillary action

Solvent tank – atmosphere saturated with solvent vapour by lining it with paper soaked in the solvent

Thin layer of powdered solid on plastic or glass plate

Spots of solutions applied here, including one or more reference compounds

Solvent

Apparatus for thin-layer chromatography

An alternative is to use a TLC plate impregnated with a fluorescent chemical. Under a UV lamp the whole plate glows, except in the areas where organic compounds absorb radiation, so that they show up as dark spots.

R_f *values* can be used to record the distances moved by chemicals in a mixture relative to the distance moved by the solvent.

thio compounds: sulfur compounds with similar formulae and structures to equivalent oxygen compounds. The thiosulfate ion, $S_2O_3^{2-}$, for example, can be thought of as a sulfate ion, SO_4^{2-}, with one oxygen atom replaced by a sulfur atom. Oxygen and sulfur are in the same group of the periodic table with the same number of electrons in the outer shells of their atoms. They therefore form compounds with similar formulae and structures.

tin (Sn) is a shiny metal with a long history. It was mined in Cornwall from the Roman times until the last mine closed in 1998. Tin was valued as an ingredient of a range of *alloys* including pewter (with antimony), solder (with lead) and *bronze* (with copper). The main use of tin today is for coating the steel for tin cans. The layer of tin stops the iron corroding. Tin's Sn^{2+} ions are not toxic. Small traces of dissolved tin contribute to the characteristic taste of some tinned foods such as canned fruit and tomatoes.

There are two *allotropes* of tin. At room temperature tin has a metallic structure but below the *transition temperature* 13.2°C the stable form is grey tin which has the diamond structure.

$$\text{grey tin} \underset{\text{below } 13.2°\text{C}}{\overset{\text{above } 13.2°\text{C}}{\rightleftharpoons}} \text{metallic tin}$$

It takes a long time for atoms in a solid to rearrange themselves. So metallic tin has to stay cold for a long time before it shows the symptoms of 'tin plague' as it gradually alters to the crumbly, brittle, grey form.

Tin, with the *electron configuration* $[Kr]4d^{10}5s^25p^2$, comes below germanium but above lead in *group 4* of the *periodic table*.

titanium (Ti) is a very strong metal which is much less dense than steel. It melts at the very high temperature of 1675°C . It does not corrode because, like *aluminium*, it is protected by a thin layer of oxide on the surface of the metal. The difficulties of titanium extraction mean that the metal is not as widely used as might be expected considering its properties.

Titanium is a *d-block element* with the electron configuration $[Ar]3d^24s^2$. Titanium forms compounds in the +2, +3 and +4 oxidation states but only the +4 state is common. Titanium(IV) chloride is a colourless liquid formed as an intermediate in *titanium extraction* and in the manufacture of *titanium(iv) oxide*.

Smoke grenades produce dense clouds of titanium(IV) oxide by the rapid *hydrolysis* of titanium(IV) chloride.

titanium extraction: a process for extracting the metal from its ores which are rutile (TiO_2) and ilmenite ($FeTiO_3$). Titanium is the fourth most abundant metal in the Earth's crust and it would be more widely used if the methods of extraction were less difficult and cheaper. In theory it should be possible to extract titanium from its oxide with carbon but in practice some of the titanium reacts with carbon forming carbides which make the metal brittle.

After the ore has been purified it is heated with carbon in a stream of chlorine gas at about 1275 K.

$$TiO_2 + 2Cl_2 + C \rightarrow TiCl_4 + CO_2$$

The titanium(IV) chloride condenses as a liquid which can be purified by *fractional distillation*.

The *reducing agent* for producing titanium metal from its chloride is a reactive metal such as magnesium, (or sodium). This is a *batch process*.

In one version of the process hot titanium(IV) chloride passes into a steel reactor inside a furnace containing molten magnesium in an atmosphere of argon at 800 K. A series of exothermic reactions heats the reactor to 1100 K producing titanium(III) and titanium(II) chlorides. Magnesium slowly reduces these chlorides to the metal as the furnace is raised to a temperature of 1300 K. Overall:

$$TiCl_4 + 2Mg \rightarrow Ti + 2MgCl_2$$

After 36–50 hours the reactor is allowed to cool for at least four days. The solid product is crushed and then *leached* with dilute hydrochloric acid to remove magnesium chloride. The titanium metal is purified by vacuum distillation at a high temperature.

Titanium is used mainly to make aircraft engines and airframes. Other major uses are the production components of chemical plants, such as *heat exchangers*.

titanium(iv) oxide is a brilliant white *pigment*. In has two crystalline forms: anatase and rutile. Millions of tonnes of titanium(IV) oxide are made each year. The main use of the pigment is in *paint* but it is also a surface coating for paper, a filler in plastics and an ingredient of cosmetics and toothpaste. The oxide makes a good white pigment because it has a very high refractive index but absorbs almost no light in the visible part of the spectrum. When ground to a fine powder it is both intensely white and very opaque, so as a pigment it has excellent covering power. The oxide is also inert and non-toxic, which is why it has replaced lead oxide in many applications.

There are two processes for producing titanium(IV) oxide from ilmenite – the long established sulfate process and the newer chloride process.

titration: a *volumetric analysis* technique for finding the concentrations of solutions and to investigate the amounts of chemicals involved in reactions. Titrations are widely used because they are quick, convenient, accurate and easy to automate.

The procedure only gives accurate results if the reaction is rapid and is exactly as described by the chemical equation. So long as these conditions apply, titrations can be used to study *acid–base, redox, precipitation* and complex-forming reactions.

A measured volume of the unknown is transferred to a flask with a *pipette*; then the standard solution is added carefully from a *burette*. The *end-point* is determined by adding a few drops of an indicator or by using an instrument such as a pH meter, a *colorimeter* or a conductivity meter.

Sometimes no indicator is needed because the excess reagent itself produces a permanent colour change at the end-point. This happens in *potassium manganate(VII)* titrations.

The diagram shows the apparatus for a titration involving a solution A which reacts with solution B. Suppose the equation for the reaction takes the form:

$$n_A A + n_B B \rightarrow \text{products}$$

which means that n_A moles of A reacts with n_B moles of B.

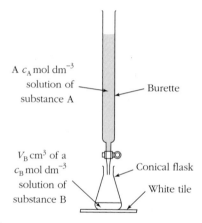

A c_A mol dm^{-3} solution of substance A — Burette

V_B cm^3 of a c_B mol dm^{-3} solution of substance B — Conical flask

White tile

Apparatus for a titration

In the laboratory, volumes of solutions are normally measured in cm^3 but they should be converted to dm^3 in calculations so that they are consistent with the units used to measure concentrations. (1 dm^3 = 1000 cm^3)

The concentration of B in the flask is c_B mol dm^{-3} and its volume is V_B dm^3.

The concentration of A in the burette is c_A mol dm^{-3}. V_A dm^3 of the solution A are added until the indicator shows that the end-point has been reached.

The amount of B in the flask at the start $= V_B \times c_B$ mol

The amount of A added from the burette $= V_A \times c_A$ mol

The ratio of these amounts must be the same as the ratio of the amounts shown in the equation:

$$\frac{V_A \times c_A}{V_B \times c_B} = \frac{n_A}{n_B}$$

In any titration, all but one of the values in this formula are known. The one unknown is determined from the results.

(See *acid–base titrations* and *iodine–thiosulfate titrations* for worked examples.)

Tollens' reagent is a test reagent used to distinguish *aldehydes* from *ketones*. Warming Tollens' reagent with an aldehyde produces a precipitate of silver, which coats clean glass with a shiny layer of silver so that it acts like a mirror.

The reagent consist of an alkaline solution of diamminesilver(I) ions, $[Ag(NH_3)_2]^+$. Aldehydes reduce the silver ions to metallic silver. Ketones do not react.

toughness: a property of materials which measures how much energy is needed to break them. Tough materials are hard to break. Metals and polymers are tough. Ceramics and glass are not tough – they are brittle.

toxic chemicals: see *hazard*. Toxic substances include poisons, *carcinogens*, mutagens and *teratogens*.

trace elements: the micronutrients which plants need in small amounts from the soil. Examples are cobalt (Co^{2+}), copper (Cu^{2+}), iron (Fe^{2+} or Fe^{3+}), manganese (Mn^{2+}) and zinc (Zn^{2+}). Plants need these nutrients in much smaller amounts than the nitrogen, phosphorus and potassium supplied by NPK *fertilisers*.

tracers: chemicals used to keep track of chemicals in the environment, in living organisms, in a chemical process and in many other circumstances where chemicals are on the move. The tracer is chosen so that it does not interfere with the changes to be studied. For this reason radioactive isotopes are often chosen as tracers because they are chemically identical with the substances being tracked but easily detected from the alpha, beta or gamma radiation emitted during *radioactive decay*. Examples of radioactive tracers include the use of:

- iodine-131 to study the behaviour of the thyroid gland
- hydrogen-3 (*tritium*) to investigate the metabolism of drugs in the human body
- phosphorus-32 to explore the uptake of phosphate fertilisers from the soil by plants.

transesterification: a process used to make *biodiesel*.

trans fats are a form of unsaturated *fat* produced when liquid *vegetable oils* are turned into solid fats through the process of partial *hydrogenation*. Naturally occurring *fatty acids* exist as the *cis* isomers. In the presence of the nickel catalyst used for hydrogenation, some of the *cis* isomers become *trans* (see *cis–trans isomers*). This means that there is a proportion of *trans* fats in products made using partially hardened oils.

The formation of *trans* fats contributes to the hardening process because a molecule of a *trans* fat is straight, rather than kinked. This means that its shape is very like the straight chain of a fully saturated fat.

Trans fats have a similar effect on blood *cholesterol* as saturated fats – they raise the type of cholesterol in the blood that increases the risk of heart disease.

transition elements are *d-block elements* which have partially filled d energy levels in one or more of their *oxidation states*. In the d-block series from scandium to zinc this definitions includes all of the metals except for the first and the last. Scandium, $[Ar]3d^14s^2$, is often excluded because it only forms compounds in the +3 state and it loses all its three outer electrons when it forms a 3+ ion. *Zinc*, $[Ar]3d^{10}4s^2$, is excluded because all its compounds are in the +2 state. Losing the outer two 4s electrons gives an ion Zn^{2+} with the electron configuration $[Ar]3d^{10}$ in which all the d energy levels are full.

Transition elements share a number of common features. They:
- are *metals* with useful mechanical properties and high melting points
- form compounds in more than one *oxidation state*
- form *coloured compounds*
- form a variety of *complex ions*
- act as *catalysts* either as metals or as compounds.

	Sc	Ti	V	Cr	Mn	Fe	Co	Ni	Cu	Zn
+7					+7					
+6				+6						
+5			+5							
+4		+4	+4		+4					
+3	+3	+3	+3	+3		+3	+3			
+2					+2	+2	+2	+2	+2	+2
+1									+1	
	Sc	Ti	V	Cr	Mn	Fe	Co	Ni	Cu	Zn

Main oxidation states of the elements Sc to Zn. Note that scandium and zinc form ions only in one state.

The chemistries of *chromium, cobalt, copper, iron, manganese, nickel, titanium* and *vanadium* provide many examples of these features.

transition state: the state of the reacting atoms, molecules or ions when they are at the top of the activation energy barrier for a reaction step. Transition states exist for such a brief moment that they cannot be detected or isolated, unlike the *intermediates* in reactions formed between two steps. The combination of reacting atoms, molecules or ions in a transition state is sometimes called an *activated complex.*

transition temperature: the temperature at which two *allotropes* of an element are in equilibrium. One form is stable below this temperature, the other is stable above the transition temperature. (See *sulfur* and *tin.*)

transmittance measures the extent to which a sample in a spectrometer absorbs radiation at a particular wavelength. Printouts from some spectrometers show transmittance on one axis.

Transmittance, T, is defined as the ratio of the intensity of the radiation leaving the sample, I, to the intensity of the radiation entering the sample, I_0.

$$T = \frac{I}{I_0}$$

For a particular wavelength of radiation, transmittance is low when a sample absorbs strongly. It is 100% if the sample does not absorb at all.

trend: a term often used by chemists to describe the way in which a property increases or decreases along a series of elements or compounds. In the *periodic table* the term can describe the variations of a property down a *group* or across a *period*. In organic chemistry the term may describe the changes in a property from one member of an *homologous series* to the next.

triglycerides are *esters* of *fatty acids* with glycerol (propane-1,2,3-triol) which is an alcohol with three —OH groups. Animal *fats* and *vegetable oils* are examples of triglycerides. Triglycerides, like other *lipids*, are soluble in organic solvents.

General structure of a triglyceride. In natural fats and vegetable oils, R, R' and R" may all be the same or they may be different.

triiodide ion: the ion that forms when *iodine* dissolves in aqueous potassium iodide.

$$I_2(s) + I^-(aq) \rightleftharpoons I_3^-(aq)$$

Iodine is only very slightly soluble in water. It dissolves in potassium iodide solution because it forms I_3^- (aq). A reagent labelled 'iodine solution' is normally $I_2(s)$ in KI(aq).

The I_3^-(aq) ion is a yellow-brown colour which explains why aqueous iodine looks quite different from a solution of iodine in a *non-polar solvent* such as hexane.

triiodomethane reaction: a reaction which produces triiodomethane (iodoform) from a compound containing either a CH_3C- group (a β-keto group) or a CH_3C- group.

The reaction conditions are to warm a drop of the compound with a solution of iodine in sodium hydroxide.

The triiodomethane reaction is used as a test for the presence of a methyl group next to a hydroxyl group, or a carbonyl group, in an organic molecule. If the result of the test is positive, triiodomethane appears as a pale yellow precipitate.

The reaction takes place in three steps for an alcohol and two steps for a carbonyl compound. The first step, oxidation to a carbonyl group, is not necessary for a carbonyl compound.

Equations for the triiodomethane reaction

Pure triiodomethane is a lemon-yellow powder which slowly decomposes in the presence of moisture to release iodine. It was once used as a *antiseptic* in wound dressings. Now its only medical use is as the antiseptic in a paste with bismuth and paraffin for packing abscesses.

Chlorine and bromine in alkali react in a similar way to produce trichloromethane (chloroform) and tribromomethane (bromoform).

365

trioxygen is the systematic name for *ozone*.

triple bond: three covalent bonds between two atoms as in *nitrogen*, *alkynes* and the cyanide ion. With three electron pairs involved in the bonding there is a region of high electron density between two atoms joined by a triple bond.

$$\overset{\times}{\underset{\times}{\text{O}}}\text{N}\overset{\times}{\underset{\times}{\text{O}}}\text{N}\overset{\times}{\underset{\times}{\text{X}}} \qquad \qquad \text{H}\overset{\times}{\underset{\circ}{\bullet}}\text{C}\overset{\times}{\underset{\circ}{\circ}}\text{C}\overset{\times}{\underset{\circ}{\bullet}}\text{H}$$

$$\text{N}\equiv\text{N} \qquad\qquad \text{H}-\text{C}\equiv\text{C}-\text{H}$$

Examples of molecules with triple bonds

The molecular orbital model for a triple bond shows that one of the bonds is a normal σ *bond* while the other two bonds are π *bonds*.

triple point: the unique combination of temperature and pressure at which the solid, liquid and gaseous forms of a substance are at equilibrium. The triple point of water is at 611 N m^{-2} (0.006 atmospheres) and 273.16 K ($0.01°$C).

tritium is the *isotope* of hydrogen, $^3_1$H. It is radioactive by *beta decay* with a *half-life* of 12.3 years.

tungsten extraction: see *extraction of metals*.

A–Z Online

Log on to A–Z Online to search the database of terms, print revision lists and much more. Go to **www.philipallan.co.uk/a-zonline** to get started.

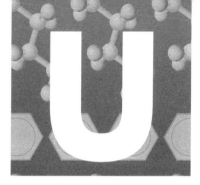

ultraviolet radiation, UV, is *electromagnetic radiation* with shorter *wavelengths* (and so higher frequencies) than the violet end of the visible spectrum. The energy of UV photons is high enough to excite the electrons in covalent bonds to higher energy levels. Excited molecules may then split into atoms (dissociate). Thus UV light can start chemical reactions by forming *free radicals*. (See *free-radical chain reaction* and *ozone in the stratosphere*).

ultraviolet spectroscopy (UV spectroscopy) is particularly useful for studying colourless organic molecules with unsaturated *functional groups* such as $C=C$ and $C=O$. The molecules absorb UV radiation at frequencies which excite shared electrons in double bonds. A UV spectrometer records the extent to which samples absorb UV radiation across a range of *wavelengths*.

In organic molecules with *conjugated systems* the UV absorption peak moves to longer wavelengths as the number of alternating double and single bonds increases. The maximum absorption by ethene is at 185 nm; this shifts to 220 nm for buta-1,3-diene. These are both colourless compounds.

The longer the conjugated chain the stronger the absorption. Beta-carotene with eleven conjugated carbon–carbon double bonds absorbs strongly with a peak shifted so far that it is not in the UV region but lies in the blue region of the visible spectrum (450 nm) so carotene is bright orange (see *coloured compounds*).

UV spectrometers make it possible to extend the techniques of *colorimetry* and *visible spectroscopy* to colourless compounds. Scientists use UV spectroscopy to check that *medicines* contain the correct amounts of drugs and that the products do not deteriorate in storage.

unit cell: the smallest unit of a crystal structure which, when piled up in three dimensions, gives a whole crystal.

The unit cell of a cubic crystal is a minute cube. An atom or ion inside the unit cell entirely belongs to that cell. An atom or ion at a corner is shared between the eight unit cells which meet in three dimensions at a corner.

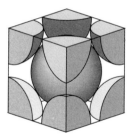

Unit cell for the caesium chloride structure

The unit cell for caesium chloride consists of one caesium ion and $8 \times \frac{1}{8}$ of a chloride ion. So the composition of the unit cell still corresponds to the *empirical formula* CsCl. (See *caesium chloride structure*.)

unidentate ligand: see *monodentate ligand*.

universal indicator is a mixture of several *acid–base indicators* which changes colour through the colours of the spectrum from red to indigo as the *pH* rises from 1 to 11. Narrower range indicators are also available which cover a smaller range of pH.

unsaturated compounds are compounds containing one or more double or triple bonds between atoms in their molecules. The term is often applied to the *alkenes* and *alkynes* which typically undergo addition reactions. The term is also commonly used to describe unsaturated *fats* and *fatty acids* which have double C=C bonds in their hydrocarbon side chains.

urea is a white crystalline solid. It is an end product of the metabolism of proteins, and is excreted as a waste product in urine.

Urea slowly hydrolyses, releasing ammonia, making it a useful nitrogen *fertiliser*. It is manufactured by the reverse process of heating ammonia and carbon dioxide under pressure.

Urea and methanal (formaldehyde) are used to manufacture a range of *thermosetting polymers* which are used as adhesives in chipboard, for producing electric sockets and plugs.

$$H_2N - C \overset{\displaystyle O}{\underset{\displaystyle NH_2}{\big\backslash}}$$

Structure of urea. Urea is the diamide of carbonic acid so it has the alternative name carbamide.

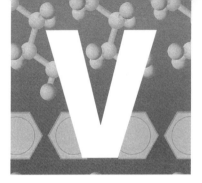

vacuum distillation: distillation under reduced pressure which makes it possible to distil liquids with very high boiling points below the temperatures at which they start to decompose. Lowering the pressure lowers the temperature at which liquids boil.

The process is used in oil refineries to separate the *hydrocarbons* in the residue from fractional distillation at atmospheric pressure. Vacuum distillation separates lubricating oils and waxes from other hydrocarbons fed to the plant for *catalytic cracking*.

Vacuum distillation is also used on a small scale to separate products of synthesis which would decompose if heated to their boiling points at atmospheric pressure.

vacuum filtration: see *Buchner flask and funnel*.

valency theory covers all the theories of chemical bonding. The outer electrons of an atom are its valency electrons which take part in chemical bonding. Electron transfer leads to *ionic bonding* or electrovalency. Electron sharing leads to *covalent bonding*.

Chemists use a variety of ways of describing or predicting the number of bonds formed by the atoms of an element. These include:

- the pattern of charges on the ions in a group of element – *group 2* metals, for example, form 2+ ions
- the number of covalent bonds which atoms normally form in molecules – carbon, 4; hydrogen, 1; oxygen, 2; nitrogen, 3; and the halogens, 1
- the *coordination numbers* in crystals and *complex ions* – giving the numbers of nearest neighbours bonded to an atom
- *oxidation numbers* – which provide a formal code for keeping track of the numbers of electrons taking part in bonding.

Dot-and-cross diagrams are an expression of the Lewis theory of chemical bonding. The *VSEPR* theory uses the Lewis theory to account for the *shapes of molecules*. The Lewis theory is not quantitative and cannot account for *bond enthalpies* and *bond lengths*.

The first quantitative theory, based on *quantum theory* and *atomic orbitals*, was the valence-bond model. In order to account for the formation of four bonds by carbon atoms, and the tetrahedral shape of a methane molecule, this theory had to introduce the notion of hybridisation. A carbon atom in its ground state has the electron configuration $[He]2s^22p^2$.

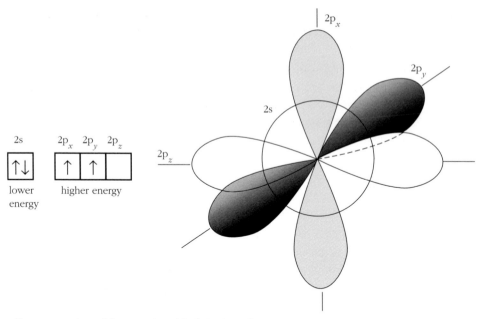

2s   2p$_x$ 2p$_y$ 2p$_z$

| ↑↓ | ↑ | ↑ | |

lower   higher energy
energy

Representation of the atomic orbitals in a carbon atom

The two electrons in the 2s orbital are paired up. In order to form four covalent bonds one electron has to be promoted from 2s to 2p giving four unpaired electrons – one in the 2s orbital and one each in the three 2p orbitals. In order to account for the shape of a methane molecule, the four atomic orbitals are then mathematically rearranged to give four equivalent, hybrid orbitals which point towards the corners of a regular tetrahedron. Overlap between these sp³ hybrid orbitals and the s-orbitals in four hydrogen atom gives rise to the C—H bonds in methane.

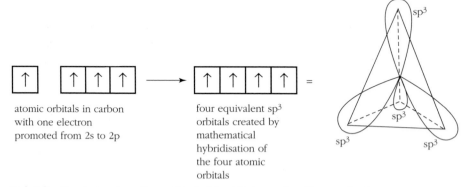

atomic orbitals in carbon with one electron promoted from 2s to 2p

four equivalent sp³ orbitals created by mathematical hybridisation of the four atomic orbitals

Hybridisation to produce four sp³ hybrid orbitals, each with one electron

The Lewis theory and valence bond theory assume that electrons are localised between the two bonding atoms. In molecular orbital theory, which dominates chemical thinking now, it is assumed that electrons belong to the molecule as a whole. In molecular orbital theory, electrons occupy *molecular orbitals* that spread through the whole molecule. Both theories deal with *sigma bonds* and *pi bonds* but the mathematical approach is different.

Molecular orbital theory introduces the idea of *delocalised electrons* which can account for the structure and bonding of molecules such as benzene as well as the properties of semiconductors and metals.

van Arkel diagram: a triangular diagram showing compounds between two elements (binary compounds), elements and alloys plotted to show their types of bonding: *metallic bonding, covalent bonding, ionic bonding* and *intermediate bonding*. Each chemical is plotted on the diagram according to the mean *electronegativity* of the atoms (horizontal axis) and the difference in the electronegativity values (vertical axis).

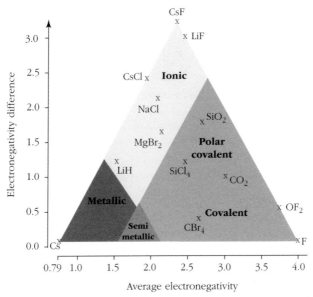

van Arkel diagram showing the regions for the main types of bonding with selected chemicals plotted

van der Waals forces are weak *intermolecular forces*. Chemists disagree about the definition of the term. Some use the term for all intermolecular forces; others exclude *hydrogen bonding*, while another group uses the term only for the weakest attraction between temporary and induced dipoles (*London forces*).

Johannes van der Waals (1837–1923) was a Dutch scientist who studied *real gases* and their deviations from *ideal gas* behaviour. He devised a modified gas equation for real gases. His equation includes two correction factors based on the *kinetic theory of gases* – one to allow for the existence of intermolecular forces and the other to allow for the fact that gas molecules have a definite volume.

van der Waals radius: the effective radius of an atom when held in contact with another atom by weak *intermolecular forces*.

Atoms do not have a definite size. The apparent size of an atom depends on the way it is bonded to a neighbouring atom, so the *ionic radius* and the *covalent radius* of an atom are not the same. Generally, the stronger the bonding the smaller the effective radius. Intermolecular forces are much weaker than covalent bonds, so van der Waals radii are relatively large.

van der Waals radii determine the effective size of a molecule as it bumps into other molecules in a liquid or gas, and when it packs together with other molecules in a solid.

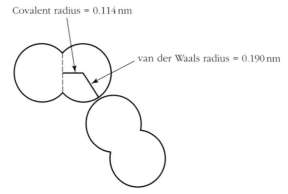

Covalent radius = 0.114 nm

van der Waals radius = 0.190 nm

Covalent and van der Waals radii for bromine

vanadium (V) is a metal, a **d-block element** with the electron configuration [Ar]$3d^34s^2$. The metal is used to make alloy **steels** which are strong and **tough**, making them suitable for machine tools and parts of engines. Vanadium is a typical **transition element** – it forms coloured compounds in several **oxidation states**, it forms **complex ions** and has compounds which can be used as **catalysts** such as vanadium(v) oxide in **sulfuric acid manufacture**.

In solution, vanadium forms ions in the +2, +3, +4 and +5 oxidation states. The +5 state is available as the yellow solid ammonium vanadate(v).

+5	VO_2^+(aq)	yellow
+4	VO^{2+}(aq)	blue
+3	V^{3+}(aq)	green
+2	V^{2+}(aq)	violet
+1		
0	V	

Oxidation states of vanadium showing the colours of the ions

Standard **electrode potentials** help to identify reducing agents to reduce vanadium(v) to the succession of lower states. (See **electrochemical series**.) Iodide ions reduce vanadium(v) to vanadium(IV).

$$E^{\ominus}/V$$

$$I_2(aq) + 2e^- \rightleftharpoons 2I^- \qquad +0.54$$

$$VO_2^+(aq) + 2H^+(aq) + e^- \rightleftharpoons VO^{2+}(aq) + H_2O(l) \qquad +1.00$$

Half-equations and electrode potentials for the reduction of vanadium(v) to vanadium(IV)

Copper metal reduces vanadium(v) to vanadium(iii).

Zinc in acid will reduce vanadium(v) all the way to vanadium(ii). The V^{2+}(aq) ion is a strong *reducing agent:*

$$Zn^{2+}(aq) \quad + \quad 2e^- \rightleftharpoons Zn(s) \qquad E^\ominus = -0.76 \text{ V}$$

$$V^{3+}(aq) \quad + \quad e^- \rightleftharpoons V^{2+}(aq) \qquad E^\ominus = -0.26 \text{ V}$$

Half-equations and electrode potentials for the reduction of vanadium(iii) to vanadium(ii)

vapour pressure is the pressure of a vapour in a closed container with its own liquid. The full term is either:

- equilibrium vapour pressure – a reminder that in a closed container a liquid and vapour reach a state of dynamic equilibrium
- saturated vapour pressure – a reminder that the atmosphere above the liquid holds as much vapour as it can at equilibrium at a particular temperature.

The vapour pressure of a liquid rises as the temperature rises. A liquid starts *boiling* when its vapour pressure equals the external pressure.

vapours are gases formed by evaporation of substances which are usually liquids or solids at room temperature. So chemists talk about oxygen gas but water vapour.

Vapours are easily condensed by cooling or increasing the pressure because of relatively strong *intermolecular forces*. Vapours therefore tend to deviate markedly from *ideal gas* behaviour.

Physicists sometimes broaden the definition of a vapour to includes gases such as butane, ammonia and carbon dioxide which can be liquefied at room temperature by increasing the pressure. (These are gases below their *critical temperature*.)

vat dyes: dyes such as indigo which are insoluble in water but can be converted to a soluble form to dye cloth. Chemical reduction converts indigo to an almost colourless, water-soluble dye which soaks into cloth such as denim for jeans. Oxidation converts indigo back to the insoluble blue form which precipitates in the fibres.

vegetable oils are liquids extracted from plants which are esters of fatty acids with propane–1,2,3-triol. They are *triglycerides* which belong to the broad class of *lipids*. The *fatty acids* in vegetable oils contain a higher proportion of *unsaturated* fatty acids than animal *fats*. Examples of vegetable oils are olive oil, sunflower oil and palm oil.

Triglycerides with unsaturated fatty acids have a less regular structure than saturated fats. The molecules do not pack together so easily to make solids, so they have lower melting points and have to be cooler before they solidify. *Hydrogenation* hardens a vegetable oil by raising its melting point.

Hydrogenation was originally used to turn oils into fats for *margarine*. The disadvantage of the process is that it produces some *trans fats*. Food scientists have found ways to turn vegetable oils into solid spreads without hydrogenation. Instead they use a process which allows triglycerides to swap fatty acid molecules. The result of this process of 'interesterification' is a product with a higher melting point. This happens because the process happens in a way that allows the molecules to pack together more neatly to make a denser material. Interesterification does not produce *trans* fats but it does produce triglycerides that are not found naturally.

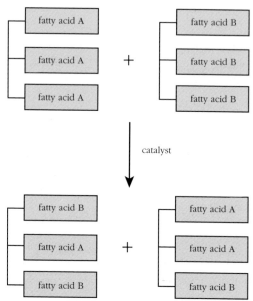

Interesterification. Two triglycerides react in the presence of a catalysts to give a mixture of products with a higher melting point. Catalysts for interesterification include sodium methoxide and the enzyme lipase.

vibration of bonds: the spring-like vibration of covalent bonds as they bend and stretch. A vibrating *polar covalent bond* is an oscillating dipole which can interact with infrared radiation. Like other energy changes involving atoms and electrons, energy is gained or lost in fixed amounts (quanta). According to *quantum theory*, $\Delta E = h\nu$. It turns out that the sizes of the energy jumps for vibrating bonds correspond to frequencies in the infrared region of the spectrum. Each type of bond has a particular value for the energy jumps so it absorbs at a characteristic frequency. This is the basis of *infrared spectroscopy*.

viscous liquids are thick and sticky liquids such as syrups and treacle. Some liquids are viscous because they consist of a tangled mass of long-chain molecules. *Lubricants* from crude oil consist of *hydrocarbons* with chains of more than 25 carbon atoms in the molecules.

Other liquids are viscous because of extensive *hydrogen bonding*. Propane-1,2,3-triol (glycerol) is a sticky liquid for this reason. It flows much more slowly than propan-1-ol. Hydrogen bonding also contributes to the high viscosity of concentrated solutions of sugars in water.

On an atomic scale, toffee and *glass* have disordered structures like liquids. They are, however, so viscous and flow so slowly that they are effectively solids.

visible radiation is *electromagnetic radiation* with *wavelengths* between 400 nm and 700 nm. This band of radiation is visible because it can bring about reversible chemical changes in cells of the retina. These chemical changes lead to electrical impulses in the nerve cells of the eye interpreted in the brain as colours.

Coloured compounds absorb radiation in the visible region of the spectrum. Chemists study these compounds using *visible spectroscopy* or *colorimetry.*

visible spectroscopy is used to determine the concentration of solutions of transition metal complex ions and other coloured compounds. The technique is based on the law that states that the absorbance of a solution is directly proportional to its concentration. The advantage of using a spectroscope rather than a colorimeter is that measurements can be taken using monochromatic light. The procedure often involves *calibration* of the instrument with solutions of known concentration

vitamins are a group of chemically unrelated organic compounds which are needed in very small amounts in the diet for healthy growth and body functions. The B vitamins and vitamin C are soluble in water. The other vitamins are fat soluble.

The B vitamins are essential for the activity of some *enzymes*. Lack of vitamin B_{12} leads to a form of anaemia. One of the triumphs of twentieth century organic chemistry was the analysis and total *synthesis* of vitamin B_{12}. Vitamin C is *ascorbic acid.*

volatile organic compounds (VOCs) include a wide variety of carbon compounds which are *atmospheric pollutants.* They are gases or liquids which evaporate easily. Motor vehicles are one source of VOCs, including *hydrocarbons* from the evaporation of fuels and from unburnt *fuel* released from exhausts. Another source is leakage from pipelines and storage tanks. The organic solvents used in the chemical industry and in chemical products are yet another source of this type of air pollution. Some *paints*, varnishes, and waxes contain organic solvents, as do many cleaning, disinfecting, cosmetic and degreasing products. One of the aims of *green chemistry* is to find new formulations of these products based on water instead of non-aqueous solvents.

volatile substances are solids or liquids which evaporate easily. A volatile substance easily turns into a *vapour. Iodine* is an example of a volatile solid. Most familiar volatile compounds are molecular liquids at room temperature. Examples are water, hexane, *ethanol* and *propanone.*

The equilibrium *vapour pressure* of a substance is a measure of its volatility at a particular temperature.

Makers of *perfumes* take advantage of differences in volatility.

voltmeters measure the potential difference between two points in a circuit in volts. Chemists use voltmeters to measure $E^{\ominus}_{cell}$ values of *electrochemical cells.*

The value needed when determining *standard electrode potentials* is the voltage when no current is flowing between the electrodes. These values are measured with an accurate voltmeter with a very high internal resistance.

volume is the amount of space taken up by a sample. The *SI unit* of volume is the cubic metre (m^3). Gas volumes are converted to m^3 before substituting values into the *ideal gas* equation.

Chemists generally measure volumes in cubic decimetres (dm^3, formerly in *litres*) or cubic centimetres (cm^3).

Measuring volumes of gases and solutions is an essential part of quantitative chemistry. (See *gas volume calculations, molar volume, reacting volumes of gases* and *volumetric analysis.*)

volumetric analysis: a very important aspect of quantitative analysis because volumetric methods can be very *accurate*. An analyst can deliver a volume of liquid from a 50 cm³ grade B burette with a *measurement uncertainty* of ±0.06 cm³. If the total volume is 25 cm³, this represents an uncertainty of about 1 part in 400 (see *titrations*).

VSEPR theory: the 'valence shell electron pair repulsion theory'. This is the theory which helps to predict the *shapes of molecules* from the number of bonding pairs of electrons and lone pairs of electrons in the outer (valence) shell of the central atoms in a molecule.

Do you need revision help and advice?

Go to pages 388–407 for a range of revision appendices that include plenty of exam advice and tips.

washing powders or liquids are complex formulations for cleaning clothes. They may include:

- *detergents* to increase wetting power, separate grease from fabrics and keep dirt in suspension
- *sequestering agents* such as citrates or *zeolites* to soften *hard water*
- *enzymes* to break down protein stains such as blood stains
- optical brighteners to keep white fabrics looking bright and white
- oxygen *bleach* such as sodium peroxoborate(III) which only acts above 60°C
- *perfume* and colour to make the product distinctive and attractive to customers.

wastes are unwanted solids, liquids and gases discarded from homes, commerce, industry, agriculture and the public services. All these can help to conserve resources and limit the amount of waste dumped in *landfill* by:

- *reduction* – the chemical industry has an extensive programme of *green chemistry* research and development to introduce new processes which give higher yields and create less waste so that the disposal problem is cut
- *re-use* – this depends on solvents and detergents which help to clean products or containers so that they can be used again
- *recycling* – a variety of physical and chemical processes recover metals, polymers and other materials so that they can be added to the raw materials used to make new products; in some industries, such as the steel industry, *recycling* is well established
- *recovery* – this includes the recovery of the energy resources tied up in materials; *incineration* not only disposes of the waste but provides energy to generate electricity. However, incineration can be hazardous if not carried out at a high enough temperature to ensure that harmful chemicals do not escape into the air.

water (H_2O) is a familiar liquid with properties which can seem anomalous when compared to other compounds. Despite the small size of its molecules, water is a liquid at room temperature because of *hydrogen bonding*. The O—H bonds in water molecules are highly polar because oxygen is so *electronegative*. Water molecules are polar because they are not linear thanks to the two lone pairs of electrons which help to determine the *shape of the molecule*.

As a *polar solvent*, water can hydrate ions and dissolve salts. Water can also dissolve some organic compounds because it can form hydrogen bonds with hydroxyl groups in alcohols, sugars or carboxylic acids and with —NH_2 groups in amines.

Water plays an important part in many reactions:

- **_hydration reactions_** – water forms aqua **_complex ions_** with metal ions
- **_acid–base reactions_** – water acts as both an acid and as a base because it is an amphoteric compound
- **redox reactions** – reactive metals such as **_group 1_** elements reduce the hydrogen in water to hydrogen gas
- **_hydrolysis_** reactions – this includes the **_hydrolysis of non-metal chlorides, hydrolysis of metal aqua ions, hydrolysis of salts_** as well as **_hydrolysis of an organic chloro compound_** and of other organic compounds such as esters (**_acid catalysis_** or **_base catalysis_**).

water cycle: the cycling of water in the environment between the oceans, the atmosphere and the crust of the Earth.

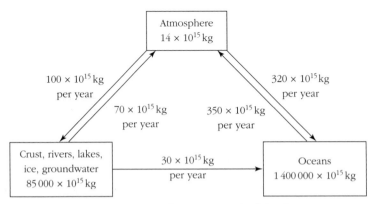

The water cycle, showing the amounts of water in each of the three main reservoirs and the flows between them

The specific latent heat of vaporisation of water is relatively large so huge amounts of energy are transferred by the water cycle. Energy from the Sun evaporates water in the tropics. Winds carry the warm moist air to higher latitudes where the energy is released as the water condenses and falls as rain.

water of crystallisation: water molecules which make up part of the crystal structure of a compound. There are five molecules of water of crystallisation for each $CuSO_4$ unit in blue copper(II) sulfate crystals, $CuSO_4.5H_2O$. The full systematic name for copper sulfate is tetraaquocopper(II) tetraoxosulfate(VI)-1-water which shows that four of the water molecules *hydrate* the copper ions; the fifth water molecule forms **_hydrogen bonds_** with sulfate ions.

Heating hydrated crystals drives off the water as steam, leaving the anhydrous salt. Stronger heating may then lead to other changes (see **_nitrates_** for example). Other salts with water of crystallisation include:

- hydrated sodium carbonate (washing soda crystals) – $Na_2CO_3.10H_2O$
- hydrated magnesium sulfate (Epsom salts) – $MgSO_4.7H_2O$

water treatment provides water for drinking, water for industry and water for medical uses.

Some of the chemical processes used to provide water for homes include:

- **_coagulation_** – the addition of aluminium(III) or iron(III) salts to add positive ions which help to coagulate negatively charged **_colloid_** particles so that finely divided solids clump together and settle out

- **adsorption** – removing organic chemicals on the surface of activated **charcoal**
- **disinfection** – use of chemicals to kill micro-organisms which might otherwise cause disease (see **chlorine water treatment**). Increasingly **ozone** or **ultraviolet radiation** are being used in place of chlorine because they are less hazardous and do not react with organic impurities to produce chlorinated hydrocarbons that may be harmful
- **water softening** – removing calcium and magnesium ions from **hard water** by precipitating them as insoluble carbonates.

Ion exchange can remove all the ions from water to give very pure water needed industrially and commercially.

wavelength: the distance between the peaks (or troughs) of a wave. Wavelengths of **electromagnetic radiation** vary from about 1000 m for radio waves down to about 10^{-3} nm for gamma rays.

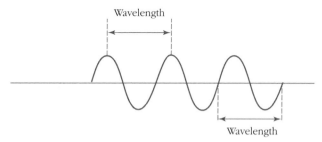

Wavelength

Wavelength

A waveform showing the wavelength

Light in the visible region of the spectrum has wavelengths from about 400 nm at the blue end to 700 nm at the red end.

All **electromagnetic radiation** travels at the same speed, c, in a vacuum. The **frequency**, ν, wavelength, λ, and speed are related by $c = \nu\lambda$.

waxes are materials which can be moulded when warm but are hard and brittle when cold. They are insoluble in water and are water repellent. Waxes are ingredients of polishes and are used to waterproof cloth and leather. Natural waxes such as beeswax from honeycombs and lanolin from wool are esters of **fatty acids** with alcohols with only one hydroxyl group. This distinguishes them from **fats** and **vegetable oils**. The formula of beeswax is $C_{15}H_{31}COOC_{30}H_{61}$. Carnauba wax, an ingredient of polishes, varnishes and lipstick, comes from the leaves of a palm tree which grows in Brazil.

Hydrocarbon wax, such as the paraffin wax used for candles, is one of the products of the **vacuum distillation** of the residue from the **fractional distillation** of crude oil.

weak acids are only slightly ionised when they dissolve in water. In a 0.1 mol dm^{-3} solution of ethanoic acid, for example, only about one in a hundred molecules reacts with water to form **oxonium ions**. The more dilute the solution the greater the degree of ionisation. The acid dissociation constant, K_a measures the strength of an **acid**.

Note the important distinction between strength and concentration. Strength is the extent of ionisation. **Concentration** is the amount of acid present in mol dm^{-3}. Note too that it takes as much sodium hydroxide to neutralise 25 cm^3 of 0.1 mol dm^{-3} of a weak acid (such as ethanoic acid) as it does to neutralise 25 cm^3 of 0.1 mol dm^{-3} of a strong acid such as hydrochloric acid.

Measuring the *pH* of a solution of an acid is not enough to show whether or not the acid is strong or weak. A solution of an acid with pH 4 might be a very dilute solution of a strong acid or a concentrated solution of a weak acid.

The salts made from weak acids and strong bases are alkaline in solution (see *hydrolysis of salts*).

weak bases are only slightly ionised when they dissolve in water. In a 0.1 mol dm^{-3} solution of *ammonia*, for example, only about one in a hundred molecules reacts with water to form ammonium ions. The base dissociation constant, K_b measures the strength of a *base*.

As with *weak acids*, it is important to distinguish between strength and concentration.

The salts made from weak bases and strong acids are acidic in solution (see *hydrolysis of salts*).

weathering is a process which breaks down rocks physically and chemically. Physical weathering cracks rocks into smaller fragments increasing the surface area exposed to chemical attack. During chemical weathering, rocks react with water, oxygen and carbon dioxide. The types of reaction involved are *hydrolysis* and *oxidation*. Weathering releases soluble ions which plants need for growth.

weight, measured in newtons (N), is the pull of the Earth's gravity on an object. A laboratory balance responds to weight but is calibrated to give readings in grams or kilograms which measure the mass of the sample. On the surface of the Earth, the pull of the Earth is proportional to mass.

weight = mass $\times$ g where the constant g = 9.8 N kg^{-1}

So chemists use 'weighing' to determine the mass of a sample. This accounts for the continuing use of concentration units such as weight/volume per cent (w/v), which should strictly be the mass/volume per cent.

wetting happens when water spreads out and covers the surface of a solid. Water wets clean glassware but breaks up into separate droplets on greasy glassware. This makes it easy to see if pipettes and burettes are clean.

Wetting is an important step in getting materials clean but water does not wet fabrics that are contaminated with greasy dirt. One of the functions of *surfactants* in *detergents* is to lower the *surface tension* of water so that it will wet dirty surfaces.

Surfactants are added to pesticides so that when sprayed onto plants the solution spreads over the leaves and is absorbed. With no surfactant the solution would break up into tiny droplets on the surface of the leaves.

word equation: an equation which describes a chemical change in words instead of symbols. Writing word equations identifies the reactants and products, so it is a useful first step towards *balanced equations* with symbols. For example:

sodium + water → sodium hydroxide + hydrogen

Word equations are useful as ways of summarising general patters of behaviour. For example:

acidic oxide + water → oxoacid

carboxylic acid + alcohol → ester + water

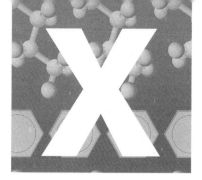

X-ray crystallography is a technique used to determine crystal structures by interpreting the diffraction patterns formed when *X-rays* are scattered by the electrons of atoms in crystalline solids. Interference between the scattered X-rays produces patterns recorded photographically. Crystallographers have worked out techniques for interpreting the patterns so that they can determine:

- the *crystal structures of ionic compounds*, the *crystal structures of metals* and the *crystal structures of non-metals*
- the *atomic radius* and *ionic radius* for atoms and ions,
- the *shapes of molecules* including *DNA* and *proteins*.

Because X-rays are scattered by electrons it is possible to draw up *electron density maps* from the diffraction patterns showing the positions of atoms. Small atoms such as hydrogen with few electrons do not show up well, so their position can be hard to determine.

The use of X-ray crystallography to determine the structures of crystals was pioneered by Lawrence Bragg (1890–1971) who developed the theory and his father William Bragg (1862–1942) who designed the instruments.

Among the great successes of X-ray crystallography were the determinations of the structures of:
- the protein alpha-helix by Linus Pauling in 1951
- myoglobin and haemoglobin by John Kendrew and Max Perutz in the 1950s
- DNA by Rosalind Franklin, Maurice Wilkins, Francis Crick and James Watson in 1953
- vitamin B_{12} by Dorothy Hodgkin in 1956.

All these scientists were awarded Nobel prizes except Rosalind Franklin who died in 1958 at the age of 37, so that she could not be considered for the prizes awarded for DNA work in 1962.

X-rays are *electromagnetic radiation* with very short wavelengths in the region 1×10^{-9} m (1 nm). X-rays have wavelengths of the same order of magnitude as the lengths of atomic bonds. As a result it is possible to use X-ray diffraction for *X-ray crystallography*.

xenon (Xe) is a noble gas which makes up less than 0.1% of the atmosphere. Xenon is separated from liquid air by *fractional distillation*. For a long time chemists thought that the noble gas elements of *group 8* were completely unreactive so they called them the 'inert gases'. This changed in 1962 when the British-born Canadian chemist Neil Bartlett was working on the properties of platinum fluorides. Based on the accidental discovery that platinum(VI) hexafluoride could combine with oxygen molecules, he used values for *ionisation energies* to predict the formation of a compound between platinum(VI) hexafluoride and xenon and produced the yellow solid $Xe^+PtF_6^-$. Since then chemists have produced a range of xenon compounds but only with the most reactive elements oxygen and fluorine. Examples are XeF_2, XeF_4 and XeF_6.

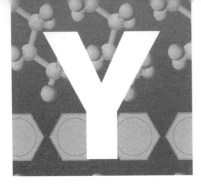

yield: yield calculations are used to assess the efficiency of a chemical process. A judgement about the overall efficiency of a chemical synthesis takes into account the *atom economy* as well as the yield.

Few synthetic reactions are completely efficient and most reactions, especially organic reactions, give lower yields. There are several reasons why the overall yield may be low:
- the reaction may be incomplete (perhaps because it is slow or because it reaches an equilibrium state) so that a proportion of the starting chemicals fails to react
- there may be side-reactions that make by-products instead of the required chemical
- recovery of all the product from a reaction mixture is usually impossible
- some of the product is usually lost during transfer of the chemicals from one container to another when the product is separated and purified.

The 'theoretical yield' is the mass of product assuming that the reaction goes according to the *balanced equation* and that there are no losses during the stages of carrying out the process. The 'actual yield' is the mass of product obtained.

The 'percentage yield' is given by this relationship:

$$\text{percentage yield} = \frac{\text{actual yield}}{\text{theoretical yield}} \times 100\%$$

Worked example

What is the theoretical yield of *aspirin* when 15.5 g of 2-hydroxybenzoic acid reacts with excess ethanoic anhydride? What is the percentage yield if the actual yield of aspirin is 7.25 g?

Notes on the method

Start by writing the equation for the reaction. This need not be the full balanced equation so long as the equation includes the *limiting reactant* and the product.

Because the ethanoic anhydride is in excess, the limiting reactant is the 2-hydroxybenzoic acid. This means that the ethanoic anhydride can be ignored during the calculation – though the correct stoichiometric amounts must be quoted.

Answer

The 'balanced' equation is:

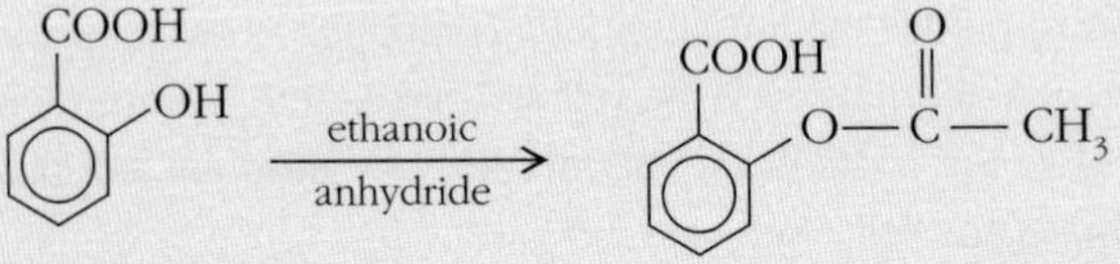

The molar mass of 2-hydroxybenzoic acid, $C_7H_6O_3 = 138$ g mol^{-1}

The amount of 2-hydroxybenzoic acid at the start of the synthesis

$$= \frac{15.5 \text{ g}}{138 \text{ g mol}^{-1}} = 0.112 \text{ mol}$$

1 mol of the acid produces 1 mol of aspirin.

The molar mass of aspirin, $C_9H_8O_4 = 180$ g mol^{-1}

The theoretical yield of aspirin = 0.112 mol × 180 g mol^{-1} = 20.2 g

Percentage yield = $\dfrac{7.25 \text{ g}}{20.2 \text{ g}}$ × 100% = 35.9%

Are you studying other subjects?

The *A–Z Handbooks (digital editions)* are available in 14 different subjects. Browse the range and order other handbooks at **www.philipallan.co.uk/a-zonline**.

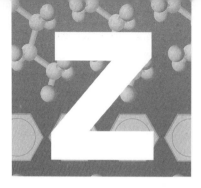

zeolites are sodium aluminium *silicates* in which the three-dimensional structures of the silicon and oxygen atoms form tunnels and cavities into which ions and small molecules and ions can fit.

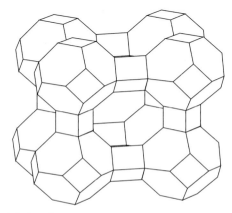

Model of a zeolite crystal structure

Synthetic zeolites can make excellent *catalysts*. They can be developed with *active sites* to favour the required reactions by acting on molecules with particular shapes and sizes. They are used for *catalytic cracking*.

Natural and synthetic zeolites are used in *ion exchange*. A synthetic zeolite is used to soften water by swapping the calcium ions in *hard water* with sodium ions. For this reason many washing powders contain zeolites.

Zeolites can be very good drying agents; they can absorb water molecules selectively because the water molecules fit into the holes in their crystal structure.

Similarly, zeolites can acts as 'molecular sieves', sorting out molecules by size. Some zeolites will absorb nitrogen from the air while letting oxygen pass through. This is one of the methods used to separate *oxygen* from the *air* on a large scale. Once the zeolite is saturated with nitrogen it can easily be regenerated by lowering the pressure and releasing the waste nitrogen back into the air.

zero order reaction: a reaction is zero order with respect to a reactant if the rate of reaction is unaffected by changes in the concentration of that reactant. The concentration term for this reactant is raised to the power zero in the *rate equation*.

$$\text{rate} = k[X]^0$$
$$= k\,(\text{a constant})\,\text{because}\,[X]^0 = 1$$

(Any term raised to the power zero equals 1.)

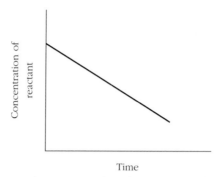

Variation of concentration of a reactant plotted against time for a zero order reaction
The gradient of this graph measures the rate of reaction. The gradient is a constant so
the rate stays the same even though the concentration of the reactant is falling.

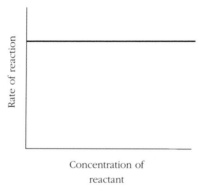

Variation of reaction rate with concentration for a zero order reaction

The rate of reaction of iodine with propanone in the presence of acid is zero order with
respect to iodine. This can be explained by proposing that the reaction takes place in
several stages including a *rate-determining step*.

Ziegler–Natta catalysts speed up the polymerisation of *alkenes* such as ethene and
propene making it possible to produce *addition polymers* at relatively low temperatures
and pressures. Ziegler–Natta catalysts are made from titanium(IV) chloride and aluminium
alkyls (such as triethyl aluminium) in a hydrocarbon solvent.

There is little chain branching in the polymers, so the poly(ethene) chains formed by
this method can pack closely together, forming the high-density form of poly(ethene).
Poly(propene) made with a Ziegler–Natta catalyst has a regular arrangement of the methyl
side chains. The molecular structure is *isotactic*. The long chains can pack together more
closely because of this regularity, allowing stronger *intermolecular forces* between the
chains and a strong polymer with the ability to form useful fibres.

N

zinc (Zn) is a metallic element; the last in the *d-block elements* of the fourth row of the periodic table. Zinc is not normally regarded as a *transition element* because both its atoms, $[Ar]3d^{10}4s^2$ and its Zn^{2+} ions, $[Ar]3d^{10}$, have full d subshells.

Zinc and its compounds show few of the typical properties of transition metals. The metal has lower melting and boiling points than transition elements. It forms only one series of compounds in the +2 oxidation state and these compounds are colourless. Examples are zinc oxide (ZnO), zinc chloride ($ZnCl_2$) and zinc sulfate ($ZnSO_4$).

Zinc oxide is an *amphoteric oxide* which means that it, and the hydroxide, will dissolve in both acids and alkalis. Adding sodium hydroxide to a solution of zinc ions produces a white precipitate of the hydroxide which redissolves either on adding acid or on adding excess alkali.

$$Zn(OH)_2(s) + 2H^+(aq) \rightarrow Zn^{2+}(aq) + 2H_2O(l)$$
$$Zn(OH)_2(s) + 2OH^-(aq) \rightarrow [Zn(OH)_4]^{2-}(aq)$$

Zinc resembles transition elements by forming *complex ions* such as $[Zn(NH_3)_4]^{2+}(aq)$. Adding ammonia solution to a solution of zinc ions produces a white precipitate of the hydroxide which redissolves in excess ammonia solution as the ammine complex forms.

$$Zn(OH)_2(s) + 4NH_3(aq) \rightarrow [Zn(NH_3)_4]^{2+}(aq) + 2OH^-(aq)$$

Some *enzymes* are only active in the presence of cofactors. The enzyme carbonic anhydrase, for example, is only effective with zinc ions bound as part of its active site. Carbonic anhydrase catalyses the reaction of *carbon dioxide* with water to form hydrogencarbonate ions and hydrogen ions. This is an important reaction for *blood buffers*.

zinc extraction produces the metal from one of its sulfide ores. The most abundant ore is sphalerite, consisting of zinc blende (ZnS). Froth flotation is commonly used to concentrate the ore which is then roasted by heating in air to convert the sulfide to zinc oxide.

$$2ZnS(s) + 3O_2(g) \rightarrow 2ZnO(s) + 2SO_2(g)$$

A neighbouring plant converts the sulfur dioxide into sulfuric acid for making fertilisers.

Most zinc (around 80%) is obtained from the oxide by *electrolysis* of zinc sulfate solution, made by dissolving the zinc oxide in sulfuric acid. The electrolysis cells have lead *anodes* and aluminium *cathodes*. Zinc forms at the cathodes.

$$Zn^{2+}(aq) + 2e^- \rightarrow Zn(s)$$

At regular intervals the zinc is stripped from the anodes, melted and cast into ingots with the metal at least 99.96% pure.

Oxygen bubbles off from the anodes where sulfuric acid reforms. Recycled acid is reused to dissolve more zinc oxide.

The main use for zinc is for protective coatings such as the layer on galvanised steel. It is also used to make *alloys*.

zwitterions are ions with both a positive and a negative charge. *Amino acids* form zwitterions when the amino groups accept protons and the acid groups give away protons.

An amino acid forming a zwitterion

In crystals and in aqueous solution, amino acids exist largely as zwitterions. Adding acid to a solution of zwitterions puts a proton on the —COO⁻ group, turning it into —COOH, so that the only charge on the ion is the positive charge on the nitrogen atom. Adding alkali to a solution of zwitterions removes the proton from the —NH₃⁺ group, turning it into —NH₂, so the ion has only a negative charge only on the —COO⁻ group (see *isoelectric point*).

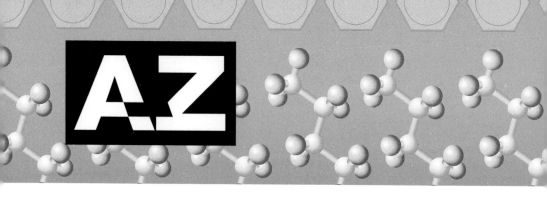

Chemistry revision lists

Using A–Z Online

In addition to the revision lists set out below, you can use the A–Z Online website to access revision lists specific to your exam board and the particular exam you are taking. Log on to **www.philipallan.co.uk/a-zonline** and create an account using the unique code provided on the inside front cover of this book. Once you have logged on, you can print out lists of terms together with their definitions, which will help you to focus your revision.

In this appendix, we have listed revision terms for the topics that feature in the A-level specifications in England, Wales and Northern Ireland (but with the exceptions indicated in the lists):

● English A-level courses: AQA, Edexcel, OCR and Salters (OCR)
● CCEA Advanced Chemistry for Northern Ireland
● WJEC Advanced Chemistry for Wales

Topics 1–31 generally appear in the first part of the courses (AS), and topics 32–50 in the second part (A2). Pick the topic you want to revise, then look up the terms listed in alphabetical order under each heading. Cross-references in the text (or on the website) will help you to build on these starting points. Note that some important terms appear more than once in the lists.

Most of the topics in this appendix also feature in the following course specifications, but they are organised in different ways:

● SQA Higher and Advanced Higher Chemistry in Scotland
● International Baccalaureate
● CIE: International GCE Advanced Chemistry and the Pre-U Certificate in Chemistry

For this reason, we have set out these revision lists on the website. You should always check what you are revising against the latest version of your course specification.

1 Atoms

Atom

Atomic number

Atomic theory

Fundamental particles

Isotopes

Mass number

Mass spectrometry

Relative atomic mass

Relative isotopic mass

Relative molecular mass

2 Radioactivity

(Salters and WJEC only)

Alpha decay

Beta decay

Nuclear fusion

Gamma radiation

Half-life (radioactive)

Ionising radiation

Isotopes

Radiation dose

Radioactive decay

Tracers

3 Electrons in atoms

Atomic emission spectroscopy

Atomic absorption spectroscopy *(Salters and WJEC only)*

Atomic orbitals

Atomic spectrum *(CCEA, Salters and WJEC only)*

Aufbau principle

Electromagnetic radiation *(CCEA, Salters, and WJEC only)*

Electron configuration

Energy levels

Hydrogen emission spectrum *(CCEA, Salters and WJEC only)*

Ionisation energy

Periodic table

Quantum numbers

4 Chemical amounts and formulae

Amount of substance

Avogadro constant

Empirical formula

Ideal gases *(AQA only)*

Molar mass

Mole

Molecular formula

Percentage composition

Relative atomic mass

Relative formula mass

Relative molecular mass

5 Chemical amounts and equations

Acid–base titration

Amount of substance

Atom economy

Balanced equations

Chemical equations

Concentration

Formula unit

Gas volume calculations

Limiting reactant

Molar mass

Molar volume

Reacting masses

Titration

Yield

6 Structure and properties

Allotropes

Changes of state

Crystal structures of ionic compounds.

Crystal structures of metals

Crystal structures of non-metals

Giant structures

Ice

Metals

Nanotechnology *(WJEC only)*

Non-metals

Physical properties

Smart materials *(WJEC only)*

7 Bonding in ionic compounds

Born–Haber cycle *(Edexcel only)*

Dot-and-cross diagrams

Electrolysis *(not in all specifications)*

Electrostatic forces

Giant structures

Hydration

Ionic bonding

Ionic radius

Ions

Lattice

Octet rule

Polarisability

Polarisation of ions

Polarising power

Sodium chloride structure

8 Covalent bonding

Bond angles

Bond length

Coordinate bond

Covalent bonds

Dative covalent bond

Dot-and-cross diagrams

Electronegativity

Giant structures

Intermediate bonding

Lone pair of electrons

Multiple bonds

Permanent dipole

Polar covalent bonds

Polar molecules

Shapes of molecules

9 Bonding between molecules

Dipole–dipole interactions

Hydrogen bonding

Ice

Intermolecular forces

London forces *(Edexcel only)*

Lone pair of electrons

Polar molecules

Van der Waals forces

10 Metallic bonding

Crystal structures of metals

Giant structures

Metallic bonding

Metals

11 Acids, bases and salts

(OCR only. Not explicit as a topic in other specifications but much of this is generally assumed)

Acids

Alkalis

Ammonia

Anhydrous salts

Arrhenius theory of acids

Bases

Double salts

Hydration

pH scale

Salts

Solubility

Water of crystallisation

12 Redox reactions

Chemical equations

Electron transfer

Half-equation

Names of inorganic compounds

Oxidising agents

Oxidation

Oxidation numbers

Oxidation states

Redox reactions

Reducing agents

Reduction

13 Periodic table

Atomic number

Atomic radius

Electron configuration

Electronegativity

Group

Ionisation energy

Periodicity

Periodicity of physical properties

Periodic table

Physical properties

Shielding

Trend

14 Group 2

Antacids

Barium

Basic oxide

Calcium

Calcium carbonate

Flame tests

Group 2

Ionisation energy

Magnesium

Shielding

Solubility

Thermal decomposition

Thermal stability of carbonates

15 Group 7

Acids

Astatine

Bleaching

Bromine

Chlorine

Chlorine oxoanions

Displacement reactions

Disproportionation reactions

Electronegativity

Fluoridation *(CCEA only)*

Fluorine

Group 7

Halide ions

Halogens

Hydrogen halides

Iodine

Iodine–thiosulfate titrations *(Edexcel only)*

Redox reactions

Silver halides

16 Energetics

Bond enthalpies

Calorimeter

Endothermic

Energy (enthalpy) level diagrams

Enthalpy change

Enthalpy change of combustion

Enthalpy change of formation

Enthalpy change of neutralisation

Enthalpy change of reaction

Exothermic

Hess's law

Specific heat capacity

Stable compounds

Standard conditions

Spontaneous reaction

System

17 Reaction rates

Activated complex

Activation energy

Catalysts

Collision theory

Heterogeneous catalyst

Homogeneous catalyst

Intermediates in reactions

Kinetic inertness

Kinetic theory of gases

Maxwell–Boltzmann distribution

Rates of reaction

18 Chemical equilibrium

Ammonia manufacture (for example of application of theory)

Dynamic equilibrium

Homogeneous equilibrium

Le Chatelier's principle

Methanol (production as an example of application of theory)

Reversible changes

Sulfuric acid manufacture (as an example of application of theory)

System

19 Production of metals

(AQA only at AS)

Aluminium extraction

Acid deposition

Carbon monoxide

Copper refining

Displacement reactions

Extraction of metals

Iron extraction

Leaching

Metal extraction

Ore

Recycling (metals)

Roasting

Sulfuric acid manufacture

Titanium extraction

Zinc extraction

20 Greener industry

(Not AQA or CCEA, details and depth of coverage vary between specifications)

Atom economy

Biodegradable materials

By-products

Catalysts

Chemical industry

Efficiency

Energy recovery

Ethanoic acid manufacture (as an example)

Green chemistry

Life-cycle assessment *(Edexcel only at AS)*

Recycling

Renewable resources

Supercritical carbon dioxide *(OCR only)*

Sustainable development

Wastes

Yield

21 Organic molecules

Cis–trans isomers

Combustion analysis

Displayed formulae

Double bond

Empirical formula

Functional group

E–Z isomers

Homologous series

Isomers

Molecular formula

Names of carbon compounds

Skeletal formulae *(not AQA or CCEA)*

Structural formulae

22 Alkanes

Alkanes

Catalytic cracking

Crude oil

Fractional distillation of oil

Free radicals

Free-radical chain reactions

Homolytic bond breaking

Hydrocarbons

Petroleum

Saturated compounds

Thermal cracking

Zeolites

23 Alkenes

Addition polymerisation

Addition reaction

Alkenes

Carbocations

Curly arrows

Electrophiles

Electrophilic addition

Heterolytic bond breaking

Hydration

Hydrogenation

Markovnikov's rule *(AQA and WJEC only)*

Monomers

Polymer

Pi bonds

Recycling (polymers)

Sigma bonds

Unsaturated compounds

24 Polymers

(Salters only)

Addition polymerisation

Copolymerisation

Plastics

Polymeric materials

Polymer solubility

Thermoplastic polymers

Thermosetting polymers

25 Halogenoalkanes

Curly arrows

Elimination reaction *(not OCR)*

Halogenoalkanes

Heterolytic bond breaking

Hydrolysis

Leaving group

Nucleophiles

Nucleophilic substitution in halogenoalkanes

Polar covalent bonds

Polar molecules

Substitution reactions

26 Alcohols

Alcohols

Aldehydes

Carboxylic acids

Dehydration

Diols

Elimination reactions *(not CCEA or Edexcel)*

Esters

Ethanol

Fehling's solution

Fermentation

Ketones

Methanol

Oxidation

Tollens' reagent

27 Fuels

(Details and depth of coverage vary between specifications)

Biodiesel

Bioethanol

Biofuels

Carbon footprint

Carbon monoxide

Carbon neutral

Combustion

Fossil fuels

Fuels

Hydrogen economy

Isomerisation

Knocking

Octane number

Petrol

Reforming

28 Pollution of the lower atmosphere

(Details and depth of coverage vary between specifications)

Acid deposition

Acid rain

Air

Anthropogenic changes

Atmosphere

Atmospheric pollutants

Biofuels

Carbon cycle

Carbon capture and storage

Carbon footprint

Carbon neutral

Catalytic converter

Climate change

Combustion

Electromagnetic radiation

Fuels

Global warming

Greenhouse effect

Greenhouse gases

Ozone in the troposphere

Photochemical smog

29 Pollution of the upper atmosphere

Atmosphere

Blowing agents

CFCs

Electromagnetic radiation

Fire extinguishers

Free radicals

HCFCs

HFCs

Ozone

Ozone in the stratosphere

30 Analytical techniques

Acid–base titrations

Breathalyser *(OCR and WJEC only)*

Electromagnetic radiation

Infrared spectroscopy

Mass spectrometry

Molecular formula *(AQA only)*

Relative molecular mass

Titration

31 Experimental skills

Acid–base titrations

Anion tests

Calorimeter

Cation tests

Errors of measurement

Flame tests

Gas tests

Hazard

Iodine–thiosulfate titrations *(Edexcel only)*

Linear relationship

Measurement uncertainty

Molar mass *(AQA only)*

Organic analysis

Organic preparations

Proportionality

Risk

Significant figures

Standard solution

Titration

Yield

32 Chemical kinetics

Activation energy

Arrhenius equation *(Edexcel only)*

Clock reaction

Enzyme kinetics *(Salters only)*

First order reaction

Half-life (chemical)

Initial-rate method

Mechanism of a reaction

Nucleophilic substitution in halogenoalkanes *(CCEA and Edexcel only)*

Rate-determining step

Rate equations

Rates of reaction

Reaction kinetics

Second order reaction

Transition state

Zero order reaction

33 The equilibrium law

Equilibrium law

Heterogeneous equilibrium *(Edexcel and WJEC only)*

Homogeneous equilibrium

K_c

K_p *(CCEA, Edexcel and WJEC only)*

Partial pressures *(CCEA, Edexcel and WJEC only)*

Partition *(CCEA only)*

Temperature effect on equilibria

34 Acid–base equilibria

Acid–base equilibria

Acid–base indicators *(not Salters)*

Brønsted–Lowry theory

Blood buffers

Buffer solutions

Diprotic acid *(AQA only)*

Enthalpy change of neutralisation *(OCR only)*

K_a

K_w

Monoprotic acids

Neutralisation reactions

pH changes during acid–base titrations *(not Salters)*

pH scale

pK_a *(not WJEC)*

pK_w *(CCEA and Edexcel only)*

Proton

Strong acids

Weak acids

35 Redox equilibria

$E^{\ominus}_{cell}$

Breathalyser *(Edexcel only)*

Corrosion of a metal *(Salters only)*

Electrochemical cells

Electrochemical series

Feasibility

Fuel cell *(not CCEA or Salters)*

Half-equation

Hydrogen electrode

Hydrogen economy *(OCR only)*

Non-rechargeable cells *(AQA only)*

Oxidation numbers

Rechargeable (or storage) cells *(AQA, OCR only)*

Redox reactions

Reference electrodes

Sacrificial protection *(Salters only)*

Standard electrode potentials

36 Thermochemistry of ionic compounds

(AQA, CCEA, OCR, WJEC)

Bond dissociation enthalpy

Born–Haber cycle

Electron affinity

Enthalpy change of atomisation

Enthalpy change of formation

Enthalpy change of hydration

Enthalpy change of solution

Lattice energy

Octet rule

(Edexcel and Salters)

Enthalpy change of formation

Enthalpy change of hydration

Enthalpy change of solution

Lattice energy

37 Thermodynamics and the direction of change

Entropy

Entropy changes and feasibility *(Edexcel and Salters only)*

Feasibility

Free energy change *(CCEA, AQA, OCR, WJEC only)*

Inert chemicals

Spontaneous reaction

Standard molar entropy

Thermodynamics

Stable compounds

38 Periodicity

(AQA and CCEA only)

Acidic oxides

Amphoteric oxides

Basic oxides

Oxides

Period 3

Periodicity

Periodicity of chemical properties

39 p-block chemistry

(Salters only)

Ammonia

Ammonium ion

Nitrogen

Nitrogen cycle

Nitrogen oxides

(WJEC only)

Aluminium

Aluminium chloride

Aluminium oxide

Amphoteric oxides

Bleaching

Boron

Boron nitride

Chlorine oxoanions

Chlorine water treatment

Dative covalent bond

Displacement reactions

Group 4

Halide ions

Hydrogen halides

Inert pair effect

40 d-block chemistry

Bidentate ligands *(not WJEC)*

Catalysts

Catalyst poisons *(AQA only)*

Chelates *(not WJEC)*

Chiral compounds *(OCR only)*

Chromium *(AQA, Edexcel and WJEC only)*

Cisplatin

Cobalt *(AQA only)*

Colorimetry

Coloured compounds *(not CCEA)*

Complex ions

Coordination compounds

Coordination number

Copper *(Edexcel only)*

d-block elements

Disproportionation *(CCEA only)*

EDTA *(not WJEC)*

Haemoglobin *(not Edexcel or WJEC)*

Heterogeneous catalyst

Homogeneous catalyst *(not CCEA)*

Ligands

Ligand substitution reactions *(not WJEC)*

Polydentate ligands *(not WJEC)*

Shapes of complex ions

Stability constants *(OCR only)*

Trace elements *(WJEC only)*

Transition elements

Vanadium *(CCEA only)*

41 Reactions of inorganic ions in solution

(AQA only as a topic)

Amphoteric oxides

Bidendate ligands

Cation tests

Chelates

Coordination number

EDTA

Hydration

Hydrolysis of metal aqua ions

Lewis acid–base theory

Ligand substitution reactions

42 Organic structures

Asymmetric molecules

Chiral compounds

Cis–trans isomers

E–Z isomers *(AQA, Edexcel, OCR, Salters, WJEC only)*

Names of carbon compounds

Optical isomers

Plane-polarised light

Racemic mixture

Stereoisomers

Structural isomers

43 Arenes and related compounds

Benzene

Delocalised electrons

Electrophilic substitution

Friedel–Crafts reactions *(not CCEA or WJEC)*

Hydrolysis of organic chloro compounds *(WJEC only)*

Methylbenzene *(WJEC only)*

Nitration of benzene

Phenol *(not AQA or CCEA)*

Phenylamine

Stability of benzene

Sulfonation of benzene *(Edexcel and Salters only)*

44 Carbonyl compounds

Addition–elimination reactions *(WJEC only)*

Aldehydes

Carbonyl compounds

Derivatives *(not AQA or Salters)*

2,4-dinitrophenylhydrazine *(not AQA or Salters)*

Fehling's solution *(not OCR or Salters)*

Ketones

Lithium tetrahydridoaluminate(III) *(CCEA and Edexcel only)*

Nucleophilic addition

Sodium tetrahydridoborate(III) *(AQA, OCR and WJEC only)*

Tollens' reagent *(not Salters)*

Triiodomethane reaction *(Edexcel and WJEC only)*

45 Carboxylic acids and related compounds

Acid anhydrides *(not Edexcel or CCEA)*

Acid chlorides

Acylation

Addition–elimination reactions *(AQA only)*

Amides

Aspirin

Biodiesel *(AQA, Edexcel and OCR only)*

Carboxylic acids

Cholesterol *(CCEA and OCR only)*

Esters

Fats

Fatty acids

Iodine number *(CCEA only)*

Nitriles *(not OCR)*

Nucleophilic substitution in derivatives of carboxylic acids *(AQA only)*

Soaps *(AQA and Edexcel only)*

Transesterification *(Edexcel only)*

Trans fats *(OCR only)*

Triglycerides

Vegetable oils

46 Organic nitrogen compounds

Acylation

Amides

Amines

Amino acids

Azo dyes *(not AQA)*

Base strength of amines *(AQA, CCEA and OCR only)*

Chromophore *(Salters and WJEC only)*

Coloured compounds *(CCEA, Salters and WJEC only)*

Diazonium salts *(not AQA)*

Isoelectric point *(OCR only)*

Nitriles *(not OCR)*

Phenylamine

Quaternary ammonium cations *(AQA only)*

Zwitterions

47 Natural and synthetic polymers

Active site *(Salters only)*

Addition polymerisation

Biodegradable materials *(not Edexcel or WJEC)*

Condensation polymers

Copolymerisation *(Salters only)*

DNA *(Salters only)*

Enzymes *(CCEA, Salters and WJEC only)*

Kevlar *(AQA, Edexcel, OCR and Salters only)*

Life-cycle assessment *(Salters only)*

Paper chromatography

Peptides

Plasticisers *(not CCEA or WJEC)*

Polyamides

Polyesters

Polymeric materials *(CCEA and Salters only)*

Polymer solubility *(Edexcel and Salters only)*

Proteins

Recycling (polymers) *(AQA only)*

48 Organic synthesis

(Not CCEA)

Chiral synthesis *(Edexcel and OCR only)*

Combinatorial chemistry *(Edexcel and Salters only)*

Combustion analysis *(Edexcel and WJEC only)*

Computational chemistry *(Salters and WJEC only)*

Drugs *(Edexcel. OCR and Salters only)*

Drug development *(Salters only)*

Medicine *(Salters only)*

Organic preparations

Organic routes

Pharmacophore *(Edexcel, OCR and Salters only)*

49 Instrumental analysis

Atomic emission spectroscopy *(Salters only)*

Carbon-13 NMR spectroscopy *(AQA and OCR only)*

Chromatography

Column chromatography *(AQA only)*

Electromagnetic radiation

Electrophoresis *(Salters only)*

Gas chromatography

GC–MS *(OCR and Salters only)*

High-performance liquid chromatography *(Edexcel, OCR and WJEC only)*

Infrared spectroscopy

Mass spectrometry

Molecular formula *(CCEA and Salters only)*

NMR spectroscopy

Proton NMR spectroscopy

Thin-layer chromatography

Visible spectroscopy *(AQA and Salters only)*

50 Investigative skills

Acid–base titrations

Anion tests

Buchner flask and funnel

Cation tests

Colorimetry

Errors of measurement

Gas tests

Hazard

Linear relationship

Measurement uncertainty

Organic analysis

Organic preparations

Proportionality

Recrystallisation

Redox titration

Risk

Titration

Yield

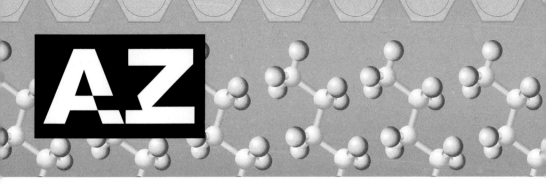

Hints for exam success

Planning a revision programme

Your revision should be systematic and based on a copy of the specification. Always check what you are revising against the latest version of your course specification.

Active revision methods

You need to make your revision active so that you maintain concentration and really test that you understand the key points while developing the necessary skills. For each revision topic, try writing the title of a topic in the centre of a sheet of paper. Then use the definitions and explanations in the main part of this book to help you build up a mind map or concept map to show how the ideas in the topic link together.

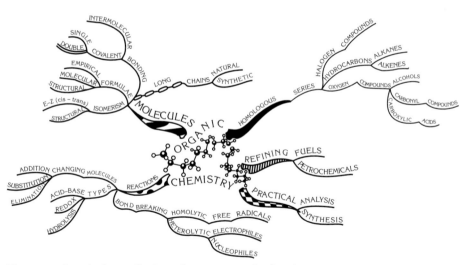

The start of a mind map for introductory organic chemistry

Suppose you are revising the chemistry of the halogens. Have a pile of scrap paper to hand and a pencil. Now, as you read make jottings, small lists, summary phrases, write equations, sketch diagrams and practise labelling them. Now tear up the paper, close the texts and notes, and write out those lists, equations, diagrams and so on. Then check to see whether you have remembered correctly.

When it comes to learning the reactions of a family of organic compounds such as the alcohols, consider using index cards. Write the equation for each alcohol reaction you have to know on one side of the card. Write the names of the reactants and the conditions for the reaction on the other side. Now you can use the cards for revision. Look at one side of the card and try to recall what is on the other side.

Practise calculations with the help of worked examples. Find questions from a book with the answers in the back so that you can check that you have worked through to the answer correctly.

Past papers

Look at past papers. The same subject matter appears, year after year, with only very minor differences in emphasis and phrasing of questions. It is because so many questions appear to be the same year after year that you should read questions with the greatest of care to note the particular emphasis, and precise nature, of a question.

Using the mark scheme

When answering questions in an exam, use the mark allocations to guide you when deciding how much to write. Three marks for part of a question probably means that the examiner is expecting three good points to be made. If asked for a chemical test, for example, one mark might be for the reagent to be used, one for the conditions and the third for describing the observations.

Some questions expect you to write at greater length, drawing together and organising relevant material. No marks are allocated for the plan you sketch out before writing your answer, but making a plan helps you to write a well-organised answer with examples and so leads to better marks.

Synoptic overview

In the second half of an A-level course you are expected to bring together your knowledge and understanding of different areas of chemistry and apply them in contexts that may be new to you. Some of the key term entries in this book are designed to give you a synoptic overview of topics so that you can make connections and link ideas. You can use key terms such as *inorganic chemistry*, *organic chemistry* and *physical chemistry* to start creating your own overview of the subject. Use the cross-references to build up charts showing how ideas link together.

Following up the cross-references from these other key terms will also help you gain a synoptic overview of parts of your course: *atomic theory, bonding, chemical industry, environmental chemistry, group, mechanism of a reaction, organic analysis, organic preparations, organic routes, periodicity, physical properties, polymers, qualitative analysis, quantitative analysis, spectroscopy, thermochemistry, valency theory.*

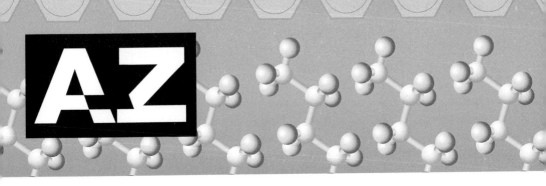

Examiners' terms

Every year too many well-prepared candidates fail to score as many marks as they should because they do not answer the question set by the examiners.

Examiners try very hard to set questions which are clear to the candidates. Even so, under examination conditions it is all too easy to rush into writing an answer before checking carefully the meaning of the question.

A useful first step is to highlight the words in the question that give instructions. Words such as 'calculate', 'describe' and 'explain'. Here is an A–Z commentary on some of the instructions often used by chemistry examiners.

calculate means that your answers will include a number (and usually units). Always show your working, then you will get almost all the marks even if you make a silly slip while calculating. Most of the marks are given for showing how you arrive at an answer.

With a calculator it may be tempting to work through several steps and then just write down the number you get. You must resist this temptation. Showing the working helps you to give an answer to the correct number of *significant figures* and with the correct units. Do not be tricked by your calculator into quoting an answer to more significant figures than are justified by the data.

Some quantities do not have units. If so, it is worth saying so in your answer to show the examiner that you have considered the point. For example, the equilibrium constant, K_p, for the reaction of hydrogen with iodine to form hydrogen iodide is a number with no units. Most equilibrium constants, however, do have units.

comment is an invitation to interpret information given in the question, or to relate the information in the question to your knowledge and understanding of the topic. This is likely to be a challenging question expecting you to write several sentences. Use the mark allocations as a guide to the number of key points that you should aim to include in your answer.

compare asks you to say something definite about each of the things to be compared. If asked to compare the action of bromine with cyclohexene and with cyclohexane, you could start by stating that cyclohexene, the unsaturated compound, rapidly decolorises a solution of bromine. You should go on to show that you also know that there is no immediate action with the cyclohexane because it is a saturated hydrocarbon. Setting out your answer in a table is a way of making sure that you say something about each of the chemicals or processes to be compared.

deduce implies that you will work out your answer with the help of information given in the question, but using your understanding of the principles involved. Your answer should show the steps in your thinking as well as your conclusions. You may have to draw on answers to previous parts of the same question.

define requires a precise statement. Using a simple example often helps to make a definition clearer. Many chemical terms have very specific definitions. This is especially true in thermochemistry. When defining standard *enthalpy change of formation*, it is not good enough to write 'the heat given out when a compound is formed from its elements'. You must specify the conditions of temperature and pressure, the standard states of the elements and the amount of compound formed.

describe asks you to outline in words, and diagrams if helpful, the main points of a topic. If asked to describe and explain the shape of an ammonia molecule you will want to draw both and dot-and-cross diagram to show the bonding electrons and lone pairs as well as a 3D diagram of the molecule with bond angles marked in. Then you will explain the *shape of the molecule* with the help of electron-pair repulsion theory.

'Describe how you would prepare...in the laboratory', asks for words and diagrams outlining the apparatus, the chemicals used and the procedure. A good answer will also include the key observations during the preparation and especially what the product is like.

determine tells you to work out a quantity, formula or equation from information given.

discuss expects you to give an argument reflecting on the key points of the topic of the question. This might be a review of the benefits and risks of using chemicals such as halogenoalkanes. Make sure your answer refers to both sides of the argument where there can be differences of opinion. The markscheme be more open for a question like this so that there can be several different ways for arriving at full marks.

explain is one of the most important words on an exam paper. Frequently you will be asked to write down some information and then give an explanation for your answer. Your explanation is likely to be based on a chemical theory. Always give reasons.

estimate probably means that you will have to calculate but that you can make simplifying assumptions to arrive at an order of magnitude answer to the question.

how would you...? means that you should describe a procedure that you could do. Imagine yourself standing at a bench in the laboratory. Now, what would *you* do? Thinking in this way will help you to avoid describing the Haber process, when asked, 'How would *you* prepare a sample of ammonia?'.

You might be asked 'How would you test for...?', if so, take care both to write down the procedure for the test and then to describe what you would observe if the test were positive. Remember that observations include smells and temperature changes as well as things that you can see, such as colour changes.

identify can generally be answered by giving the name or formula of a chemical. Check the wording of the question carefully to see whether or not you have to give your reasoning in full.

justify your answer means that you must give your reasons for an answer. This should include an explanation based on theory.

list asks for a series of points – probably just one or two words each. If asked for a certain number of examples in the list do not give more.

name is usually a request for the recommended IUPAC name. You are quite likely to be asked to give the name of an organic structure (see *names of organic compounds*).

outline tells you that the answer can be short with only the key points. Even so, you must include essential details. For instance, even in 'outline', when mentioning sulfuric acid you must state whether it is dilute or concentrated.

predict means that you cannot simply recall the answer. You must use information given in the question, or your answers to earlier parts of a question, together with your understanding of chemistry to describe what will happen under specified conditions.

sketch a diagram means that a simple outline drawing will be acceptable. Make sure that the proportions are about right and that key features are fully labelled and explained.

If asked to, 'Draw/sketch the apparatus used to…', simply take sufficient care to depict the apparatus to be used. There are very seldom any marks allocated for artistic quality, but a good, labelled drawing can save a lot of writing. Your drawing should be good enough for you not to have to label the apparatus, only the substances present in the apparatus. Small points are important. Do not draw a stopper at the top of a reflux condenser – it could be explosive.

sketch a graph means that what you draw should be the right general shape in the right proportions but the position of the line or curve need only be qualitatively correct. The axes should be labelled but need not have a numbered scale. Sometimes it is important to identify particular features to show that the line passes through the origin or cuts one of the axes at a particular value.

Take care. In an answer about rates of reaction it can be useful to sketch the distribution of molecular kinetic energies in a gas at a given temperature, and superimpose on it the distribution when the temperature is 10 degrees higher (see *activation energy*). It is all too easy to exaggerate the picture to make a point, and then when the activation energy is added it appears the reaction will be wildly explosive.

state or give both ask for short and simple, factual statements.

suggest tells you that you cannot simply recall a right answer. If you are asked for a suggestion it may be that there is more than one possible answer and that you will be given marks for any sensible suggestions. You may also be asked to suggest answers if you are given some new information which you have not met before. The examiners want you to apply your knowledge of chemistry to answer the question in an unfamiliar situation.

why? means 'explain' or 'give reasons'. Be careful not to reword the question without giving an explanation. If asked 'Why does the first *ionisation energy* of the elements in *group 2* decrease with increase in atomic number?' you have to say more than 'Because there is less attraction on the electrons by the nucleus'. This is a true statement, but not enough. The examiner expects you to show how *shielding* by inner full shells affects the pull of the nucleus on the outer electrons of atoms such as magnesium and calcium.

write, or construct, an equation means that you should describe a reaction with a *balanced equation* complete with chemical formulae.

Sometimes it is very important to include *state symbols*. This is especially true in thermochemistry and electrochemistry. In general it is better to include state symbols whenever possible, especially when aqueous reagents react to form a precipitate or a gas.